AN ARCHAEOLOGY OF THE SACRED:

Adena-Hopewell Astronomy and Landscape Archaeology

William F. Romain

The Ancient Earthworks Project

Olmsted Township, Ohio

Copyright © 2015 William F. Romain

All Rights Reserved.
No part of this book may be reproduced or utilized in any form or by any means, electronic or mechanical, including photocopying, recording, or by any information storage and retrieval system without permission, in writing, from the copyright holder.

Unless otherwise credited, all photographs and LiDAR images are by William F. Romain

Inquiries about reproducing material from this work should be addressed to William F. Romain at: romainwf@aol.com

First edition, July 2015 published by:

The Ancient Earthworks Project
   Olmsted Township, Ohio

www.ancientearthworksproject.org

ISBN 10:0692492267 (The Ancient Earthworks Project)
ISBN 13:978-0692492260 (European Article Number)

Library of Congress Control Number: 2015911389
 The Ancient Earthworks Project, Olmsted Township, Ohio

ACKNOWLEDGEMENTS

It is the rare book that does not build-upon the work and intellectual accomplishments of others. This volume is no exception. Looking back, I find that my thinking about prehistoric religion has been influence by many, some of whom I know personally; others whom I know only through their written works.

My sincere thanks to Tim Pauketat for his insights and commentary over the years. Thomas E. Emerson has influenced my thinking about ancient religions and has been a good friend over the years. Thank you Tim and Tom.

More generally, I have drawn upon the theoretical approaches of Tim Ingold, Christopher Tilley, Robert Hall, Irving Hallowell, Graham Harvey, Christopher Vecsey, Vine Deloria, and Martin Heidegger.

My efforts in the field of archaeoastronomy were fine-tuned by my Ph.D. advisor Clive Ruggles at the University of Leicester. I promised Clive that I would write-up my Hopewell findings. Well, Clive, here it is....a decade or so in the making - but done.

My M.A. advisor at Kent State University was Olaf Prufer. It was from Olaf that I learned about Hopewell. If Olaf is watching from above, I hope he is smiling at what his teachings have wrought. I will always remember our afternoon beverages at the Brown Derby.

The reader will find that this book is heavily dependent on LiDAR imagery. I first learned about LiDAR from my friend, Jeffrey Wilson. This book would have been considerably different if Jeff had not enlightened me about LiDAR. Thank you, Jeff.

Subsequent to learning about LiDAR from Jeff, my early LiDAR efforts were made possible through assistance provided by The Ohio State University, Newark Earthworks Center. For help and support I owe a debt of thanks to NEC Director Richard Shiels.

Mike Mumansky at Applied Imagery and Jeff Smith with the Ohio Geographically Referenced Information Program spent considerable time teaching me how to make the best use of LiDAR data. Thank you, Mike and Jeff.

Several ideas in this book owe their genesis to workshops I attended. Two of the most informative concerned ancient cosmologies and were sponsored by the Santa Fe

Institute, School for Advanced Research. My sincere thanks to the Santa Fe Institute for inviting me. My understandings were further enlightened by participation in a third Santa Fe workshop, sponsored by Anna Sofaer, Director of The Solstice Project.

Many thanks are extended to Jarrod Burks and Brent Eberhard. Both gentlemen provided geographical coordinate data needed to locate some of the lesser-known earthworks.

Among the many other individuals who provided valuable assistance, photographs, permissions, and information were: Stanley Baker, Jeff Carskadden, Wesley S. Clarke, Richard Conway, Robert Converse, Rob Cook, Dwight Cropper, Paul Gardner, Martin Gray, Angela Haines, Jim Henry, Ed Herrmann, Karen Leone, Brad Lepper, J. Huston McCulloch, Richard Moats, G. William Monaghan, Albert Pecora, Jennifer Pederson, Matthew Purtill, Robert Riordan, Bret Ruby, John C. Rummel, Geoffrey Sea, Wendi Schnaufer, Mark Seeman, Tim Schilling, Al Tonetti, Jean Yost, and Gail Zion. To all, thank you.

I wish to extend a special thanks to my mother. Mom has always been there to support and encourage me. Thanks mom! Thanks also to Lee Rothenberg for your encouragement.

To my wife, Evie, I extend a very special thank you. I could not have written this book without your patience, love, and support over the years. Thank you.

To all, my sincere thanks. I am solely responsible for the contents of this book.

CONTENTS

CHAPTER 1: INTRODUCTION 1

CHAPTER 2: BACKGROUND AND METHODS 20

CHAPTER 3: MUSKINGUM RIVER WATERSHED 39

Fredericktown Earthworks	51
Newark Earthworks	55
Flint Ridge	79
Glenford Earthworks	83
Marietta Earthworks	87

CHAPTER 4: HOCKING RIVER WATERSHED 97

Rock Mill Works	101
Wolf Plains Group	103

CHAPTER 5: SCIOTO RIVER WATERSHED 108

Circleville Earthworks	114
Stitt Mound	117
Dunlaps Earthwork	118
Mound City	119
Shriver Circle	124
Hopeton Earthworks	126
Adena Mound	130
Mount Logan Mound	133
Works East Earthwork	134
High Bank Earthworks	136
Liberty Earthworks	140
Johnson Works	143
Seal Township Works	145

CONTENTS (cont.)

CHAPTER 5: SCIOTO RIVER WATERSHED (cont.)

Tremper Earthworks	148
Portsmouth Earthworks	150

CHAPTER 6: PAINT CREEK WATERSHED **162**

Junction Group	169
Steel Group	170
Story Mound	172
Chillicothe Country Club Mound	173
Anderson Earthwork	174
Hopewell Mound Group	176
Frankfort Works	179
Bourneville Circle	181
Spruce Hill	182
Baum Earthworks	184
Seip Earthworks	186

CHAPTER 7: BRUSH CREEK WATERSHED **191**

Trefoil Works	195
Fort Hill	196
Serpent Mound	201

CHAPTER 8: LITTLE MIAMI RIVER WATERSHED **205**

Cedarville Earthworks	209
Pollock Earthworks	210
Williamson Mound	211
Fort Ancient	212
Turner Earthworks	216
East Fork Works	217

CONTENTS (cont.)

CHAPTER 9: GREAT MIAMI RIVER WATERSHED **219**

    Calvary Cemetery Enclosure 222

    Alexandersville 223

    Miamisburg Mound 224

    Hill-Kinder Mound 226

    Great Mound 226

    Carlisle Fort 227

    Fairfield Township Works 229

    Butler County Fortified Hill 231

    Pollock Wilson Mound Group 232

    Miami Fort 233

CHAPTER 10: DISCUSSION **238**

NOTES **266**

REFERENCES **297**

INDEX **321**

CHAPTER 1
INTRODUCTION

About two thousand years ago, people of the Eastern Woodlands of North America built an entire world of inter-related and sophisticated monumental structures. Some were gigantic earthen mounds; others were geometrically shaped enclosures. Many dwarf such monuments as Stonehenge. Built by people of the so-called Adena and Hopewell cultures, the earthworks span a time period from about 400 B.C. to A.D. 400. The earthworks we are concerned with are found mostly in south and central Ohio.

When Western explorers first entered the Ohio territory there were more than 10,000 mounds and 600 geometric earthworks in that region alone. Today, most are gone – having fallen victim to farming and urban development. Still, enough remain to stir the imagination and compel us to ask, why people would toil at the mind-numbing task of moving thousands of tons of earth over the course of hundreds of years to build these monumental structures? What could possibly have motivated them?

In this book, I offer one possible answer. Along the way I chronicle what remains of the earthworks. In essence, I present a case that religion was the motivating force behind the Adena-Hopewell phenomenon and its earthwork expressions. In this regard it is useful to keep in mind that there are distinctions between what drives the construction of an earthwork, the principles underlying its design and siting, what an earthwork meant to the people who built it, and how it may have been used.

Many mounds were used for burial purposes. Others probably marked locations where mythic or real events occurred. Some earthworks may have been places inhabited by spirits. Some may have been locations for the gathering of special plants, minerals, or medicines. Others marked locations where solar or lunar azimuths intersected with the land. Many earthworks and mounds were locales where people went to connect with the spirit world. Certain places were for healing; others were vision quest sites. Perhaps some were territorial or route markers. No doubt many earthworks were places for rituals and ceremonial activities. Perhaps some were destinations for religious pilgrimages. No doubt many earthworks were meeting places for a variety of social, political, and economic activities.

In short, earthworks and mounds were the focal points for a wide range of spiritual, social, political, and economic activities at multiple levels. Ultimately, I argue, earthwork design, location, and use were fundamentally shaped according to religious considerations. Ironically, perhaps, this same conclusion was anticipated years ago by antiquarian scholars Ephraim Squier and Edwin Davis (1848:47) who stated: "We have reason to believe that the religious system of the mound-builders...exercised among them a great, if not controlling purpose."

Adena-Hopewell religion was not an orthodox or formal institution in the Western sense. Adena-Hopewell religion was not a commandment-based religion with a monolithic deity acknowledged on Sunday mornings. Nor was it a component of society amenable to academic study as a stand-alone system. Rather, I argue that Adena-Hopewell religion comprised cultural-level cosmological understandings integrated with actor-based perceptual experiences that, together, informed a characteristic Adena-Hopewell way of being-in-the-world. This way of way of being-in-the world emerged as a relational field that entangled individual sensory experiences and interactions with a variety of non-human agents and forces, cosmic realms, and numinous earth, sky, and water phenomena. In this relational field, religious practices were connected to the landscape; and earthworks and mounds were places where the trajectories of human and non-human agents and forces intersected. For whatever their purposes, by virtue of their location and form, mounds and earthworks were focal points for intersecting celestial and terrestrial alignments, temporal convergences between past and present, liminal places between states of being including living and dead, experiential intersections between human perception and numinous phenomena, and places for interactions between human and non-human agents.

Through their incorporation of cultural-level design principles, earthworks and mounds mediated between realms and in some cases, enabled and guided hierophantic experiences that informed the phenomenological aspect of Adena-Hopewell religion. Architectural design principles imparted form to the earthworks that, in turn, provided the material setting for experience and action. Thus design principles (understood and held in common by widely separated groups of people) positioned and oriented people in such a way as to guide an intentional range of sensuous experience. To a limited extent, some of these experiences can be re-created and relived today.

With respect to any object, there are certain properties that define its physical nature. For Adena-Hopewell mounds and earthworks, the most obvious properties – and the ones we consider here, are place, shape, size, and orientation. For Adena-Hopewell earthworks, culturally-defined, normative practices relevant to place, shape, size, and orientation were combined in unique ways that resulted in site-specific configurations that made each earthwork unique, impacted the senses and experiences of the people who built and used them, while at the same time, imparting a sense of relatedness among geographically separated earthworks. Among the ways that place, shape, size, and orientation find material expression in the earthworks are through location, geometry, measurement, and alignments. In the embodied human world, location provides for a position relative to other things and serves as a focal point for assigned meanings. Geometry gives shape to the formless and establishes relationships between things. Measurement provides sequential and predictable order. Alignments, especially to the heavens, but also to terrestrial phenomena, provide a connection to the cyclic flow of time, bring structures into harmony with cosmic realms, and establish a link with forces that affect the individual and community. Of course objects have additional properties (e.g., mass, color, texture, duration). For the earthworks, however, place, shape, size, and orientation are among their most visually defining properties and, for that reason, serve as a good point of departure for our discussion.

Place and Location

We begin with *place* because it is *place* that provides the physical setting for everything else that follows. Adena-Hopewell earthworks were situated in special places (Romain 1993b, 2000, 2009; Seeman 2004). There are few guarantees of veracity for any proffered interpretation with respect to what may have made a place special. Rather, as archaeologists what we posit are 'plausible accounts produced in and for the present' (Thomas 1996:88-89). With that in mind, however, there are certain landscape features that have not changed significantly since Adena-Hopewell times. The locations of hills and valleys relative to the rising and setting of the sun and moon are essentially the same today as they were two thousand years ago. Likewise, the capabilities and constraints of human perception have not changed. All humans, for example, perceive things relative to near-far, front-back, left-right, up-down, and so on (Tilley 2008; Tuan 1977). Thus, as Christopher Tilley (1994:74) suggests: "…these most basic of personal spatial experiences, are shared with prehistoric populations in our common biological humanity. They provide tools with which to think and work."

Sometimes the features or relationships that ostensibly made a place special seem obvious to us – such as a location that provides a vantage point for a solstice sunrise along the course of a river valley, or over a prominent mountain peak. In such instances it is helpful if our initial conjecture about why a place might be special is supported by similar occurrences within the target culture as well as by relevant ethnographic data. Other times, whatever made a place special is not obvious, may be based in idiosyncratic reasons and perceptions, and may be beyond our ability to recognize. In all cases, however, if an earthwork is situated in a particular location, it is because someone in the past, i.e., a "place-maker" (*sensu* Basso 1996:5), intentionally designated the place as special.

But what exactly is meant by *place*? The question is relevant because it cuts to the very heart of experience. There are multiple ways of defining *place* (e.g., Bowser and Zedeño 2009). Archaeologist Ruth Van Dyke (2003:180) describes *place* as "the intersection of time, space, and self." This definition contains within it the implication that *self* is interwoven with place. Taking this a step further, the meaning of a place is necessarily contingent upon a lived awareness or consciousness of it. Nicole Boivin (2004: 64-65) puts it this way: "Far from being inscribed on a passive material substrate, meanings emerge out of specific encounters between embodied minds and the physical properties of the material world."

In other words, meaning is not intrinsic to a place; rather, meaning emerges as people engage with place. It is also the case that places are capable of evoking multiple meanings and those meanings are subject to change. Further, there is an emotional component to all this (Gosden 2004). In our apprehension of place, it is our emotional evaluations that determine our feelings about a particular place, influence our thinking and ultimately, drive our physiological response. Indeed, by virtue of their capacity to evoke emotional responses, special places bring some people closer to the transcendental.

Following upon this, meaning as it relates to place and the Moundbuilder earthworks necessarily draws upon landscape, which, for purposes of the present discussion includes earth, sky, and water realms. The basic premise is articulated by architect Christian Norberg-Schulz (1980:166): "The 'meaning' of any object [or place] consists in its relationships to other objects, that is, it consists in what the object 'gathers'." In his essay "Building Dwelling Thinking," philosopher Martin Heidegger (1993 [1954]:354) uses the example of how a bridge "gathers" its surroundings:

> "The bridge does not just connect banks that were already there. The banks emerge as banks only as the bridge crosses the stream. The bridge expressly causes them to lie across from each other. One side is set off against the other by the bridge. Nor do the banks stretch along the stream as indifferent border strips of the dry land. With the banks, the bridge brings to the stream the one and the other expanse of the landscape lying behind them. It brings the stream and bank and land into each other's neighborhood. The bridge *gathers* the earth as landscape around the stream."

In the world of anthropology and archaeology, the concept of gathering has found recent expression by Hodder (2006, 2011) in the way of material entanglements, lifeworld webs for Ingold (2007, 2011), networks for Latour (2005), bundles by Pauketat (2013a, 2013b), and complex objects for Zedeño (2008, 2013). By whatever term we wish to use, however, the underlying concept is basically the same.

For the Adena-Hopewell, the ceremonies and ritual movements that gave expression to their way of being-in-the-world was inexorably tied to special, or sacred places in the landscape. To the extent that it is possible, in the following chapters I explore the relationships between Adena-Hopewell monumental earthworks as situated in a particular *place*, and selected earth, sky, and water features *gathered* by such places, for it is those gathered, multi-sensual and related features and relationships that allow for the creation of meaning as part of lived experience.

It can be said that the gathered features of a place relative to self emerge in a relational field (Hodder 2011; Ingold 2006, 2011; Pauketat 2013a, 2013b; Thomas 1996; Zedeño 2009; also see Watts 2013). A useful explanation for the term *relational field* is provided by Baires, Butler, Skousen, and Pauketat (2013:3-4): "A relational field…is an entanglement of experiences comprised of multiple actions, persons, objects, natural elements, and spirits. Importantly, the relationships and entanglements are never fixed, but are always becoming, made and remade contingent on the various movements in space." Therefore, places, phenomena, humans and non-humans do not exist in isolation. Rather, they are associated, connected, and entangled in complex ways. Further, relational fields are animate, dynamic, continuously evolving, and mutually constitutive (Harvey 2006; Ingold 2006, 2011; Pauketat 2013a, 2013b; Thomas 1996). Early on, archaeologist Christopher Tilley (1994:10) recognized the essential elements of this concept when he explained:

> "A centered and meaningful space [place] involves specific sets of linkages between the physical space of the non-humanly created world, somatic states of the body, the mental space of cognition, and representation and the space of movement, encounter, and interaction between persons and between persons and the human and non-human environment."

Depending upon one's beliefs, therefore, relational fields can entangle all manner of actors to include gods, angels, demons, ghosts, spirits, souls of the living as well as the dead, and other non-human agents. For purposes there, I take the broadest definition and consider non-human agents as entities, forces, or phenomena that have the capacity to affect human experience. Accordingly, non-human agents can include plants, animals, sun, moon, stars, mountains, lakes, rivers, other earth, sky, and water things, as well as earthworks and mounds themselves. In many cultures, different kinds of non-human agents can have consciousness, self-awareness, sentience, act with deliberate intention and in some sense may be considered alive and person-like (e.g., Hallowell 1975 [1960]; Harvey 2006; also see Jones and Cloke 2008).

The point is that, as characterized by Emerson, Alt, and Pauketat (2008:218): "...the landscape, material things [to include earthworks], and human bodies are parts of relational fields that would be differentially experienced, interpreted, and engaged by the people moving through these fields." Among the many ways relational fields involving landscape can be experienced are through sensual awareness, contemplation, mindful sitting, ritual circuits, directed movements, processions, and pilgrimages.

For the Adena-Hopewell, their lifeworld-relational field was what we might call religious. However, as already implied, their life-world relational field is perhaps better described as a way of being-in-the-world, where the spirit dimension was intrinsic to everyday life. Moreover, for the Adena-Hopewell, monumental earthworks, landscape features, human and non-human agents, and lived experience were interwoven in a fabric of meaning that was itself sacred.

At the same time, the Adena-Hopewell worldview included cosmological concepts of a vertically tiered cosmos, quadripartite division of space in the horizontal, vertical *axis mundi* that connected the Upperworld, Lowerworld, and This World, a cyclic notion of time, and related ideas (Carr and Case 2005a; Charles and Buikstra 2006; Hall 1997; Penny 1985; Romain 2000, 2009). At their deepest levels these understandings appear based in how the embodied human mind-brain is constrained to operate in the particular kind of world we find ourselves in. Subject to individual interpretations mediated by culture, these cosmological understandings informed the Adena-Hopewell way of being-in-the-world.

The Adena-Hopewell way of being-in-the-world was not limited, however, to a set of core beliefs taught by religious experts and carried around in peoples' heads to be recited on special occasions. Although the cosmological concepts just noted appear to have been held in common by Native Americans across a wide geographic area and deep in time, there was more...much more. Specifically, there was an experiential aspect to the Adena-Hopewell way of being-in-the-world. One of the basic concepts underlying traditional Native American religions is the idea that the spirit realm is intimately connected to all life (Deloria 1975, 1991; Vecsey 1995; Walker 1995). From archaeological evidence (e.g., Byers 2004; Carr and Case 2005a; Charles and Buikstra 2006; Hall 1979, 1997; Penny 1985; Romain 2000, 2009) it seems certain that people of the Adena and Hopewell cultures were of the same mind-set.

As to the nature of the spirit realm thus conceived, religious scholar Christopher Vecsey (1995:14) explains it this way: "this realm...indicated by terms like *manito*, *wakan*, *orenda*, and the like, is a numinous, normally unseen, qualitatively superior realm that reveals itself. It is a great mystery that manifests itself, makes itself felt, supports human existence, and toward which ritual is directed." In describing this realm, the term *supernatural* is not appropriate, as that word suggests a distinction between sacred and profane, or between the natural world and something beyond. For Native Americans of traditional belief, the spirit realm is everywhere and is a part of everything. In other words, the so-called sacred is found in all things.

The implications of this relational worldview are noted by Deward E. Walker, Jr. (1995:102). They include:

1. "A body of mythic accounts explaining cultural origins and cultural development as distinctive peoples.
2. A special sense of the sacred that is centered in natural time and natural geography.
3. A set of critical and calendrical rituals that give social form and expression to religious beliefs and permit the groups and their members to experience their mythology.
4. A belief that harmony must be maintained with the sacred through the satisfactory conduct of rituals and adherence to sacred prescriptions and proscriptions.
5. A belief that while all aspects of nature and culture are potentially sacred, there are certain times and geographical locations that together possess great sacredness.

6. The major goal of religious life is gaining spiritual power and understanding necessary for a successful life, by entering into the sacred at certain sacred times/places."

If, as Walker suggests, a major goal of Native American religious life is to access the sacred, then, as Walker (1995:104) further explains, the way this is done is by "actually entering sacredness rather than merely praying to it or propitiating it."

It is difficult to pin down a precise definition for the term "sacred." Toward this end, however, Christina Pratt (2007:409) explains that: "sacred means where the spirit occurs. Experiences and things are sacred when they allow us to establish conscious relations with the Transcendental."

Although not necessarily a prerequisite for "entering the sacred," Native American religious practitioners often facilitate that encounter by identifying and engaging with places in the landscape where the sacred is tangibly manifested in a way that can be sensually experienced. Thus as explained by J. Donald Hughes (quoted in Suzuki and Knudson 1992:152), "Sacred space is where human beings find a manifestation of divine power, where they experience a sense of connectedness to the universe. There in some special way, spirit is present to them." At these special places, the spirit realm intersects with the ordinary world, and so at these places people are more easily able to interact with non-human beings and forces, often for the purposes of gaining information, power, or assistance, or simply to give thanks. As a consequence, Native American religions are typically more closely tied to the landscape than, for example, Abrahamic religions, which are more concerned with a conveying a particular "message."

As will be shown, for the Adena-Hopewell, special landscape places included certain mountains, promontories, viewscapes, river confluences, waterfalls, springs, gorges, places where concentrations of special rocks or minerals occur, and especially, where the trajectories of celestial events and topographic features intersected with each other. These places were marked by mounds or earthworks. Sometimes these special locations were further consecrated by human burials. I will have more to say such places in the following chapters. The important point for now is that, depending on the person and relevant phenomenon or feature associated with a sacred place, the effect might be subtle with quiet contemplation required to bring into conscious awareness the special quality of the location. Other times the effect might be dramatic as a particular phenomenon reveals itself. An example of the former might be the location where ancestors or relatives fought and died. An example of the later might be a precipice overlooking a visually impressive waterfall, or the solstice sunrise along the course of a river valley.

Sacred places can incorporate a temporal aspect as well. Walker (1995:104) refers to such locations as "sacred time-places." For the Adena-Hopewell, I present evidence that sacred time-places included locations where solar or lunar events intersected with either the local topography or constructed earthwork in a way that a visual manifestation of the sacred was presented. Mircea Eliade (1987) referred to such intersections as "hierophanies" – where numinous phenomena are manifested in ways that can be directly experienced in transcendental ways, beyond common experience.

By physically situating monumental earthworks in sacred time-space places, or constructing earthworks that themselves created sacred time-space places, the Adena-Hopewell facilitated access to the world of spirit. Moreover, since these sacred time-places were embedded in a relational web that included not only objects and events but also stories, mythologies, and histories that were created, told, and re-told, these places no doubt evolved profound meanings that contributed to individual and collective memories, cultural identity, and the spiritual life of people. For the Adena-Hopewell, the landscape was alive with spirit, power, and memory that was experienced and engaged. In this, the landscape was integral to Adena-Hopewell religion.

Shape and Geometry

For the Adena-Hopewell, once a place was designated as special and appropriate for the siting of an earthwork or mound, the next step in bringing an earthwork into existence was probably to decide on its shape. Shape gives form to substance. Shape defines the physical parameters of a thing and enfolds within it future potentials and constraints.

The interesting thing about geometric shapes is that they are often used to evoke and recall symbolic meanings, especially as related to cosmological themes. In their symmetry and balance, for example, man-made geometric shapes can symbolize the spatial and directional order of the earth realm, or the patterned movements of the celestial realm. In both cases, through the use of geometric shapes, a link is established that connects people to an aspect of the cosmos. Further, shape can influence actions as it provides cues for culturally sanctioned behaviors. Enter into a sacred circle earthwork, for example, and certain behaviors are acceptable, whereas others are not. Shape therefore, can incorporate cultural-level symbolic meanings; but it can also guide thinking and influence lived experience.

Moundbuilder earthworks were of several shapes including amorphoric shapes as found in hilltop and promontory enclosures, single geometrically shaped enclosures, earthwork complexes made-up of two or more geometric enclosures, and mounds that include conical, loaf-shaped, flat-topped platform, and occasional effigy shapes. The Adena culture is best known for conical mounds, small circle earthworks, and a few geometric enclosures resembling something in-between a circle and square. Several Adena earthwork complexes are known. Typically these are comprised of small circle earthworks situated in close proximity to each other.

People of the Hopewell culture built conical mounds, loaf-shaped mounds, platform mounds, and circle earthworks. Additionally, they built a number of large geometrically shaped enclosures in the form of circle, squares, octagons, ellipses, and more esoteric shapes (note 1). Geometric earthworks appear modeled after ideal shapes. Sometimes these shapes are executed with great precision on the ground. Other times, actual earthwork shapes are less than perfect – with lopsided circles, squares having corners that depart from right angles, or sides that are not quite straight. In any case, the Hopewell in particular, utilized a number of additional geometric concepts including arcs, spirals, parallel, bisected, and perpendicular lines, and isosceles, equilateral, and right triangles.

Earthworks and earthwork complexes made by the Hopewell are generally larger and more intricate than those of the Adena. Typically, Hopewell complexes are comprised of two or more geometric enclosures such as a circle and square, circle and octagon, two squares, oval and circle, or other combinations.

Two additional kinds of Moundbuilder structures are hilltop and promontory enclosures. Hilltop enclosures have walls that extend around the perimeter of a hilltop. Promontory enclosures have one or more walls extending across the narrow width of a promontory or bluff, thereby separating the bluff from the tableland. Given their topographic settings, hilltop and promontory enclosures are generally not geometrically shaped, although sometimes they include smaller geometric enclosures and mounds within their perimeters.

Size and Mensuration

Having established the location and shape for an earthwork, the next thing that needed to be decided was size. Adena and Hopewell earthworks were not made to random sizes. In every case I have looked-at, Adena-Hopewell earthworks incorporate a standard unit of length, or iteration thereof (note 2).

This standard unit of length is equal to about 1,054 feet (321.26 meters) (note 3). Because the 1,054-foot length is found in so many Hopewell earthworks (e.g., Hively and Horn 1982, 1984; Romain 1992a, 1996, 2000; Romain and Burks 2008a, 2008b, 2008c), I refer to it as the Hopewell Measurement Unit or simply, HMU. Greater multiples of the HMU that were sometimes used, especially in Hopewell earthworks include: 7 x 1 HMU, 6 x 1 HMU, and 2 x 1 HMU. Commonly occurring lesser, or sub-multiples include: 1/2 HMU (527 feet), 1/4 HMU (263.5 feet) and 1/8 HMU (131.8 feet).

We do not know what standard was used to establish the HMU or sub-multiples. It may be based on the distance between two natural features, or more likely, is based on a length related to the human body, such as an arm length (Romain 2000). Of interest is that sub-multiples of the 1,054-foot length – i.e., 1/4 HMU and 1/8 HMU are used for the diameters of Adena mounds and Adena circle earthworks. Based on this, it seems probable that the HMU had its origins in Adena culture. It is also the case that some earthworks use iterations of the HMU that are based on the relationships between the sides of right triangles. Multiple instances of this phenomenon will be found in later chapters.

Standard units of measure – such as the HMU, are cultural-level phenomena that reflect agreed-upon understandings among people. From an experiential perspective, however, the use of a standard unit of measure provides a visual reference to a known quantity and a way of establishing fixed relationships between things. In a world made-up of mostly fluid and changing relationships, measurements provide an element of stability.

Orientation and Alignments

Hopewell earthworks typically incorporate alignments to the sun or moon (Hively and Horn 1982, 1984, 2010, 2013; Romain 1992c, 1993a, 1995, 2000, 2004, 2005, 2009, 2014b) (note 4). Additionally, many Hopewell earthworks and earthwork complexes are simultaneously aligned to topographic features such as the course of rivers, river terraces, the lay of the land, or trajectory of a line of hills (e.g., Romain 2004; also see Hively and Horn 2010). Adena earthworks are less-often aligned to celestial events. Where they occur, Adena celestial alignments are to the winter or summer solstices; less common but not unknown for Adena, are lunar alignments.

Alignments can be along the major axis of a multi-component earthwork complex, or through a single earthwork component. Alignments of square earthworks can be along the major or minor axis of that shape, or diagonally across an earthwork from corner to corner. Celestial alignments vary in their accuracy. Most are accurate to within about one degree of arc.

The important thing about celestial alignments is that once incorporated into monumental architecture, a visual link is established between that structure and the cosmos. Of course the other thing is that, once linked to the cyclic movements of celestial bodies and events, the regular movement of those bodies lends a temporal dimension to human experience manifested as days, months, and years. Through celestially-aligned monuments therefore, human engagements are situated in a temporal context that provides for a meaningful sequence.

Together, place, shape, size, and time situate people in a sensual relationship to other things. For the Adena-Hopewell, earthen monuments mediated relationships between earth, sky, and water realms, non-human agents, and ancestors. The resulting relational web was invested with meaning with one of the consequences being the material remains that we see today. Again the point is that Adena-Hopewell religion was not limited to a system of beliefs; but rather, was a way of being-in-the-world comprised of imagination, memory, sensory experience, emotion, and material practices focused in part by earthworks and mounds.

Having laid out aspects of the Adena-Hopewell experience, a legitimate question that arises is: Why, should anyone care about any of this? Of what possible relevance in today's modern world are the religious motivations of a long-dead people?

My answer is that, through such inquiries we are reminded that religion is not a static component of cultures, prehistoric, or otherwise. It is not an epiphenomenon or incidental effect. Rather, religious belief can and does, shape history. Today, we are witness to the efforts of millions of people actively seeking to change the world based on religious beliefs. Religion matters today and religion mattered in the past. Religion has the power to alter the course of civilizations; and as a driving force, it can be as potent as any set of environmental, political, or economic factors. Case studies such as this one are important therefore, if, as archaeologists, our goal is to document the history of our species and understand who we are.

There are several unique things about this book. First the reader will notice that heavy emphasis is placed on illustrations. This relates to the fact that archaeoastronomy, landscape archaeology, and architectural design – all of which comprise a central focus of this volume, are visually oriented subjects. Our understanding is mostly informed by visual perceptions. Thus, hundreds of illustrations to include three-dimensional images, photographs, maps, and diagrams are included, while narrative text is kept to a minimum.

Further, the reader will find that the graphics rely heavily on a technology known as LiDAR. Details are provided in the next chapter. Briefly, however, LiDAR uses near-infrared laser beams generated from an aircraft-mounted laser to scan the ground below. The result is an accurate representation of the surface topography without interference from buildings, trees, or vegetation. The application of LiDAR to the study of prehistoric architecture offers exciting new possibilities. In this volume, more than 100 earthworks and mounds are assessed using LiDAR imagery supplemented by other technologies. To my knowledge, this is the first major study of its kind to use LiDAR imagery so extensively.

The danger of course is that LiDAR, but also GIS (Geographic Information System), GPS (Global Positioning System), aerial and satellite photography, and geophysical techniques can be abused, with the result being the identification of previously undocumented sites and subsequent looting or vandalism; which brings me to another point. Please keep in mind that the sites documented in this book include burial sites and earthworks that are still considered sacred by living Native Americans. When visiting these places, kindly respect the sites, their creators and descendants, by doing no further damage. I should also mention that not every site in Ohio is documented in this book. Many are not. Basically the assessments herein are limited to those for which I have good LiDAR data.

Of course it is the case that a thorough site assessment needs to go beyond a superficial analysis of photographs, maps, and other imagery and should include on-the-ground evaluation – for it is through the physicality of embodied experience that we gain a sense of place and in the case of prehistoric sites, come to appreciate the features and relationships that may have affected experiences in the past (Tilley 2008). Thus the site assessments that follow are informed by personal visits, including in most cases, multiple visits at different times of the year under varying weather conditions.

On another matter, the reader will notice I sometimes collapse the cultural designations *Adena* and *Hopewell* into the more straightforward designation, *Adena-Hopewell*. Where the cultural identity of a specific earthwork or mound is known with certainty, I note that. Often times, however, that information is not known. So too, it is often difficult or even impossible, to distinguish cultural affiliation based solely on earthwork morphology. This is especially true for smaller earthworks and mounds that have not been excavated. Even in cases where excavations have been made, cultural identity is not always certain. So too, radiocarbon dating of earthworks can be problematic – mostly due to issues related to the dating of mound fill, which may include organic materials pre-dating earthwork construction. Other times an earthwork complex, like Marietta for example, will incorporate older mounds and earthworks, thus commingling Adena and Hopewell.

Sometimes a reasonable guess can be made as to cultural affiliation based on location. Many Adena mounds, for example, are situated on ridges or on the edges of promontories. At the same time, however, in the valleys below, both Adena and Hopewell mounds and earthworks are sometimes found together.

In an effort to address these and related issues, the approach I have taken is to consider Adena-Hopewell earthwork expression as part of a larger web of relationships linked through the design principles discussed. Thus what matters for the present assessment is not so much whether a site is Adena or Hopewell, but rather, how various design principles are expressed in each earthwork.

A related matter is that what we see on the ground today reflects the end stage of earthwork construction. Further, we know that earthworks were sometimes added to, or otherwise modified over generations. Thus the question becomes: Is what we are looking at the result of a master plan envisioned by the designer of a particular earthwork, or is what we are looking at the cumulative result of episodic construction that presents the illusion of a planned design?

For small earthworks and non-accretional mounds it is reasonable to think that many were built pursuant to a master plan or mental template. So too, for large earthwork complexes, a grand design or intentional final form seems indicated by the seamless integration of multiple design shapes using celestial and topographic alignments and standard units of length. At the same time, however, a master plan would not necessarily have needed to include every earthwork feature, or component, as long as additional constructions by later builders were consistent with the basic design principles first employed. Whether built all at once pursuant to a grand scheme, or built in a modular fashion over time, the end results would appear the same. Imagine, for example, a structure built from Lego blocks. Given that the Lego blocks are comprised of standard shapes and sizes, it is impossible to determine when looking at the end product if the structure was built all at once, or in a series of episodes over days, months, or even generations.

In any case, there is accumulating evidence that some earthwork complexes such as the Newark complex, for example, were constructed very rapidly (Lepper 2010). Increasingly, similar findings are being reported for earthwork sites elsewhere in the

Eastern Woodlands including Mound A at Poverty Point (Ortmann and Kidder 2013) and Monks Mound at Cahokia (Schilling 2013). Very likely, many earthworks reflect a combination of rapid building following a master plan combined with additional features added-on in a modular fashion, following commonly-held design principles transmitted through time.

With reference to this last point, the reader will notice that my theoretical perspective tacks between the notion of shared, regional-level design principles incorporated in the earthworks and an experiential approach that considers actor-level sensory perceptions and experiences at specific sites. In this, I posit that cultural-level design rules and constraints were operationalized in earthwork designs. Within the parameters of these rules there was considerable freedom of expression resulting in local variation. But there was a proper way of doing things dictated by tradition. For example, the size of an earthwork could be whatever the designer wanted - but it needed to incorporate either the HMU or an iteration thereof. Cultural-level protocols did not allow for the use of arbitrary or idiosyncratic units of length. So too, earthworks could be aligned to sky events in different ways - e.g., through a major or minor axis, or along a diagonal; but alignments for the most part needed to be to the sun, moon, or special constellation and not some arbitrary celestial phenomenon. Quite simply, there were cultural-level design protocols that guided individual experience and account for similarities among earthworks.

At the same time, the phenomenological tack taken here is focused on local-scale appreciation of the sensory experience of a particular place or earthwork as interpreted by the mind-body. Contributing to this are the methods of archaeoastronomy which, as Lionel Sims (2009) points out, bring an element of quantification to the phenomenological perspective. The important point is that, through the earthworks, cultural-level design rules and local-scale sensuous experience were simultaneously engaged and enmeshed in a relational web created and mediated by human minds. Moreover, this relational web was ever-changing as its constituent parts to include human and non-human entities morphed in various ways.

The book is arranged in the following way. Chapter 2 provides an introduction to the Adena and Hopewell cultures, a brief discussion of LiDAR, and explanations for several key archaeoastronomical concepts. The heart of the book is found in chapters 3 – 9. In these chapters, relationships of the Moundbuilder earthworks to the landscape are considered, along with assessments of how design principles related to shape, size, and orientation are incorporated in each earthwork.

For organizational continuity the earthworks are organized according to major river basins. Figure 1.1 shows the relevant areas. Chapter 3 includes sites located in the Muskingum River Basin, Chapter 4 presents sites found in the Hocking River Basin, and so on, through Chapter 9. Chapter 10 provides a few summary points and concluding remarks.

Figure 1.1. Map of Ohio Showing Distribution of Earthworks

Above: William C. Mills (1914:Pl. XI) map of Ohio showing documented earthworks (annotation added). In addition to the above map, Mills's, Archaeological Atlas of Ohio includes detailed county maps. In total, the maps show 5,396 sites, some of which include multiple earthworks. The Atlas was the result of 20 years of survey work and field verification by staff of the Ohio State Museum (Dancey 1984).

CHAPTER 2
BACKGROUND AND METHODS

Important to any discussion is that we agree on how certain terms are used. Crucial to any scientific explanation is that we document how we arrived at our findings so others can replicate or assess our results. The purpose of this chapter is to explain several terms and methods used throughout the book. We begin with a brief explanation of what is meant by *Adena* and *Hopewell*.

Adena

There are taxonomic difficulties with the terms *Adena* and *Hopewell* (Applegate and Mainfort 2005; Brown 1992; Otto and Redmond 2005; Swartz 1971). No one knows what names Moundbuilder peoples call themselves. That information is lost to time. Both *Adena* and *Hopewell* are names that were applied by twentieth century archaeologists to describe groups of people who shared similarities in earthwork and artifact morphologies. However, people of the so-called Adena and Hopewell cultures were never integrated into distinctive polities. Rather, as archaeologist Orrin Shane (1971:142) explained years ago: "you really cannot deal with cultures in the sense of an Adena [or Hopewell] culture....What you have to look at are local sequences that have developed in ecologically meaningful areas, and the development of local cultures."

Archaeologist David Hurst Thomas (1993:81-82) offers a different perspective: "Neither a particular culture nor a political power, Hopewell became the first pan-Indian religion, stretching from Mississippi to Minnesota, from Missouri to West Virginia; for the first time, people sharing neither language nor culture were drawn together by a set of beliefs and symbols." McCord and Cochran (2005:359) opine that, at least for east-central Indiana, Adena and Hopewell "...represent different parts of a single ceremonial system." Archaeologist Mark Seeman (1995:122) considers Hopewell to be an "ideological system."

Building upon Seig and Hollinger (2005), my view is that Adena and Hopewell are best thought-of as archaeological horizons – where *horizon* refers to "...a unit of time bracketing the rapid spread, often over an enormous area, of a technical procedure, set of artifacts, art style, *religious cult*, or other archaeological phenomenon that endures for only a short time" (emphasis added) (Gibbon 1998:374). In the case of Adena and Hopewell, what I believe spread across the Eastern Woodlands were socio-religious concepts and practices bundled (*sensu* Pauketat 2013a) as religious movements. To greater or lesser extents, parts of these religious movements were adopted by social groups who lived in dispersed ecological settings associated with specific river valleys. As these groups were spread-out across the Eastern Woodlands, it is likely that many spoke different languages, some of which were mutually unintelligible (Seeman 1995:124). What did get communicated, however, were basic ideas, perhaps as communicated by later prophets such as Black Elk, Handsome Lake, Sitting Bull, Wovoka, Tecumseh, and others in historic times.

From our top-down perspective, the result appears to be a culture or at least a worldview shared by people across a wide geographical area, expressed in similar representative and flamboyant ways. But Adena and Hopewell were not monolithic polities or cultures. Rather, I would opine that what we call Adena and Hopewell are simply archaeological horizons that reflect the material outcomes of religious movements. As with all religious movements, some last only a few years, others – like Adena and Hopewell lasted hundreds of years.

Some archaeologists believe that Adena extends as far back as 800 B.C. Others posit a beginning date of about 500 B.C. Most agree that the end of Adena occurred at about A.D. 100. Adena is mostly situated across southern Ohio, southeastern Indiana, northern and central Kentucky, parts of West Virginia and southwestern Pennsylvania.

Hopewell flourished from about A.D 1 to A.D. 400. There are several regional expressions of Hopewell across the Eastern Woodlands. The phenomenon that archaeologists consider 'classic Hopewell', however, is concentrated in south-central Ohio. Based on the parameters just noted, it is apparent that there is some geographic and temporal overlap between Adena and Hopewell.

As to Adena, archaeologist Mark Seeman (1986:566) provides a useful list of diagnostic attributes:

> "Adena" ...is that rather amorphous Early Woodland cultural manifestation in the Ohio Valley identified with burial mound construction, elaborate mortuary ceremony, and such specific artifact classes as tubular pipes, semi-keeled gorgets, quadriconcave gorgets, expanded-center gorgets, Adena ovate-based projectile points, hematite celts, copper bracelets, and Adena Plain, Fayette Thick, and Montgomery Incised ceramics...."

To this list we can add a distinctive style of stone tablets known as Adena tablets.

Adena people were hunters and gatherers. Their meat sources included deer, turkey, rabbit, waterfowl, and fish. They also gathered wild fruits, berries, and nuts such as acorns, walnuts, and hickory nuts. Among the plants they utilized were tobacco, squash, goosefoot, knotweed, maygrass, little barley, sumpweed, and sunflower. Few Adena habitation sites are known. But it appears that Adena people lived in small, seasonally occupied sites, although a few, large-scale specialized workshop sites are known (e.g., the Peter Village near Lexington, Kentucky – see Clay 1986).

More visible in the archaeological record are Adena earthworks. As Ohio Archaeologist editor Robert Converse (2003:176-177) notes, "...whether or not Adena borrowed or originated the construction of burial mounds, it certainly was the first culture to extensively practice over a wide area the erection of artificial mounds for the burial of their dead."

Adena mounds are generally conical in shape and range from barely noticeable rises on the landscape, to enormous structures reaching roughly 70 feet in height and more than 200 feet across. Some mounds include burials, others do not. Where burials are found in Adena mounds, it is often the case that more than one individual is interred. In the simplest case there is a primary burial situated on or below the floor of the mound. Often times this burial is found in a log, bark-lined, or stone tomb. The primary burial is then covered-over by a low mound, with additional burials and layers of earth added over time resulting in an increasingly larger mound.

Some Adena mounds cover the remains of circular wooden structures. It may be that these structures were used for preparing the dead for burial. Whatever their purpose, after some period of use the wooden structures were dismantled and/or burned, with a mound built over the decommissioned structure.

Another kind of earthwork built by people of the Adena culture were so-called sacred circles. These structures range from more than two hundred feet in diameter to much smaller. Most are simple earthen circles with an interior ditch and single opening through their perimeter wall. Some have a burial mound in the center of the circle. Often times, Adena circle earthworks are found in groups.

Hopewell

In many ways, Hopewell appears to be an elaboration of Adena; but there are noticeable differences. Hopewell earthworks are generally larger and more complicated than Adena. So too, Hopewell submound mortuary structures are different from Adena (Clay 1986). Hopewell artifacts are of a greater variety and complexity than Adena. Distinguishing artifact classes for Hopewell include: rocker-stamped pottery, platform smoking pipes, flint bladelets, cymbal-shaped ear spools, large ceremonial obsidian points, mica cutouts, and panpipes.

Within Hopewell there are differences between groups that lived in different river valleys, particularly with regard to burial customs. At the same time, what seems to unite *Hopewell* is that people of the so-called Hopewell culture shared a worldview that was expressed in a similar and uniquely flamboyant way in their earthworks, but also in smaller expressions such as smoking pipes, copper breastplate designs, headdresses, projectile points, bladelets, earspools, pottery, and other crafted items.

Many of the objects made by Hopewell people were crafted from exotic or hard-to-find substances including: silver, gold, meteoric iron, copper, obsidian, mica, galena, garnet, crystal quartz, alligator, shark, and bear teeth, freshwater pearls, and marine shell. Some of these materials were obtained from distant sources (Seeman 1979). Mica, for example, came from the Appalachian Mountains. Obsidian came from the Far West; while marine shell and shark teeth came from the Atlantic or Gulf coasts. Likewise, alligator and grizzly bear teeth came from distant sources, as did quantities of exotic flints.

Like the Adena, Hopewell people were hunters, gatherers, fishers, and horticulturalists. Their primary source of protein came from white-tailed deer. Other animals and birds that were utilized included elk, bear, beaver, rabbit, raccoon, geese, duck, and turkey. Fish and freshwater mussels were also part of the Hopewell diet. Fruits and berries were gathered to include honey locust, raspberry, elderberry, and strawberry. Oak trees, which produce massive quantities of acorns are abundant in Ohio and acorns were eaten by Hopewell people, as well as hickory, black walnut, and hazelnut.

Like the Adena, seed plants utilized by the Hopewell included goosefoot, marshelder, sunflower, knotweed, maygrass and little barley. Squash, gourds, and tobacco were also grown. Compared to Adena, horticultural efforts intensified during Hopewell times. There is some evidence that small amounts of maize were grown (Riley, Edging, and Rossen 1990).

Hopewell people lived in small hamlets, typically comprised of two or three dwellings; although a few settlements were bigger and can be considered small villages (Griffin 1996). Settlements were located along major river valleys, usually near a geometric earthwork (Converse 2003:248-249; Dancey and Pacheco 1997; Prufer 1965).

It is difficult to know the details of Adena and Hopewell social and political life. At the domestic level, nuclear and extended families were the fundamental units in both cultures. To build very large earthworks – like the Adena

Miamisburg and Grave Creek mounds, or the large geometrically shaped earthworks of the Hopewell, some form of social and political organization was needed. Labor forces had to be recruited and assembled at certain times and locations. Specific tasks needed to be accomplished, whether carrying basket loads of earth, engaging in survey work, cooking food, or bringing water. To coordinate these activities it is likely that someone was in charge, at least for each particular local building episode. By implication, it seems likely that there were local leaders among the Adena and Hopewell who coordinated and directed mound building activities.

In the case of the Hopewell, where earthwork designs could be quite complicated, social and political leaders may have been advised by specialists. It is not likely that everyone had the requisite knowledge to build a large, complicated earthwork. The knowledge needed to correctly align an earthwork, for example, would have required exacting information about the movements of the sun and moon relative to the surrounding landscape over time. To build a large earthwork would also have required intimate knowledge of the earth to include soil characteristics, drainage, slope, and flooding potential, as well as competence in the equivalent of geometry, mensuration, and ground surveying. People having this kind of knowledge combined with ritual knowledge were likely specially trained. Other Hopewell ritual specialists probably included healers as well as psychopomp shamans with expertise in conducting complex mortuary rituals.

Hunters and warriors have a long tradition of high social standing among Native American peoples. From the exceptional craftsmanship exhibited by Adena and Hopewell projectile points, as well as impressive collections of predator teeth found in Hopewell contexts, there is reason to believe that hunters and warriors were held in high regard.

Craftsmen of various specialties would have been important. Among such specialists were expert knappers capable not only in working with flint, but also obsidian and quartz crystal. Others possessed the ability to shape raw copper into a variety of intricate designs; while yet others were skilled in carving stone into smoking pipes of exquisite detail. Hopewell pottery and woven textiles suggest specialist artisans and craft-persons.

Moundbuilder Cosmology

In Chapter 1 I suggested that Adena-Hopewell religion can be thought-of as a relational web comprised of core beliefs and lived experience. Among the beliefs found in all religions are those relating to cosmology, where cosmology refers to how the cosmos is structured and its origin. Often times cosmological beliefs are incorporated in monumental architecture. Indeed, the predilection to incorporate cosmological beliefs into monumental architecture is found cross-culturally and includes such examples as Stonehenge, the Egyptian pyramids, ziggurats of Mesopotamia, and Angkor Wat; as well as Cahokia, Chaco Canyon, and the Inca, Mayan, and Olmec centers in the New World. From this, we can expect that cosmological beliefs are likewise incorporated in Adena-Hopewell monumental earthworks. Presumably, through consideration of their design principles we can begin to identify some of these cosmological beliefs.

From accumulated evidence (e.g., Buikstra, Charles and Rakita 1998; Byers 1996; Carr and Case 2005b; DeBoer 1997; Hall 1979; Penny 1985; Romain 2000, 2004, 2009) it appears that Adena and Hopewell cosmological beliefs were similar in many respects to those of historically known Eastern Woodlands peoples. Undoubtedly the details of some concepts changed over time and certainly it is the case that basic concepts were expressed in a variety of ways among different people. Based on a wide range of sources, however, (e.g., Emerson 1989; Hudson 1976; Lankford 2007d; Penny 1982; Reilly 2004; Romain 2009), several of the most basic concepts found across the Eastern Woodlands can be described as follows.

The cosmos is divided vertically into earth, sky, and water realms. The sky realm is associated with the Upperworld. It is the realm of the sun and Thunderbird beings. The Lowerworld is associated with water and is the domain of the Great Serpent, or Underwater Panther. Balanced between the two realms is This World, where people live. Sometimes, This World is described as a flat disk, turtle, or island, floating on the primordial waters. Often the Upperworld and Lowerworld are thought to have multiple levels; but sometimes the cosmic structure is simply reduced to Earth and Sky realms. Typically, an *axis mundi* connects the three realms, where an *axis mundi* is a vertical trajectory that extends in a perpendicular fashion to the flat earth thereby connecting the cosmic levels. The *axis mundi* concept can be expressed by a special mountain, a particular tree or upright pole, column of smoke, earthen mound, or other ways.

In the horizontal plane, the cosmos is generally believed to be divided into quarters, with world quarters delineated by the cardinal directions, solstice directions, or lunar standstill directions. World quarters usually have their own set of associated spirit beings, animals, colors, or other features. In the ways just mentioned, the deeper notion that the cosmos is made-up of complementary dualities is expressed vertically and horizontally and in multiple ways.

Other cosmological beliefs typical to the Eastern Woodlands include the idea of center at the intersection of the *axis mundi* and four quarters, a cyclic view of time, the existence of hidden dimensions or realms not normally visible, the idea that phenomena are animated by a life force or soul, the capability of people, animals, and spirits to change their physical appearance through metamorphosis and transformation, and belief that the spirit world can be experientially engaged, usually through altered states of consciousness.

An important set of beliefs found across the Eastern Woodlands and apparently having great time depth concerns the fate of the soul after death. There are many variations but essentially these beliefs can be summarized as follows. People have more than one soul (note 1). A useful and simple distinction is made by Lankford (2007a:213) - i.e., one soul is the life-soul, the other is the free-soul. After death, the life-soul remains with the corpse and is associated with the bones of the deceased. It is this soul that is most often feared as it can cause bad fortune, illness, and death. For this reason, different kinds of strategies or ghost-thwarting devices and symbols are often employed to ensure that this soul stays in its grave (Bacon 1993; Carr and Case 2005b; Hall 1976; Romain 2009:120-121).

It is the free-soul that leaves the body after death and journeys to the Land of the Dead, where it is reunited with the Ancestors. The free-soul travels to the Land of the Dead by following a path known to different tribes as the Path of Souls, Pathway of Departed Spirits, Spirit's Path, Spirit Way, Spirit's Road, or Ghost's Road (Miller 1997). In each instance and by whatever name, this Path is often identified by Native people as the Milky Way band of stars.

Along the way the soul encounters challenges, tests, or obstacles (Lankford 2007c). One common challenge is in the form of a somewhat unpleasant, adversarial and/or judgmental entity that blocks the way forward. Sometimes this entity is a dog, sometimes an old woman or man, sometimes an eagle, or other creature. Often this being is said to be located at the place where the Milky Way splits into two paths. Casual observation of the Milky Way easily locates the place. Known as the Great Rift, the star Deneb marks the beginning of the Rift (Lankford 2007c). If the soul successfully meets the challenge or is deemed worthy by the judging entity, it is allowed to move forward on the path to the Land of the Dead. If not, the soul is sent along the other fork in the Milky Way leading to oblivion, or other bad outcomes.

The Land of the Dead is located at the end of the Milky Way Path. In many versions of the myth, the Milky Way terminus and the Land of the Dead are guarded by a Great Serpent (Lankford 2007c). This Serpent rules the Lowerworld. Generally, the Great Serpent is associated with darkness, night, water, death, plant and animal plant fecundity, and the female aspect of a dualistic cosmos. Although it is a Lowerworld entity, the Great Serpent has flight capabilities and is sometimes seen in the night sky as a fire-dragon, meteor, or star constellation. Among the Pawnee the star constellation Scorpius is a cognate for the Great Serpent (Lankford 2007b). Most Eastern Woodlands tribes explain that the Lowerworld Great Serpent and Upperworld Sun or Thunderbirds are engaged in an eternal struggle with each other.

For Adena and Hopewell, their core beliefs were likely transmitted across the generations by oral teachings, mythologies, narratives, legends, songs, dances; ritual behaviors to include mortuary activities, votive deposits, and sacrifice; representational forms to include pottery and textile making, figurines, masks, and rock art; and important to the discussion here, through earthwork design and location. Where the cosmological model of the universe just outlined becomes relevant is in the understanding that, if people of the Adena and Hopewell cultures held similar beliefs about the cosmos as those expressed by their descendants, then it is likely that Adena-Hopewell ritual practices were concerned with balance and maintaining a harmonious relationship with the cosmos. As I explain in the chapters that follow, Adena-Hopewell earthworks reflect a serious interest in maintaining balance and harmony with the cosmos using geometry, alignments, mensuration, and locational design principles. Additionally, however, many of the most important Adena-Hopewell earthworks appear to have been used to bring into the physical world, expressions of their mythologies, including the journey of the soul to the Land of the Dead.

Methods

If we wish to know how close the Moundbuilder earthworks are to ideal geometric shapes, what their dimensions are, or how closely they are oriented to celestial events or terrestrial features, then we need accurate maps and survey data. Unfortunately, even though people of the Adena and Hopewell cultures built thousands of mounds and a significant number of geometrically shaped enclosures, few intact structures remain. Most have been severely damaged, excavated, or totally obliterated by urban development, gravel mining, and farming operations. In those instances where earthwork remnants remain, it is often the case that they are not easily recognizable at ground level. In such cases, however, geophysical surveys, GPS ground surveys, LiDAR imaging, or some combination of these techniques can sometimes be used to retrieve useful information. Of the techniques just-mentioned, LiDAR imaging is especially useful.

LiDAR

LiDAR is an acronym for *Light Detection and Ranging.* Airborne LiDAR scans the earth from an aircraft using near-infrared laser light. As the aircraft flies over the terrain, thousands of laser pulses per second are beamed toward the earth's surface. These laser pulses strike buildings, trees, vegetation, and the ground. The laser light is reflected back to the aircraft as a series of returns. First returns are reflected off the highest object struck, such as a tree limb. Second returns are reflected off the next highest object, such a rooftop. Other returns are reflected off lower objects to include the ground. Most LiDAR systems identify between four and six returns per pulse. Based on how long it takes for a return to bounce back to the aircraft, the distance to an individual target, or coordinate point can be determined by a computer on-board the aircraft. Also on-board the aircraft is an inertial navigation unit and GPS (Global Positioning System) instrumentation. The inertial navigation unit monitors the orientation of the aircraft relative to yaw, pitch, and roll. The airborne GPS unit in combination with one or more ground-based differential GPS stations provides accurate location data for the aircraft and where the laser pulses strike. The on-board computer processes the laser, GPS, and inertial navigation data with the result being a set of three-dimensional coordinates for any given target point. Vertical precision or 'resolution' for the LiDAR imagery in this book is plus or minus 6 inches. A typical LiDAR model is generated from millions of coordinate points, the result being an image that shows in exquisite detail, features not otherwise visible at ground level, or in aerial photographs. Depending on the requirements of the analysis, returns can be filtered so that images include trees, buildings, or only the bare earth surface.

For archaeology, the value of LiDAR is that it provides an accurate, three-dimensional view of the landscape (Romain and Burks 2008a, 2008b, 2008c). Moreover, many LiDAR software programs enable vertical exaggeration of ground features, color-coding of elevations, and the ability to vary lighting conditions. As I demonstrate in later chapters, these capabilities are useful for bringing-out earthwork walls and other features.

One of the most impressive features of LiDAR is its ability to penetrate ground cover such as trees and vegetation. This is accomplished in two different ways. First, some of the beamed laser light photons will penetrate some leaves if they are not too thick and/or dense. Photons that reach the ground are then reflected back to the aircraft. Second, when the aircraft flies over an area, overlapping swaths are flown so that the laser pulses have multiple opportunities to strike the ground from different angles. Given the hundreds of thousands, or millions of pulses that are sent out by the laser, some will find un-impeded trajectories to the ground surface.

LiDAR offers a number of analytical capabilities. One of the most useful is the ability to accurately measure linear distances. This means that the dimensions of earthworks can be accurately determined. To measure distance, for example, the mensuration tool built into many LiDAR imaging software programs is used to draw a line between two points of interest. The software identifies the LiDAR coordinate points closest to the designated start and end locations and locks onto those points. The program then determines the horizontal distance between the start and end points.

Other capabilities that some LiDAR software programs offer include the ability to generate contour maps, profiles of the terrain, determination of the highest point in an area, and line-of-sight and viewshed data.

Perhaps the most dramatic feature of LiDAR, however, is that it offers the ability to manipulate images in virtual three-dimensional space. Once a LiDAR image is created, that image, or model can be rotated in any direction and the apparent altitude of the viewer can be changed. In the case of the Moundbuilder earthworks, this allows us to view the earthworks from entirely new perspectives.

Of course, there are times when, no matter how we manipulate the imagery, LiDAR will not reveal the presence of an earthwork that has been totally obliterated. LiDAR cannot reconstruct what once was. In cases where earthworks have been completely plowed-down, built-over, or dug out, the coordinate point data needed for LiDAR processing are lost.

In any case, LiDAR is visually oriented technology. Once a few basic concepts are understood, interpretation rests with the viewer. For black and white LiDAR images, shades of gray show differences in elevation. Also provided for many of the images in this book are distance scales. However, since many of the three-dimensional images rely on perspectives other than top-down, these scales are accurate only for the foreground area where they appear. As mentioned, during the image-making process certain aspects of the image can be changed such as perspective, lighting conditions, and vertical height. In most cases I have used low angle lighting to help bring-out features. So too, all images have been vertically exaggerated to better illustrate earthwork details.

True North and Grid North

One of the things to be aware of when using LiDAR is that data are typically referenced to State Plane Coordinate systems. For archaeoastronomic analyses the problem is that, State Plane Coordinate systems are tied to grid north. Grid north is not the same as true north. True north (or astronomic north) (note 2) is basically the point in the sky that the circumpolar stars appear to rotate around. Grid north is north as indicated by a particular map projection. The difference between the true north and grid north lies in the fact that the earth is spherical in shape. To fit a three-dimensional world onto a two-dimensional piece of paper, or computer screen, map projections are used. Depending upon the location, the difference between grid north and true north can be more than one degree in horizontal azimuth. Since archaeoastronomic analyses seek to represent what was seen from a particular place on the surface of the earth, we need to know where true north is and likewise, our computed sightlines need to be referenced to true north.

In Ohio, the Lambert Conformal Conic Projection is the basis for the Ohio State Plane Coordinate System and the LiDAR data provided by the State of Ohio through the Ohio Geographically Referenced Information Program is therefore referenced to State Plane, or grid north. In the Lambert Conformal System the difference between true north and grid north is called the theta angle (θ). The θ angle varies depending on longitude and other factors. Specifically, the theta angle increases the further one is east or west of the central meridian for the state. Depending upon location, a correction for the theta angle will likely need to be applied to LiDAR azimuths to establish true north. At Newark and Fredericktown, for example since those locations are close to the central meridian (which is 82° 30') the correction is negligible. At Marietta, however, the correction is 0°.66.

The following formula is used to convert grid north to true north (adapted from United States Department of Commerce 1978:3):

true north = grid azimuth + θ - λ

where, θ is the mapping angle for the longitude of the station, and λ is the arc-to-chord, or "second-term" correction. The value for theta (θ) is found by reference to the appropriate table (United States Department of Commerce 1978:table II; also see Dracup 1974). Theta correction values can be plus or minus, depending upon whether the site is east or west of the central meridian, respectively. Further, the signs for theta are reversed and applied accordingly when converting the other way - i.e., from true north to LiDAR, or grid north. In the chapters that follow, all necessary theta corrections have been applied.

Calculated alignments presented here are likely accurate within a range of error as potentially influenced by several factors including: differences in actual tree height from that presumed here; differences in horizon elevation due to unknown extent of forest clearing; refraction values different from those used herein resulting from non-standard atmospheric conditions at time of observation, and unique locational factors such as whether the original observer was standing at the base of a mound or at its top. Considered all these factors together, posited alignments are likely accurate to within plus or minus one degree, except where otherwise noted.

Archaeoastronomic Concepts

In the assessments that follow, we are interested in determining how close the trajectory of an earthwork feature or sightline intersected with a celestial body two thousand years ago. To help explain how that assessment is done, it is useful to explain some basic archaeoastronomic terms and methods.

We begin with the term, *sightline*. Sightline refers to the line of sight for an observer as viewed from a particular location. A sightline can extend along a wall, from corner to corner across an earthwork, from gateway opening to gateway opening, between two mounds, or along the major or minor axes of an earthwork. A given sightline will have an azimuthal value or direction as measured from true north.

Calculating Solar and Lunar Azimuths

Knowing the orientation of a sightline relative to true north is only part of the information we need to determine if an earthwork feature is aligned to a celestial event. We also need to know the *azimuth* of the celestial body we are interested in. *Azimuth* simply means the horizontal angle measured clockwise in degrees from north. Thus east has an azimuth of 90 degrees, south has an azimuth of 180 degrees, west is 270 degrees, and so on. Two azimuths are therefore involved in our determinations – i.e., the azimuth of the sightline for the earthwork or feature we are assessing and the azimuth of the celestial body we are interested in. The closer the two azimuths are to each other, the closer we can say the alignment is.

To determine the azimuth of a celestial body, spherical trigonometry is used. The situation is complicated, however, by the fact that the sun, moon, stars, and planets are in a constant state of motion. Moreover, their relative positions in the sky have changed over time due to changes in the obliquity of the ecliptic and in the case of the stars, precession. As a result, our calculations need to use position data for the sun, moon, stars, and planets appropriate to the time and location being studied. Data needed to make the necessary calculations include site latitude, declination of the celestial body of interest for the approximate year, and horizon elevation.

The declination of a celestial body can be thought-of as that body's latitude as projected into the sky. It is measured in degrees north and south of the celestial equator. For example, if a star has a declination of +40°, it means that if the star were brought down to earth it would fall directly on the latitude of 40° North. The declination of the sun changes throughout the year and on a larger scale, has changed over the past thousands of years. Currently, the sun's declination ranges from +23°.4 at the summer solstice to -23°.4 at the winter solstice. The moon's declination also changes but in a more complicated way. The range of the moon's declination is often described in terms of major and minor standstills. Archaeoastronomer Clive Ruggles (2005:166-167) explains the concept:

> "The rising point of the moon moves up and down the eastern horizon (and similarly the setting point on the western horizon) between limits that are reached once every month. These limits themselves vary over a cycle of 18.6 years, so that in every nineteenth year the range of rising (and setting) positions of the moon is at its widest, while nine and a half years later it is at its narrowest. The outer and inner 'limits of the limits,' where the moon rises and sets at times that are (rather misleadingly) referred to as its major and minor 'standstills,' form eight lunar horizon targets...."

The exact azimuths where these rise and set events will be visible on the horizon are a function of date, latitude, horizon elevation, and various correction factors to include refraction and parallax.

The term *horizon elevation* refers to the vertical angle of the distant horizon relative to a flat plane as observed from a particular point. It is measured in degrees. So, for example, looking out across an ocean, the horizon elevation is 0° (without dip). Looking toward a distant mountain range, the horizon elevation might be 2° or 3° – depending on the height of the mountains and distance of the observer from the mountains.

The sun, moon, and most stars rise and set at an angle relative to the horizon. As a result, the apparent rising and setting azimuths of these bodies will change as a function of horizon elevation. Looking west for example, if the horizon elevation is 0° - i.e., perfectly level, then the azimuth of the summer solstice sunset might be 300°. If, however, the horizon elevation is 2° then the summer solstice sunset might be 298°.

Further, the apparent position of a celestial body, to include its rising and setting positions is affected by refraction. Refraction is caused by the bending of light through the atmosphere. A correction for refraction must therefore be subtracted from a celestial body's apparent altitude. The appropriate correction is found by reference to standard tables (e.g., Wood 1978:fig. 4.5) (note 3).

In the case of the moon, an additional correction for parallax needs to be applied. The value for this correction is +0°.95 (note 4).

Lastly, since our initial calculations are made with reference to the center of a celestial body, we need to apply a correction if we wish to use either the tangency of the upper or lower limb of the sun or moon with the horizon as the moment of rise or set. This correction is -0°.25 for upper limb tangency and +0°.25 for lower limb tangency. For the analyses here, lower limb tangency is nearly always used.

In summary, for the present study, the azimuths for solstice and lunar standstill events were calculated according to the following formula (McCormac 1983:345):

$$\cos A = \frac{\sin \delta - \sin \varphi \sin h}{\cos \varphi \cos h}$$

where, A stands for the azimuth, h is the horizon elevation, φ (phi) represents the latitude of the site, and δ (delta) is the declination of the sun or moon.

For the sun, the following declination values for A.D. 100 were used: summer solstice sun = +23°.67; winter solstice sun = -23°.67 (note 5). For the moon, the following declination values for A.D. 100 were used: north maximum =

+28°.83; north minimum = +18°.53; south maximum = -28°.83; south minimum = -18°.53.

As indicated, a date of A.D. 100 was used for most calculations, although this date should not be interpreted as the date of construction for any particular earthwork. Since Adena-Hopewell earthwork radiocarbon dates range from about 500 B.C. to A.D. 400, with most of the analyses herein concerned with Hopewell, the A.D. 100 date is a convenient approximate midpoint to use for calculations. The obliquity of the ecliptic has been decreasing at approximately 0°.013 per century. Taking this rate of change into account, we can establish that, if all other values are held the same, the difference in the sun's setting azimuth between, for example, 100 B.C. and A.D. 400 was only 0°.08. This difference is so small that it would have been virtually undetectable by the human eye. For all practical purposes, the sunset in A.D. 400 occurred at the same place on the horizon as it did in 100 B.C., or even 500 B.C.

Horizon elevations were determined by observation using either a handheld Suunto clinometer, or total station; by analysis of LiDAR data, or digitized United States Geological Survey (USGS) 7.5-minute series maps. Where USGS maps or LiDAR imagery were used, 80 feet were added to the foresight location to account for the prehistoric presence of trees (note 6). Where map data or LiDAR imagery were used, horizon elevations were determined according to the formula:

$$\tan z = VD/HD$$

where, VD means vertical distance, HD means horizontal distance, and

$$VD = C - A$$

where, C is the foresight location (highest elevation), and A is the backsight location (lowest elevation).

These formulae provide an initial value for the apparent horizon. The earlier mentioned corrections for refraction, parallax, and limb tangency are applied to this apparent horizon value, with the result being a corrected horizon elevation value used for calculation purposes.

Lastly, the reader will note that I have limited my astronomical analyses mostly to the sun and moon; and even in those instances, I have concentrated on solstice and lunar standstill alignments. The Serpent Mound, Mound City, and Great Hopewell Road alignments to stellar phenomena are exceptions. The problem with stellar alignments is that because there are so many stars in the sky, some feature of an earthwork will undoubtedly line-up with some star or constellation. In such cases the question of intentionality becomes an issue. It is, for example, an easy matter to plot every possible combination of alignments incorporated in an earthwork and then find something in that mix that purportedly accounts for the earthwork's orientation. Unfortunately, that 'shotgun' approach is sometimes taken by some researchers. The difficulty is that not every conceivable celestial alignment was necessarily recognized or intentionally built into a particular earthwork. To the extent possible I have tried to avoid this difficulty and minimize the possibility of reading too much into posited alignments by taking a minimalist approach. In what follows, not every conceivable alignment is plotted. In general, what I look for are patterns in the data - i.e., repeated instances of solstice or lunar alignments - which in turn, support the likelihood that any one such alignment is intentional. Thus, for example, if we find repeated instances of lunar alignments in a number of different earthworks at the Newark Earthworks complex, then the likelihood is that any one such posited alignment is intentional is increased.

CHAPTER 3
MUSKINGUM RIVER WATERSHED

The Muskingum River Watershed drains roughly 8,000 square miles of eastern and southern Ohio and extends across 27 counties. The major river is the Muskingum, which flows roughly north to south into the Ohio River. Tributaries that feed into the Muskingum include the Kokosing and Licking rivers. Major archaeological sites in the Muskingum Watershed include the Fredericktown, Newark, and Marietta earthworks (Figure 3.1).

Fredericktown Earthworks

The most northern group of earthworks considered in this book are the Fredericktown Earthworks. Nestled in the northwest corner of the Muskingum Watershed, the Fredericktown Earthworks overlook the North Branch of the Kokosing River. There are several mounds in the general area. However, the Rowley, Stackhouse, and Kandel mounds and earthworks appear inter-related. Viewed from the Rowley Mound, the summer solstice sun will rise between the Stackhouse and Kandel earthworks. Both Stackhouse and Kandel are, in turn, oriented to the solstice sunrise through their diagonal and major axes, respectively. Thus the summer solstice rise is emphasized in different ways and at two different scales.

There are no documented reports of any of these three earthworks having been excavated; although an unconfirmed report claims that an expanded-center bar gorget - typical to the Adena culture, was found in the Rowley Mound in the 1800s. That these three earthworks were intended as part of an integrated design is suggested by their relative placement in terms of how Stackhouse and Kandel flank the solstice sightline viewed from the Rowley Mound; and the emergent isosceles triangle that situates all three relative to each other.

The diagonal solstice alignment of Stackhouse is interesting in that, if it is an Adena construction, then it anticipates similar diagonal alignments found among many of the large Hopewell squares. The geometry of Stackhouse and

Kandel is interesting in that both resemble a shape that is neither circle, nor square, but something in-between.

Newark Earthworks

The Newark Earthworks are situated in the Newark-Licking River Valley, in the western part of the Muskingum River Watershed. The valley is located in the Glaciated Plateau physiographic region of Ohio. Several unique features combined to make the Newark location special. To begin with, Newark is surrounded by hills. These hills reach to several hundred feet in height. As Figure 3.7 shows, several hills are situated at locations that would allow observers standing on their summits to view all four solstice rise and set events over corresponding hills, across the valley (presuming that the trees on the various observation points were cleared so as to allow these kinds of views). Researchers Ray Hively and Robert Horn (2013) were the first to make this observation.

The hill to hill solstice alignments are, of course, fortuitous as they result from natural phenomena, not human intentions. Nevertheless it is understandable how the intersection of all four solstice directions with landscape features might be interpreted as evidence for Newark as a special place where earth and sky meet and where spirit manifests itself in a visible way.

Enhancing this hierophantic experience is the following. Of the solstice alignments that crisscross the valley, the most impressive involve hilltops H1 and H4 (Figure 3.7). Looking from hilltop H1, the summer solstice sun will rise over hilltop H4. Looking the other way, from H4 to H1, the winter solstice sunset will occur over H1. The H1-H4 sightline extends longitudinally along the center length of the entire 8-mile long, Newark Valley. Support for the idea that this particular relationship between earth and sky phenomena was noticed by the Newark Earthwork builders is found in the location of earthworks where solstice sightlines intersect with the landscape and along valley trajectories. Fredericktown, the Johnson Works, and Fort Hill are three examples.

The complementary opposite of the H1-H4 sightline is the sightline between H2 and H3. Viewed from H2, the winter solstice sun will rise over H3. Viewed from H3, the summer solstice sun will set over H2.

As Figure 3.7 shows, the H1-H4 and H2-H3 sightlines intersect. The place where these sightlines intersect is at the center of the Newark Valley, at a point nearly equidistant between the Octagon-Observatory Circle and Great Circle (Hively and Horn 2013). In other words, the intersection of the four solstice directions marks the center-place for the Newark Earthworks.

The Moundbuilders had no role in determining the location of the hills surrounding Newark and the solstice alignments that resulted. They did, however, create their own, complementary alignments. Most are lunar; but some are solar and stellar. The details of the man-made alignments are presented elsewhere in this chapter. Important to recognize, however, is that the Newark Earthworks are not aligned along the hilltop to hilltop solstice sightlines. Rather, the centers of the major earthworks are situated where lunar sightlines as viewed from the surrounding hilltops intersect. As Figure 3.7 shows, the center of each major earthwork is situated where two lunar azimuths intersect. It remains a mystery as to how these long sightlines were established on the ground. Nevertheless they are an objective fact and similar instances of precise alignments across even greater distances are documented elsewhere in this book.

Of the lunar alignments shown in Figure 3.7, there is one in particular that is especially interesting - and that is the moon maximum north rise sightline from H1. What make this sightline interesting is that it extends along the same azimuth as the river terrace that flanks the west side of the Octagon and Observatory Earthwork. In other words, the moon azimuth, the lay of the land as referenced to the river terrace, and the earthwork are all oriented in the same direction. In this, the earth, moon, and earthwork were brought into a synchronous and harmonious relationship. Many more examples of this kind of convergence are found elsewhere in this book.

While crisscrossing celestial and earth alignments were of considerable importance in establishing the location and special nature of Newark, there was more. Appropriate to a dualistic view of the cosmos, in addition to the sky, water played an important role. As Figure 3.7 shows, three major water sources flow into the area - i.e., the North Fork Licking River, the South Fork Licking River, and Raccoon Creek. These waterways meet at Newark, where they form the main Licking River. The Newark Earthworks are located at this river confluence.

The North Fork of the Licking River flows from north south. Interestingly, however, the South Fork of the Licking River flows from south to north. Thus the Newark confluence marks the place where two opposite flowing rivers meet. Add to this the fact that Raccoon Creek joins from the west and the potential for directional symbolism is evident.

The observation that Newark is situated at a river confluence is interesting. Where the matter begins to take-on potential cosmological significance is the observation that, if we add Ramp Creek to the mix - which is situated a couple of miles south of the earthworks, then the result is that the main Newark Earthworks are essentially surrounded by waterways (Romain 2005c). If a circle is drawn around the Newark Complex, watercourses are found to enclose about 320 degrees, or 90% of that circle. The result is something of an earth island, reminiscent of later Eastern Woodlands mythologies that describe the earth as an island surrounded by primordial waters (e.g., Beauchamp 1922; Connelley 1899; Count 1952; Hewitt 1918; Hudson1976; Lankford 2007). It is probably no coincidence that watercourses - including natural, enhanced, and entirely man-made are often found separating earthworks from the surrounding topography. This is a common attribute of Adena and Hopewell earthworks at multiple scales ranging from small Adena circles with interior ditches, to large geometric earthworks like the Newark Great Circle, to huge hilltop enclosures like Fort Ancient (Connolly 1996; Romain 2004a).

Another water feature that likely made Newark special was the lake or lake-bog situated in the middle of the earthwork complex. This lake-bog is shown on old maps of the Newark Earthworks (see e.g., Figure 3.29). A notation on Atwater's map of 1820 indicates at that time, the lake was about 150 acres in extent. This is what the Salisburys (1862:29) said about the lake:

> "The modern history of the Lake is somewhat singular. It is said that, when the first settlers reared their cabins in this vicinity, the basin was so dry, smooth and unobstructed that the people of the surrounding country found it a convenient place for a race course. Afterwards it gradually filled with water and became a noted place for duck hunting....The proprietors endeavored to drain it....finally a deep and extensive ditch accomplished the desired object....The extensive network of ditches upon its surface, exhibit a deep stratum of peaty soil, at the base of which is some marl and many shells."

There are a couple of interesting bits of information in this passage. First, it seems that the water level of the lake was subject to variation - sometimes being dry, while other times forming a respectable lake. So too the discovery of a deep peat layer indicates that various times, the lake may have been a peat bog. Given this, I would not be surprised if, during Adena-Hopewell times this feature was a combination lake-bog. In Ohio, bogs provide a diversity of unusual flora and fauna. Among some of remarkable life forms found in Ohio bogs are the massasauga swamp rattlesnake, sphagnum moss, poison sumac, and two species of carnivorous plants - i.e., the pitcher plant and sundew plants. Although probably not the most important of all the things that made Newark special, surely these peculiar life-forms contributed to the special nature of the place.

Today the area is drained and the lake-bog is gone. The existence of the lake-bog during the florescence of the Newark Complex, however, is suggested by the fact that no parallel walkways lead directly into or across the lake-bog area. Quite to the contrary, the parallel-walled walkway that connects the Octagon to the Wright Square deviates around the area.

Adding to the intrigue of the Newark lake-bog is another Salisbury observation that suggests ancient manipulation of the lake-bog's water levels. The information is contained in the Salisburys' description of the parallel walled passageway leading from the Ellipse Earthwork to the Wright Square. According to the Salisburys (1862:5):

> "The north wall extends S. 40° W. about 850 feet where there is a break or inset of the wall of 111 feet., a little beyond which appears to have been the outlet of the ancient lake Y. The channel is bridged with earth between the walls, except within 15 ft. of each, under which and the passage, was probably a tunnel for the passage the water. A little to the west, appears to have been a low dam extending across the neck of the swamp, and on a little spur of dry land extending within this dam, is a small clay ring *m*."

Based on this description as well as the Salisbury map showing these features, it seems possible that the Moundbuilders were intentionally controlling water levels in the lake-bog by channeling and damming. When water levels were low, the dam was closed; when too high, the dam and channel were opened and water drained toward the South Fork Licking River.

In summary, what is striking about Newark is the complex and inter-related nature of earth, sky, and water relationships expressed in multiple ways and at a range of scales. Newark was, quite simply, the most sophisticated earthwork complex built by the Hopewell. It may be the most sophisticated and intricate earthwork complex ever built, anywhere in the world.

In cosmological terms, I would propose that Newark was quite literally, a cosmic center (*sensu* Brown 1997). Perhaps Newark was considered a portal to the Otherworld. I offer this interpretation based in the observation that, with its surrounding watercourses and hills, Newark is a place where landscape features define its spatial boundaries in a very visible way. Add to that, crisscrossing solstice azimuths and the result is Newark as a metaphorical earth island and center place oriented to the cosmic quadrants, at the intersection of earth and sky. Within this center place the Moundbuilders likely performed the most sacred of their rituals – i.e., rituals intended to maintain cosmic balance, ensure plant and animal abundance, maintain health, and facilitate in death, the transition of the soul from the land of the living to the land of the dead.

As to Newark as a portal to the Otherworld, I would point out that the Milky Way Path of Souls and its earth analogue, the Great Hopewell Road begin in Newark. Add to that, a lake-bog in the middle of an earthwork complex that was used at least in part for burials and having associations with the moon and Milky Way and a bigger picture comes into view. One can imagine a low-lying fog hovering across the valley, with perhaps the distant outline of an occasional earthwork poking up through the mist. Centered in this Otherworldly place was the lake-swamp-bog, flanked here and there by the dead, forever interred in their burial mounds. As the Milky Way came into view across the summer sky, and the earthworks were crisscrossed by solar and lunar trajectories, maybe a doorway was believed to open-up - i.e., a doorway to the Otherworld. It was through this doorway that souls of the deceased began their journey to the Land of the Dead.

We will continue with exploration of that idea in later chapters. For now, there is another earthwork feature that may be related. Look closely at the Wyrick and Salisbury maps and what is seen is that low walls surround the earthworks. These walls are seldom commented upon and are mostly now obliterated. A few reconstructed remnants, however, are seen near the Great Circle. In the 1860s the Salisburys said the walls were about 1 1/2 feet in height. Given their low height, they were certainly not designed to keep people or animals out.

In Figure 3.29 I have highlighted the interior areas contained by these perimeter walls, as well as the main earthwork walls. The resulting boundary has a few breaks in it. But enough of the complex is surrounded to suggest that the space within was intentionally distinguished from outside space. Maybe what was being created were boundaries and pathways not so much for the living, but for spirits of the dead. Maybe these barriers and causeways were meant to help guide the free-soul (perhaps accompanied by mourners, relatives, and psychpomps) to the entrance to the Great Hopewell Road, and from there, along the Milky Way Path to the Otherworld.

Flint Ridge

Flint Ridge is the source for Flint Ridge flint. It was one of the most valuable and unique resources the Hopewell had. Flint Ridge flint is very colorful and highly prized by flint-knappers. Among the many implements the Hopewell made from this flint were exquisite projectile points, blades and bladelets.

Flint Ridge is about 9 miles in length, 3 miles wide, extending from east to west. Its western end comes to within about 10 miles of Newark. The flint is found in a geologic bed ranging from 2 - 10 feet thick. At several locations flint outcroppings are visible and large chunks of flint are found on the surface. At other places the flint beds have been mined by the Hopewell and others. Where it has been mined, hundreds of prehistoric quarries are in evidence along the ridge. The quarries are pit features, some as much as 20 feet deep. LiDAR imagery shows that most of these pits are clustered near the present-day Ohio Historical Society Flint Ridge Museum.

Given the importance and proximity of Flint Ridge to Newark it comes as no surprise to find that the Newark Earthworks are connected to Flint Ridge in a couple of different ways. As Figure 3.31 shows, physical access to Flint Ridge from the area of the Newark Earthworks is provided by following the South Fork of the Licking River to Swamp Run, up Claylick Creek to the Hazlett Mound. The Hazlett Mound is about 1/2 mile from the western edge of the main quarry area.

Of special interest is the lunar association between the Newark Earthworks and Flint Ridge. As Figure 3.31 shows, a line drawn from the center of the Newark Octagon along the azimuth of the moon's minimum south rise (i.e., 116°.2 degrees), intersects Flint Ridge, at a distance of 10.5 miles, at the very heart of where the heaviest concentration of prehistoric flint quarries are located. (This same sightline also intersects the near-center of the Wright Square.) One might be inclined to think that this alignment is fortuitous - and perhaps it is - except for something else. As explained below, a lunar crescent symbol is found at the Hazlett Works.

There are several mounds located on Flint Ridge. The most impressive earthwork feature found at Flint Ridge, however, was the Hazlett Works. Today little remains of the Hazlett Works. Originally, the Hazlett Works consisted of three mounds surrounded by an enclosure wall made at least in part of flint blocks. The main Hazlett Mound was of earth and was about 85 feet in diameter and 13 feet in height (Mills 1921:255). A tomb structure was found within the mound. This square-shaped tomb measured 37 feet by 37 feet on its outside and was comprised of flint blocks. Two skeletons were found in the tomb. Accompanying one of the skeletons were two earspools, a large copper gorget, shell beads, and part of a wolf jaw. The Hazlett Mound is attributed to the Hopewell (Carskadden and Fuller 1967; Mills 1921:161).

As mentioned, one of the things that makes the Hazlett Works special is its location. Specifically, it overlooks the approach into Flint Ridge from Newark. Additionally, however, as Figure 3.32 shows, a circle enclosure was originally part of the Hazlett Works. This circle earthwork was located due west of the Hazlett mound. Of special interest is that this circle enclosure had a crescent-shaped embankment wall within. The east-facing opening of the Hazlett Crescent (and also the circle itself) is consistent with similar crescent-in-circle earthworks at Newark and the Yost Earthworks to be discussed next. If, as I believe, the crescent-inside-a-circle design was meant as a lunar symbol, then what we have in the Hazlett Crescent Earthwork is a moon symbol at the near-terminus of a lunar sightline that extends from the lunar-oriented Newark Octagon, to Flint Ridge. In this a symbolic association is drawn between the Newark Earthworks, the moon, Flint Ridge, and flint.

As might be appropriate for the burials within the Hazlett Mound, the mound is associated with the summer solstice sunset and Newark through two summer solstice sunset azimuths - the first from the Hazlett Mound to Coffman Knob (Figure 3.32), and the second from Coffman Knob to hilltop location H5 in Newark (figure 3.7).

About 8 miles south of Flint Ridge are a remarkable series of earthworks and mounds comprised of the Yost Earthworks, Glenford Hilltop Enclosure, Roberts Mound, and Wilson Mound. Each of these mounds and earthworks is unique. LiDAR assessment indicates a geometric relationship between these four earthworks.

One of the interesting features associated with the Glenford Hilltop Enclosure is that a pathway extends from its southeast entrance. This pathway extends along the summit of a ridge and leads directly to the Wilson Mound. Moreover, near the southeast entrance to Glenford, this pathway intersects a small square-oid shaped earthwork. This earthwork is shown on an old Salisbury map and although largely destroyed, remnants of its walls are visible in LiDAR imagery. From the way they are physically linked, it appears that the Glenford Hilltop Enclosure, Small Squaroid earthwork, and Wilson Mound were part of an interrelated site complex.

The Yost Earthwork deserves special mention. Much of the earthwork has been destroyed by plowing. Old maps (see Moats 2012) indicate it was originally much larger than what is visible today. In any case, one of the most unique aspects of Yost is the crescent-in-circle earthwork still extant at the site. Originally, a considerable number of similar crescent-in-circle earthworks were found at Newark. Since the destruction of the Newark circles, the Yost Circle and Crescent appears to be the sole remaining exemplar of this particular kind of earthwork. As suggested earlier in this chapter, the crescent-in-circle earthwork may be a lunar symbol. And as proposed by Moats (2012) the Yost Earthwork likely has lunar alignments incorporated into its design.

Marietta Earthworks

Marietta is located in the Unglaciated Allegheny Plateau region of southern Ohio, about 67 miles southeast of Newark. The region is characterized by steep, rugged terrain. The Muskingum River Valley at Marietta is relatively narrow, averaging only about one mile in width. Steep, high ridges parallel both sides of the Muskingum River. Notably, the Marietta Earthworks are located at the confluence of the Muskingum and Ohio Rivers. As such, Marietta controlled access into the deeper areas of the Hopewell region to include the Newark earthworks. Notably too is that high quality clays are found in the immediate vicinity of the Marietta Earthworks. During the 1800s these clays were used to make bricks (Squier and Davis 1848:76).

One of the interesting things about Marietta is that it integrates a major Adena earthwork – i.e., the Conus Mound, with mounds and earthworks that are Hopewell. Based on its representative conical shape, surrounding embankment with internal ditch, and typical Adena dimensions (i.e., 1/4 HMU), the Conus Mound is very likely an Adena mound. At the same time, diagnostic materials recovered from the nearby Capitolium Mound establish that earthwork as Hopewell (Pickard 1996). Astronomic assessment shows that the Capitolium Mound is astronomically tied to the Conus Mound along a solstice alignment.

As a matter of fact, there are a considerable number of astronomic alignments in evidence at Marietta and similar to Newark, Marietta is situated at the intersection of celestial sightlines. As shown by Figure 3.41, the Conus Mound (itself a likely *axis mundi* symbol) is located at the intersection of the summer solstice sunrise and summer solstice sunset sightlines. From that point of origin, the rest of the Marietta Complex can be laid-out using straightforward iterations of the HMU, and shapes based on squares, oriented to the winter solstice sunset. Further, the complex extends parallel to the lay of the land established by the southeast to northwest trending river terraces of the Muskingum River. Thus the Marietta Complex is at the intersection of earth, sky, and water realms...a special place, indeed.

Figure 3.1. Map of sites in Muskingum River Watershed

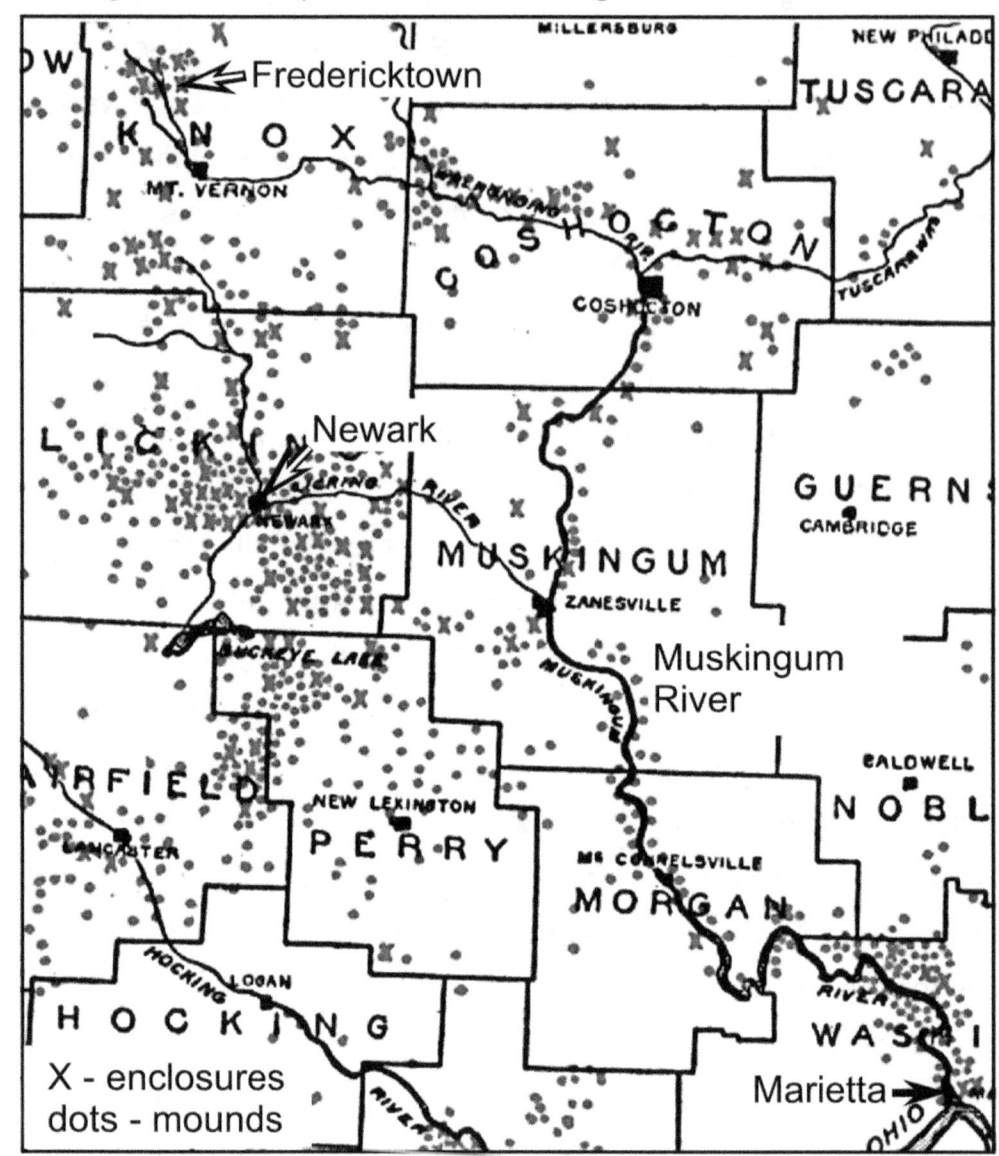

Above: Detail of Mills (1914:Pl. XI) map showing earthworks and mounds in the Muskingum River Watershed (annotation added).

Figure 3.2. Fredericktown Earthworks

The Fredericktown Earthworks are located in Knox County, about 30 miles north of Newark overlooking the North Branch of the Kokosing River. The three earthworks of interest are: Stackhouse, Kandel, and the Rowley Mound.

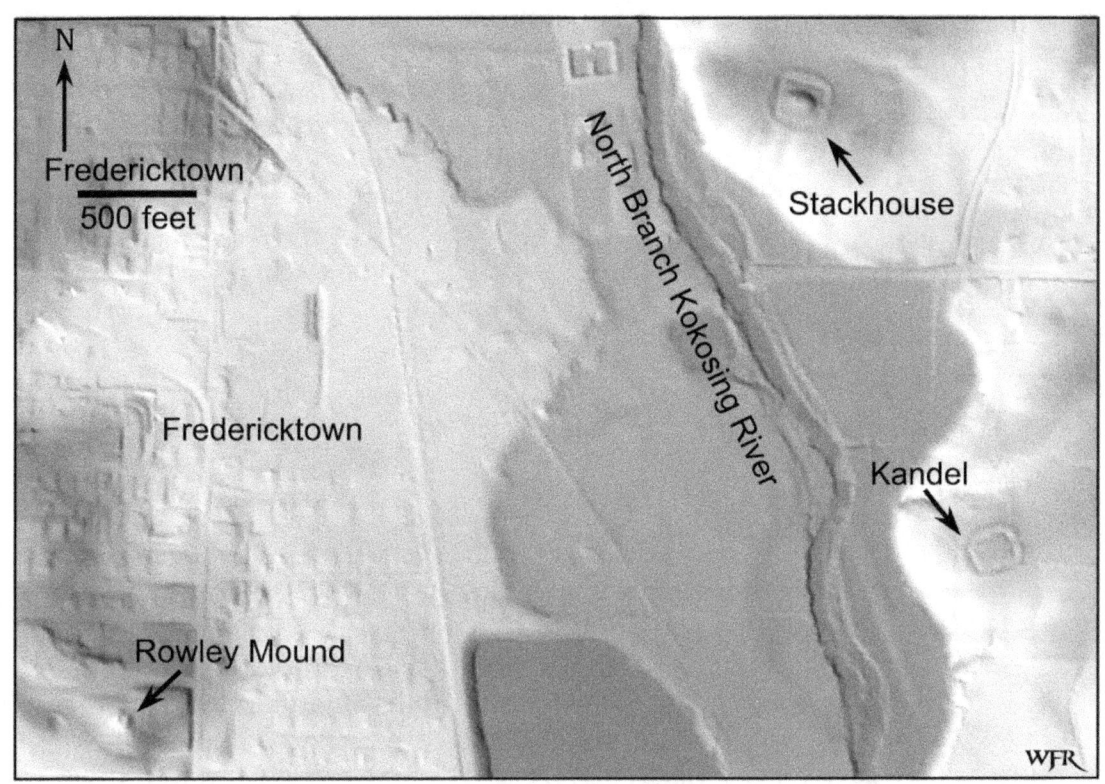

Above: LiDAR view of the Fredericktown Earthworks.

Stackhouse Mound (also known as the Braddock Mound). Photo courtesy of John C. Rummel.

Rowley Mound (also known as the Raleigh Mound). Photo courtesy of John C. Rummel.

Figure 3.3. Fredericktown: Stackhouse Earthwork

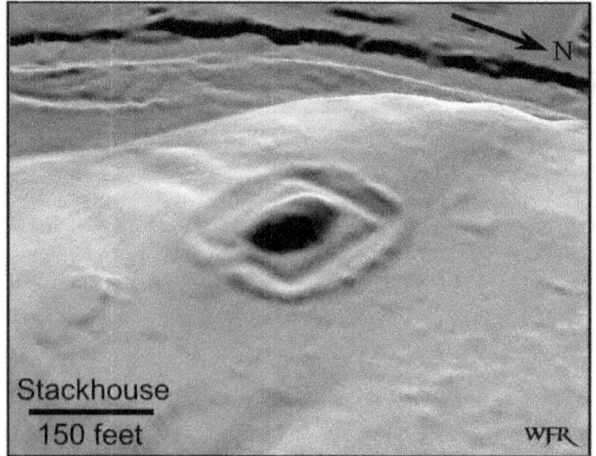

Left: LiDAR image of the Stackhouse Earthwork. The center mound has not been excavated.

Below: LiDAR image of similarly shaped earthwork part of the Davis Group in Hocking County. A log-entombed human burial was found in the center of this mound (Thomas 1894:448).

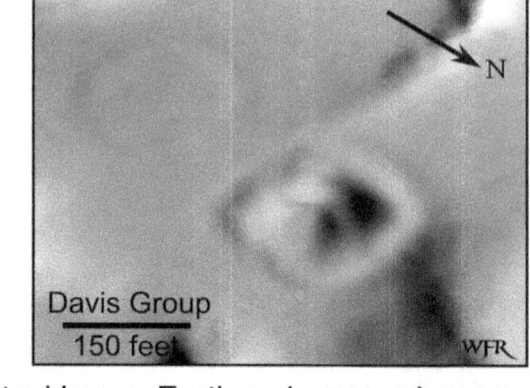

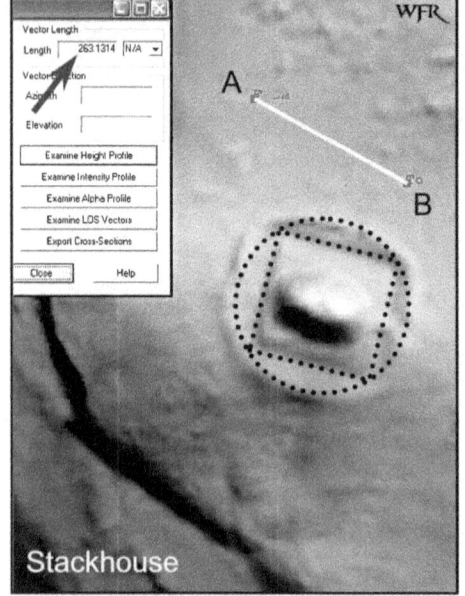

Left: the Stackhouse Earthwork uses a lesser multiple of the 1,054-foot length (1 HMU) in its design. In this case the earthwork is based on the length of 263.5 feet, or 1/4 HMU. In the image to the left, the LiDAR program has generated a line 263.5 feet in length (line A-B) (note 1). That line is used to draw a 263.5-foot diameter circle and matching square. As shown, the 1/4 HMU circle and square correspond to the earthwork's dimensions.

LiDAR analysis reveals that Stackhouse is aligned to the summer solstice sunrise along its diagonal axis (note 2). In the image to the right, the calculated summer solstice azimuth (white line) has been superimposed over a LiDAR image. Point A is where the horizon elevation has been measured from.

Figure 3.4. Fredericktown: Kandel Works

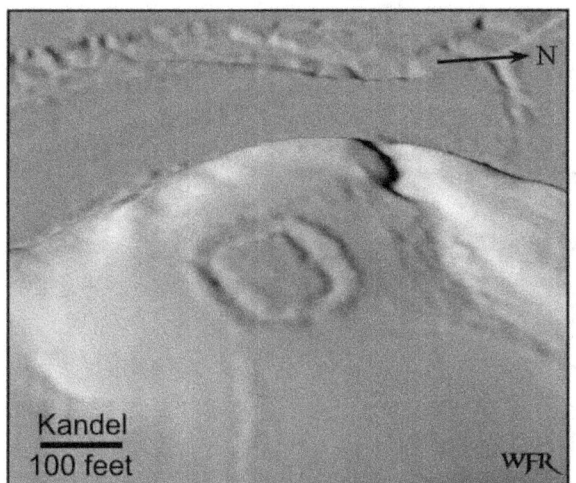

The Kandel Earthwork does not have a center mound. Similar earthworks are found elsewhere, however. The LiDAR images below, for example, show the Newlove and Jackson Square earthworks located in Clark and Jackson counties, respectively.

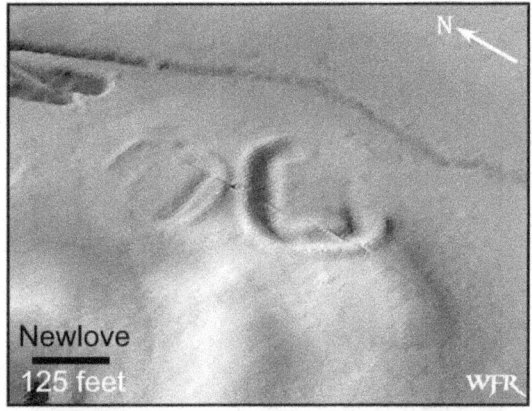

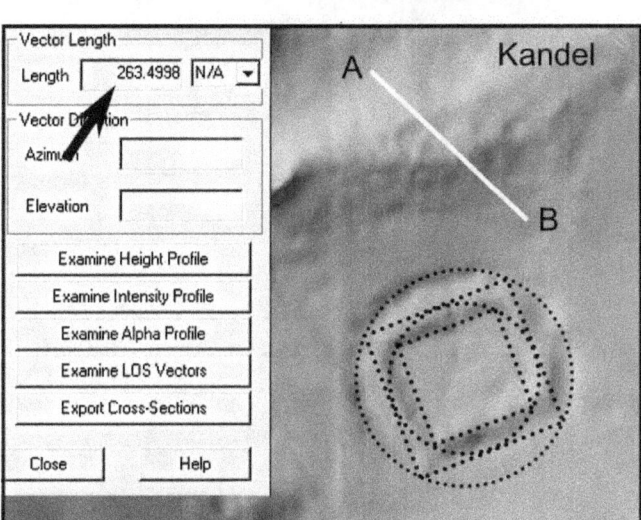

Above left: like Stackhouse, the Kandel Works uses the 1/4 HMU in its design. In the above image, the LiDAR program has generated a 263.5-foot line (A-B) (1/4 HMU). That line is used to make a 263.5-foot diameter circle, matching square, and nested figures. As shown, the 1/4 HMU-derived figures closely define the earthwork's dimensions.

Kandel is aligned to the summer solstice sunrise along its major axis (note 3). In the image to the right, the calculated summer solstice azimuth has been superimposed over a LiDAR image. Point A is where the horizon elevation has been measured from.

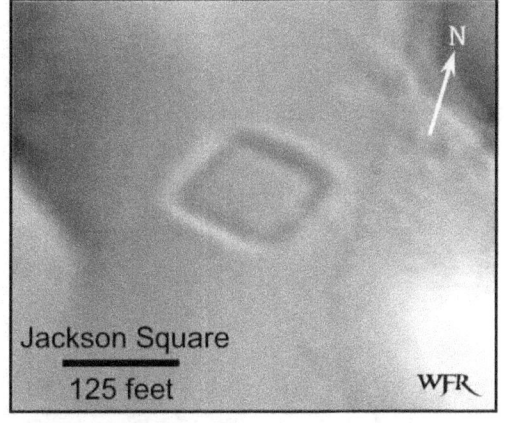

summer solstice sunrise calculated az = 65°.9

Figure 3.5. Fredericktown: Earthwork Relationships

The Fredericktown Earthwork Complex has attributes typical to both Adena and Hopewel. Of interest are the inter-site relationships.

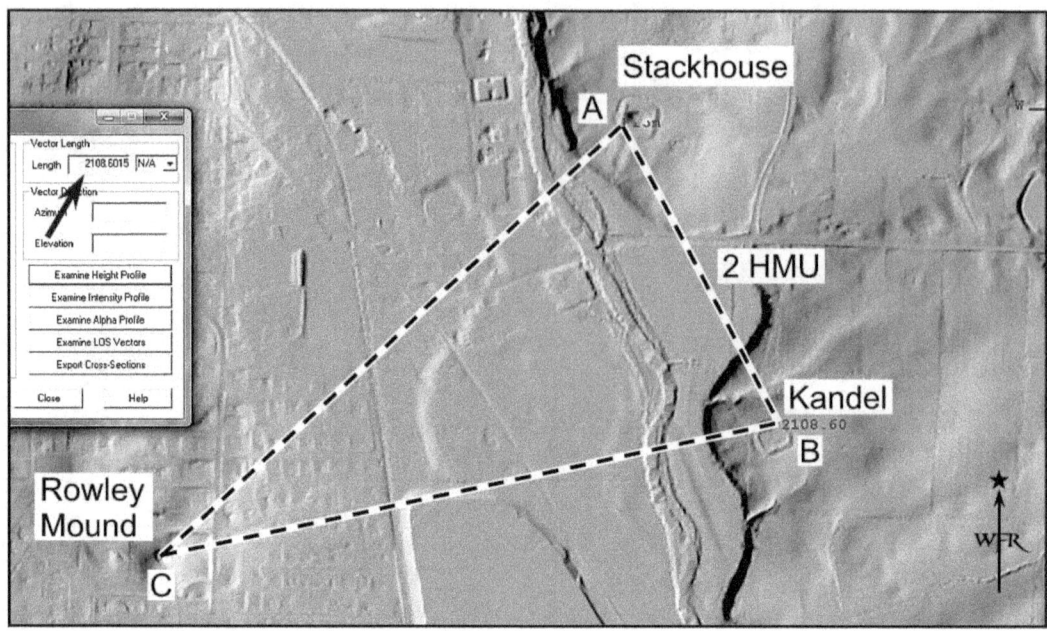

Above: the distance from Stackhouse point A to Kandel point B is 2,108 feet - or 2 HMU. Point A is the southwest corner of the Stackhouse inner platform. Point B is the northeast corner of the Kandle inner platform. Lines A-B-C form an isosceles triangle.

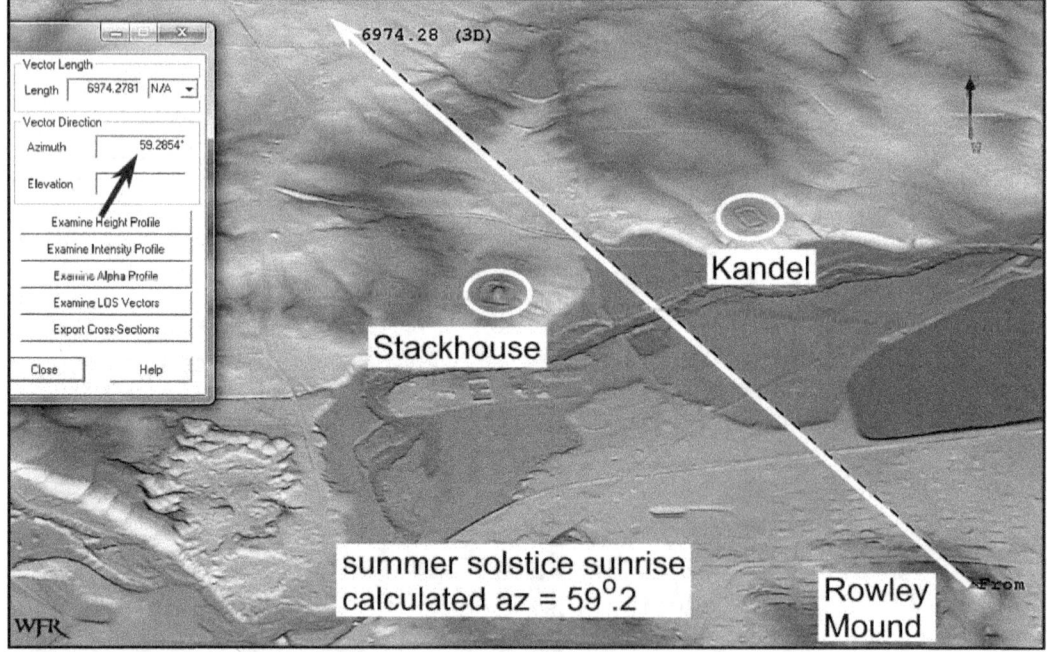

Above: viewed from the Rowley Mound, the summer solstice sun will rise in alignment over the small valley between Stackhouse and Kandel (note 4).

Figure 3.6. The Newark Earthworks

The Newark Earthworks Complex is the most sophisticated and intricate earthwork complex of its kind in the world. The site covers more than four square miles of a broad river valley situated at the confluence of Raccoon Creek and the North and South Forks of the Licking River (note 5).

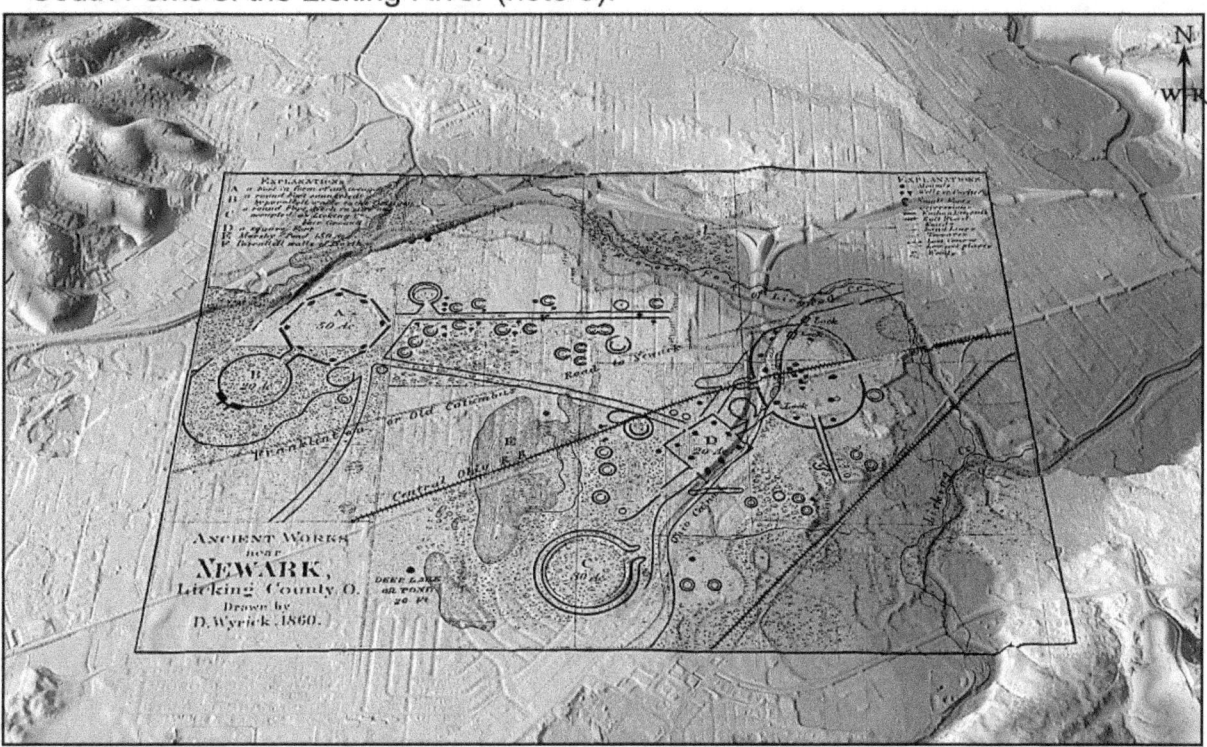

Above: Wyrick map from 1860 superimposed over LiDAR image of the Newark Valley.
Below: Wyrick (1860) map of Newark (annotation added).

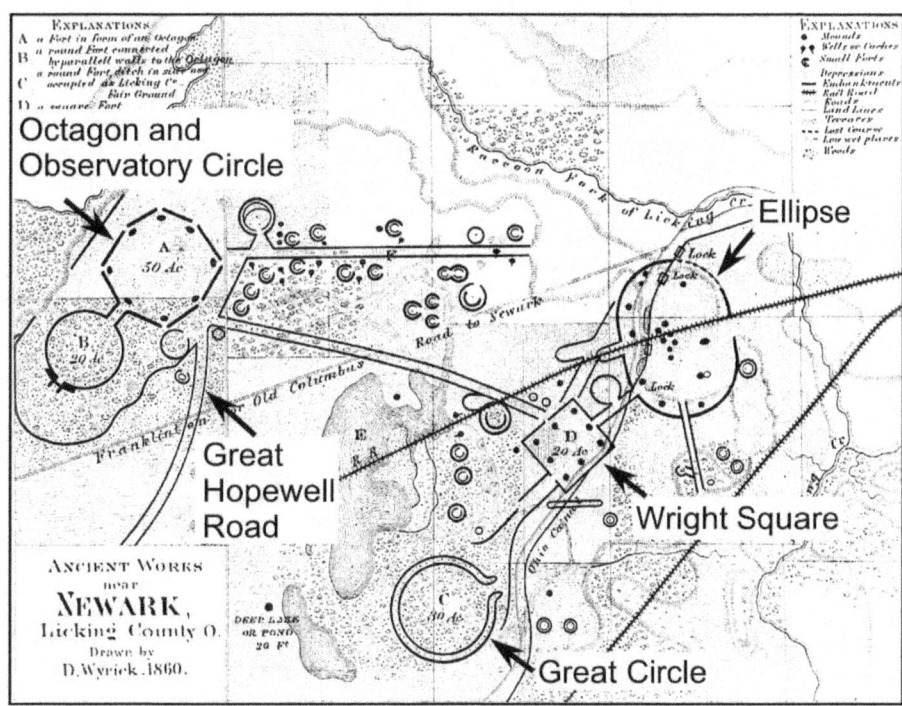

Figure 3.7 Newark Large-Scale Alignments

There are two sets of large-scale alignments at Newark - natural occurring solar alignments and man-made lunar alignments (note 6). The natural intersecting solar alignments make the Newark Valley special. The lunar alignments emphasize that.

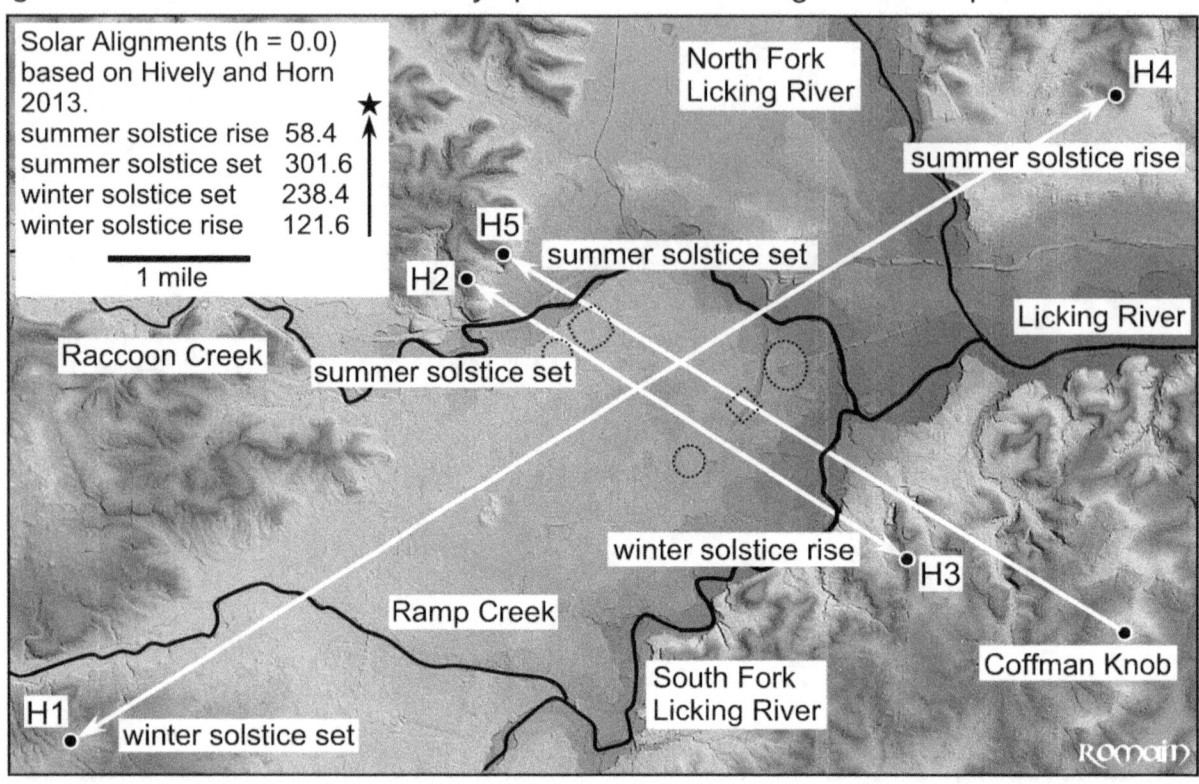

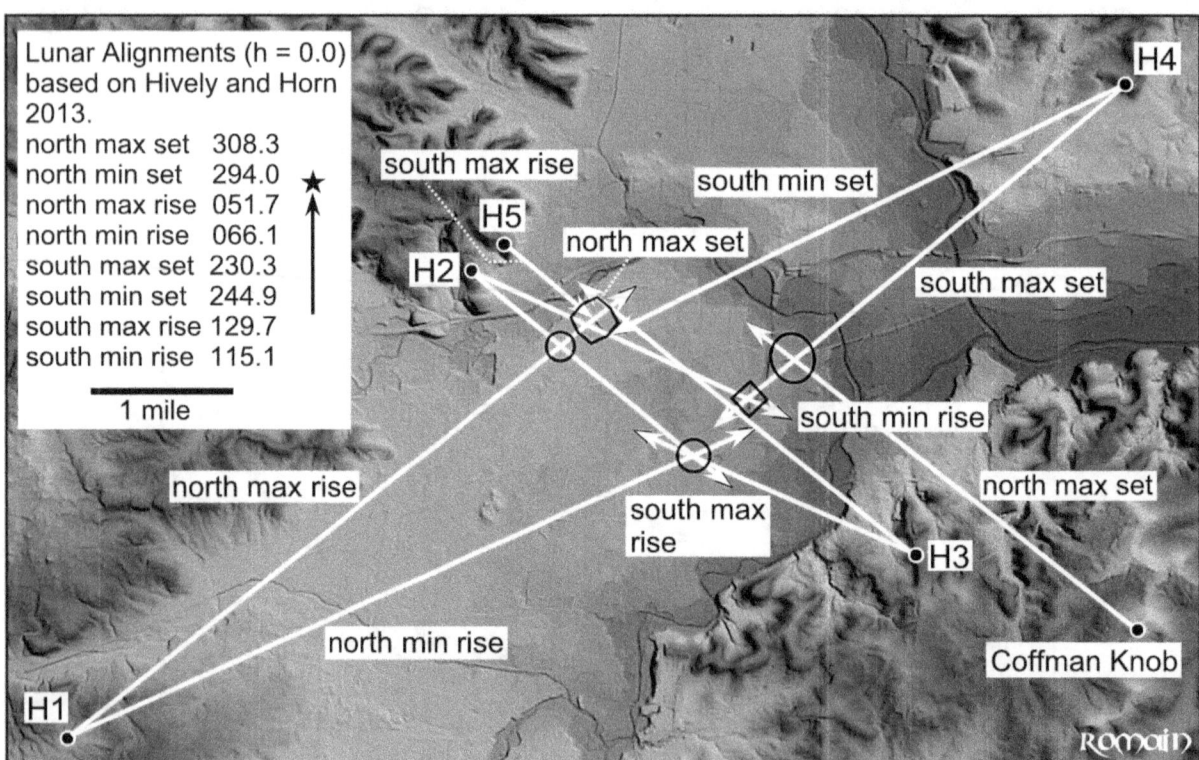

Figure 3.8 Newark Hilltop Vantage Points H1 - H4

Important to understanding the relationships between the earthworks and surrounding hills is the matter of visibility - from one hill across the Newark Valley to the opposite hill. In the figures below, the shapes and relationships for several of the hills are shown.

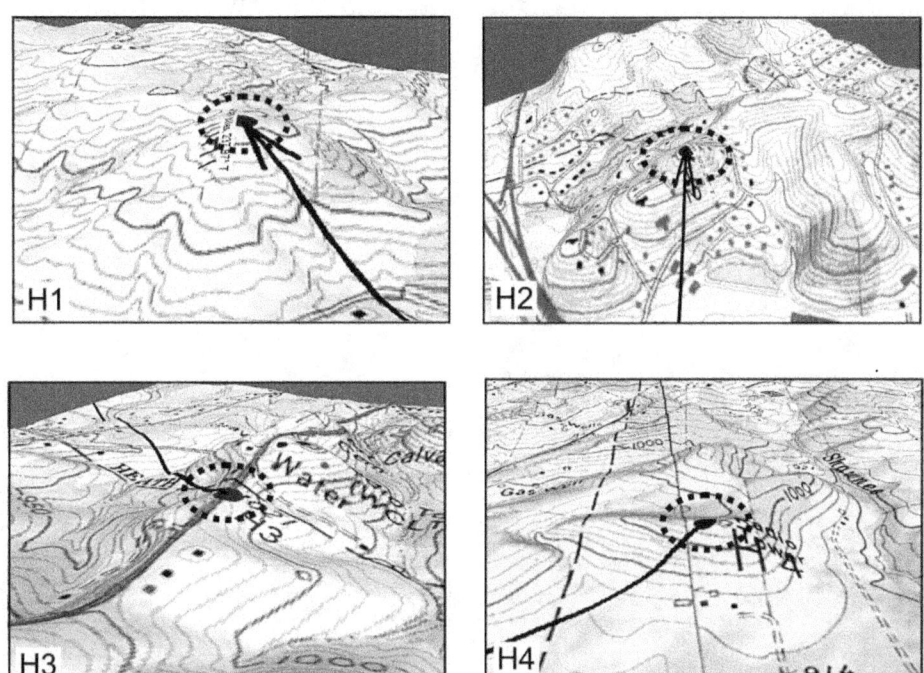

Above: three-dimensional map views of the H1 - H4 vantage points. Arrows show plotted solstice azimuths. Maps by MyTopo, annotation added.

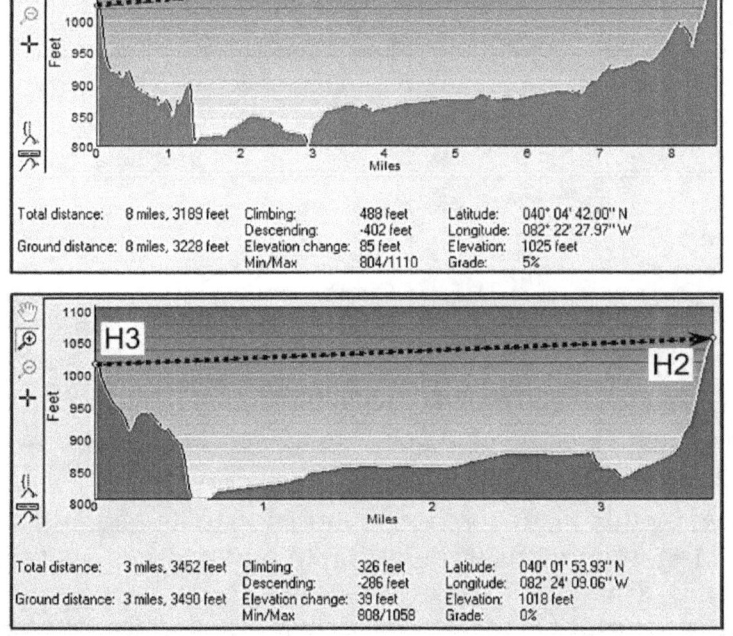

Above: line-of-sight profiles for H1 - H4 and H2 - H3. Maps by MyTopo. annotation added.

Figure 3.9. The Newark Earthworks: Octagon and Observatory Circle
The Newark Octagon and Observatory Circle is one of the most impressive of the Hopewell earthworks. The site has been partially restored. It is used as a golf course.

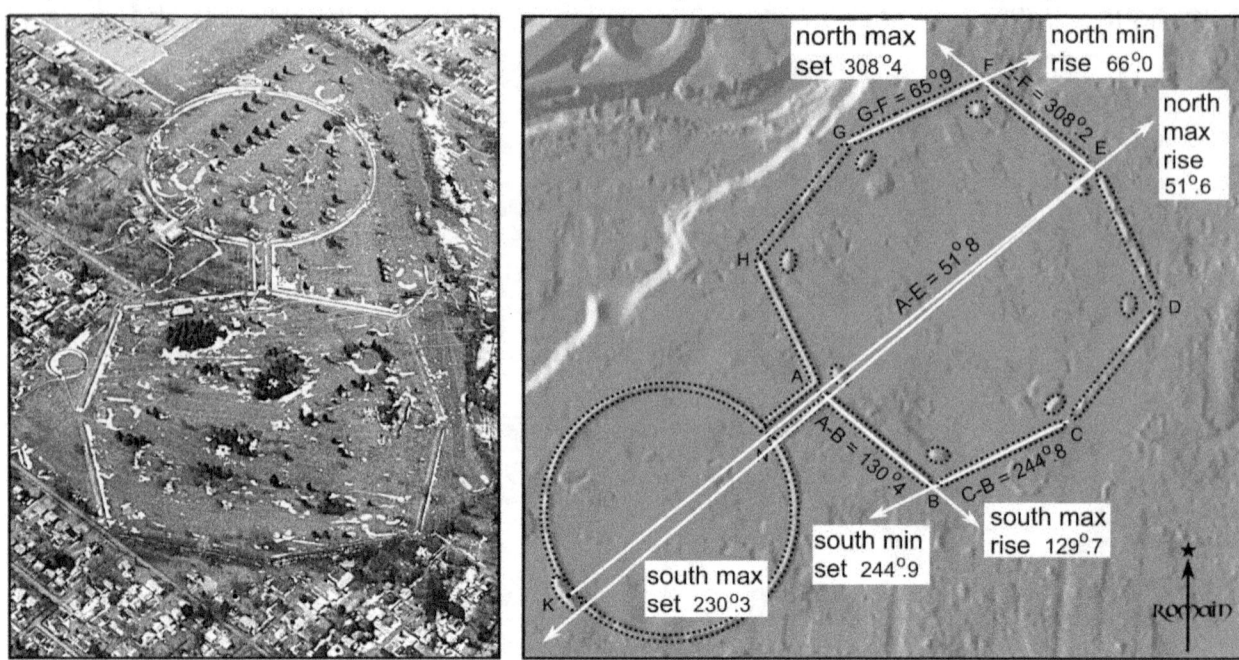

Above left: aerial view of the Octagon and Observatory Circle.
Above right: Multiple lunar alignments are incorporated in the design of the earthwork. In this LiDAR image, earthwork walls are outlined by superimposed dotted lines. LiDAR azimuths for each wall are shown in black. White lines and values show calculated lunar azimuths relative to the walls (note 7). True north shown by arrow with star. Core lunar alignments based on Hively and Horn 2013:figure 2.

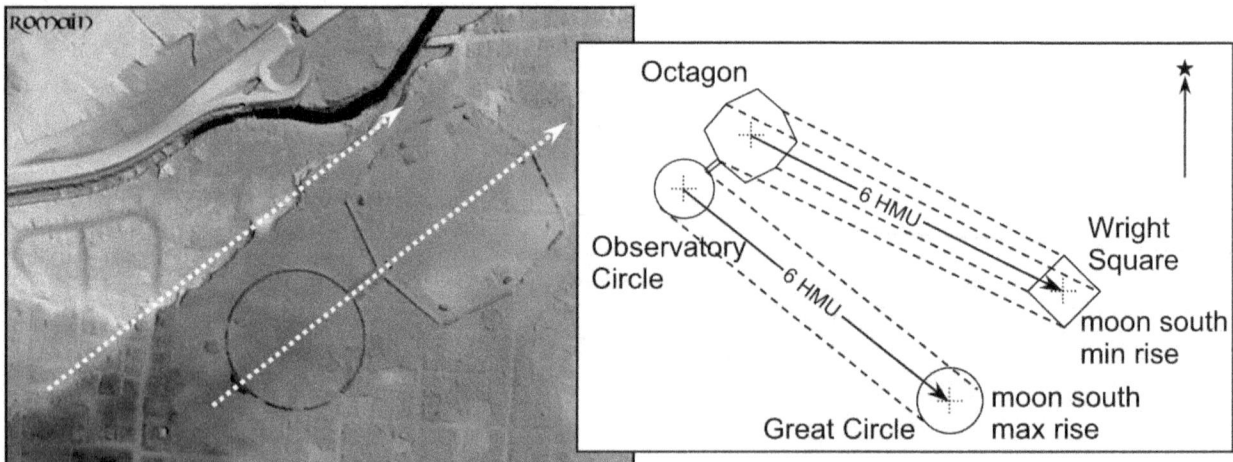

Above left: as shown by this LiDAR image, in addition to its alignment to the moon, the major axis of the Octagon and Observatory Circle earthwork is aligned parallel to the terrace edge formed by Raccoon Creek.
Above right: the Octagon and Observatory Circle are linked to the Great Circle and Wright Square earthworks along lunar azimuths and also by multiples of the HMU. Redrawn after Hively and Horn 2013: figure 3.

Figure 3.10. The Newark Earthworks: Octagon and Observatory Circle HMU
Adena and Hopewell geometric earthworks incorporate specific units of length in their design. These lengths are multiples and sub-multiples of a 1,054-foot length.

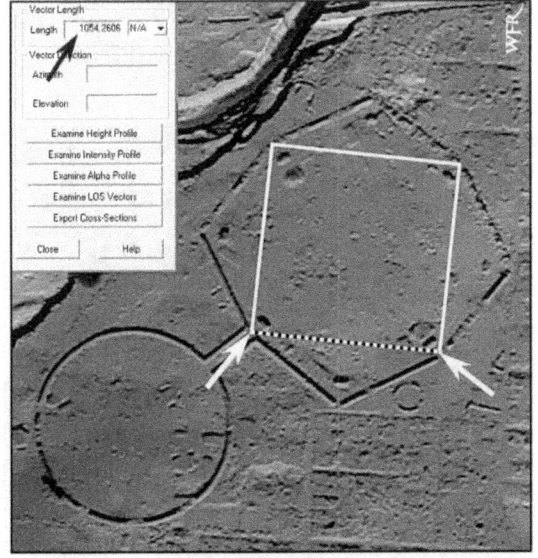

Hopewell Measurement Unit

1 HMU (1,054 feet)
1/2 HMU (527 feet)
1/4 HMU (263.5 feet)
1/8 HMU (131.8 feet)

The Hopewell Measurement Unit (or HMU) is equal to 1,054 feet. In addition to this unit of length, greater and lesser multiples of the HMU are found in the earthworks.

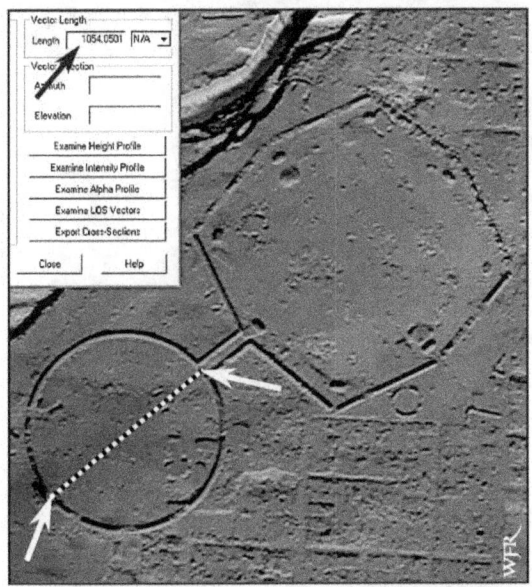

Left: LiDAR images showing how the dimensions of the Newark Octagon and Observatory Circle are based on the 1,054-foot unit of length (after Thomas 1894:464; Hively and Horn 1982.) In these figures, the LiDAR computer program has locked-on to ground coordinate points. The program then determines the distance between the coordinate points. This distance is shown in the accompanying data boxes.

Above: one of the 'gateway' mounds situated at the vertices of the Octagon. Coring into two of these mounds revealed only sterile soil (Romain 2005b). Photo by Evie Romain.

Figure 3.11. The Newark Earthworks: Wright Square

Except for a restored corner preserved in a park, the Wright Square has been destroyed by urban expansion. Based on the restored section and Thomas's (1894:466) data, however, the location and orientation for the earthwork can be reliably established. The below explanation for the location and orientation of the Square is adapted from Hively and Horn (2013).

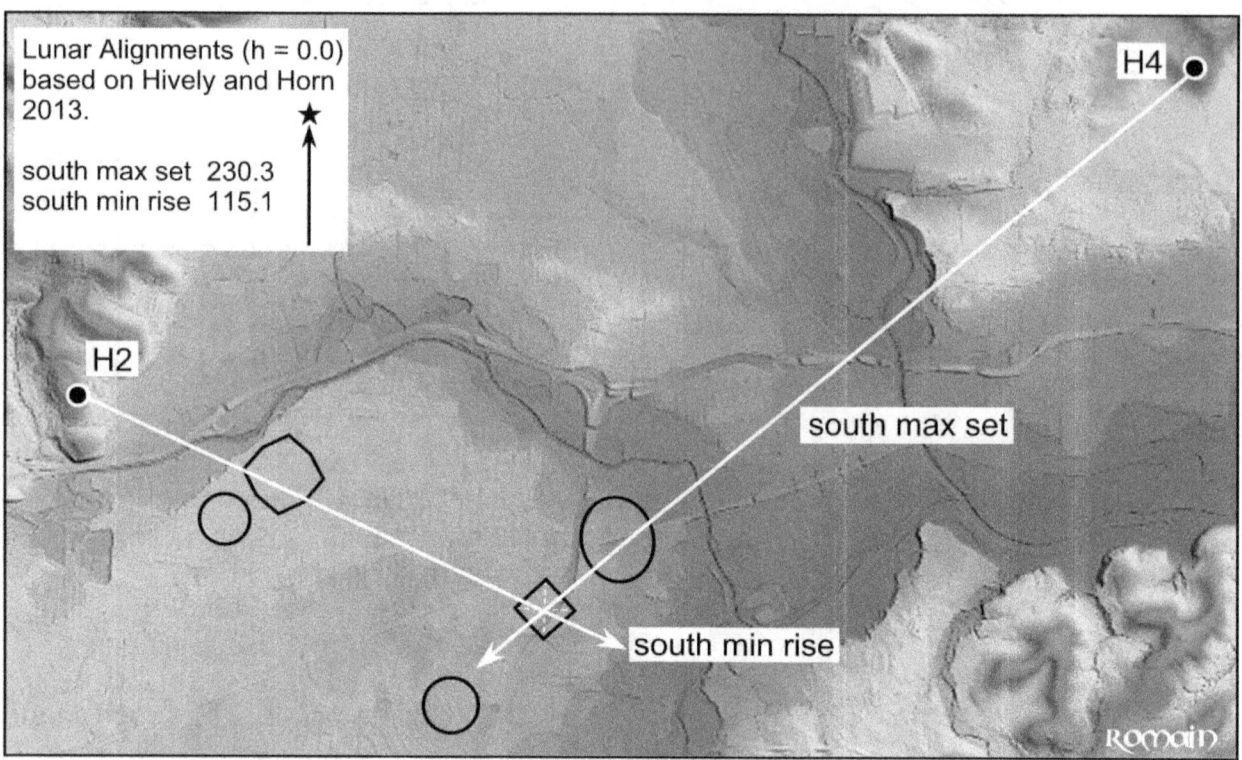

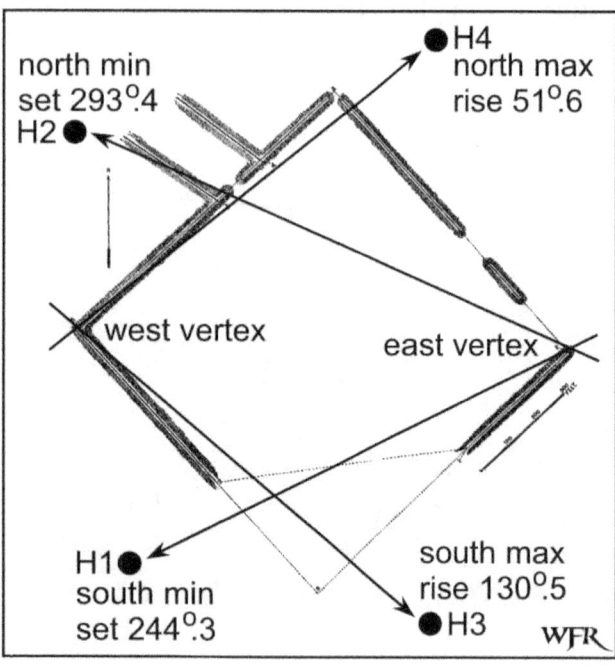

Above: the center of the Wright Square is situated at the intersection of the moon's south maximum set and south minimum rise azimuths.

Left: the orientation of the Square is established by lunar sightlines to the hilltop locations noted. Where these sightlines intersect marks the east and west vertices of the Square. Drawing of Wright Square by Thomas (1894:Pl.34), annotation added. Azimuth values from Hively and Horn 2013: table 4.

Figure 3.12. The Newark Earthworks: Great Circle

The Newark Great Circle is one of the largest of the Hopewell circle earthworks.

Above: aerial photo of the Newark Great Circle.

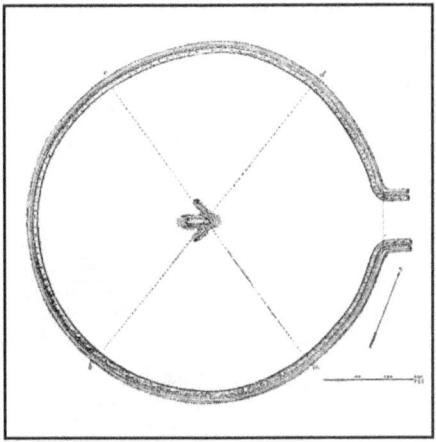

Thomas (1894:Pl. 31) map of the Newark Great Circle.

Great Circle
150 feet

The so-called "Eagle Mound" in the center of the Great Circle was found to contain structural postholes, burned bone, and worked copper (Greenman 1928).

Above: Thomas's (1894:Pl. 31) map of the Great Circle superimposed over LiDAR image.

Right: entrance into the Great Circle. A deep ditch extends along the inside of the perimeter wall. Atwater (1820:127) reports this ditch was half-filled with water when he viewed it.

Figure 3.13. The Newark Earthworks: Great Circle - Wright Square

The Newark Great Circle incorporates the 1,054-foot unit of length in its design in the following way. In the below LiDAR image, a 30-60-90 degree triangle is superimposed. If one side of this triangle is 1 HMU (or 1,054 feet) in length and the other side is 1/2 HMU (or 527 feet), the resulting hypotenuse will be 1,178 feet. The actual Great Circle deviates slightly from being a perfect circle (note 8). However, the dotted white line shows how a circle equal to 1,178 feet in diameter closely matches the Great Circle.

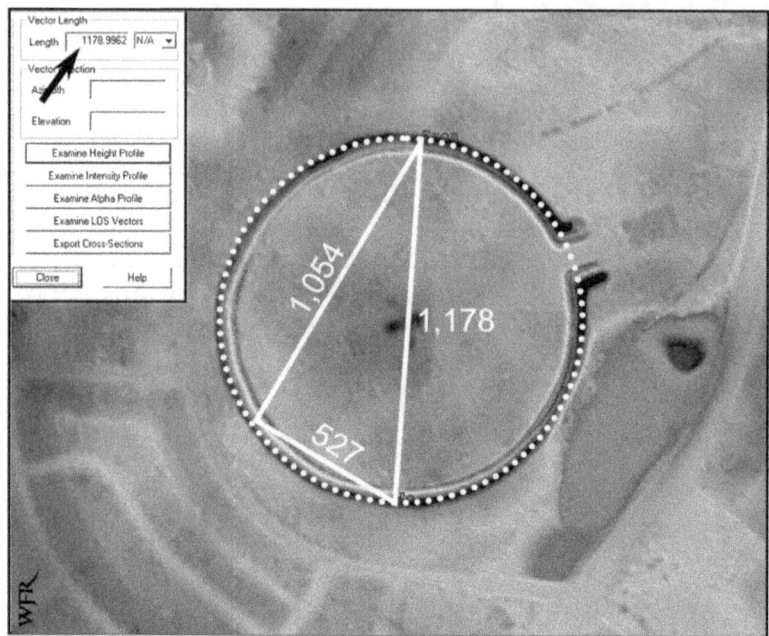

Above: experimental plot of Chenopodium sp. (goosefoot) being grown at the Newark Great Circle earthwork.

There are many interesting relationships among the earthworks that make-up the Newark Earthworks Complex. For example, the circumference of the Great Circle is equal to the perimeter of the Wright Square to within one percent. Calculations based on an ideal circumference of 3,700 feet for the Great Circle and perimeter of 3,713 feet for the Wright Square (also see Romain 2000:40).

Below: Thomas's 1894 maps over LiDAR image showing the spatial relationships between the Great Circle, Wright Square and waterways to the east.

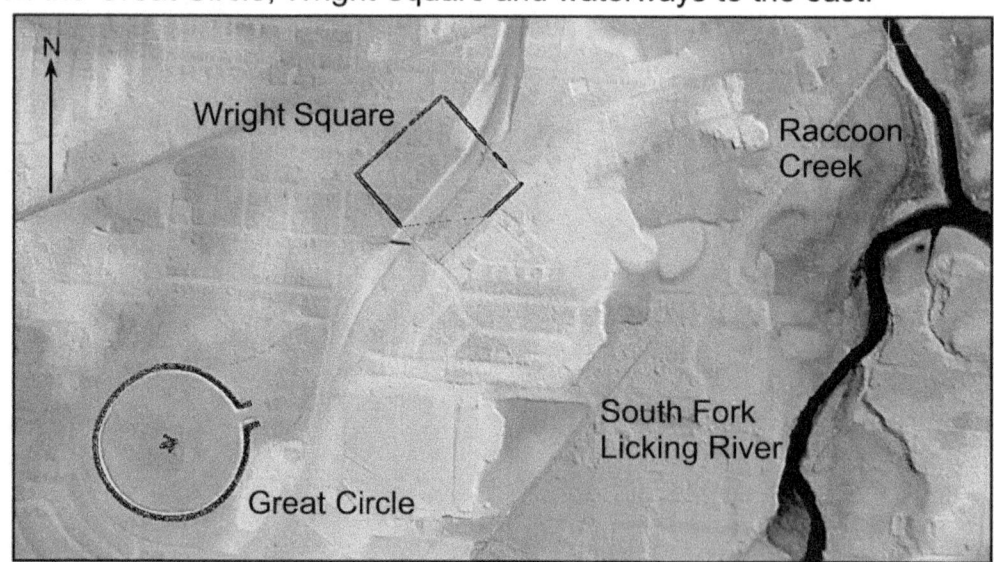

Figure 3.14. The Newark Earthworks: Great Circle Size and Alignment

The Great Circle is not quite a perfect circle. It is a bit 'lopsided.' This can be accounted for if the Great Circle was constructed using the perimeters of two different circles - equal in size, but each centered on a different point. It may be that one-half of the earthwork was laid-out using a length of rope. But when it came time to lay-out the other half, the original point of origin was not accurately located. The result was another 1/2 circle of proper size, but not coincident with the trajectory of the first half - hence a lopsided circle. In figure A, the solid white circle defines the outside edge of the north perimeter wall. The dashed line circle in figure B defines the outside of the south wall. Both the solid and dashed-line circles are 1,178 feet in diameter. Figure C shows how when combined, the two 1,178-foot circles define the peculiar shape of the Great Circle. The lopsidedness of the earthwork would not have been noticed at ground level until it was time to orient it to the moon's minimum north rise position.

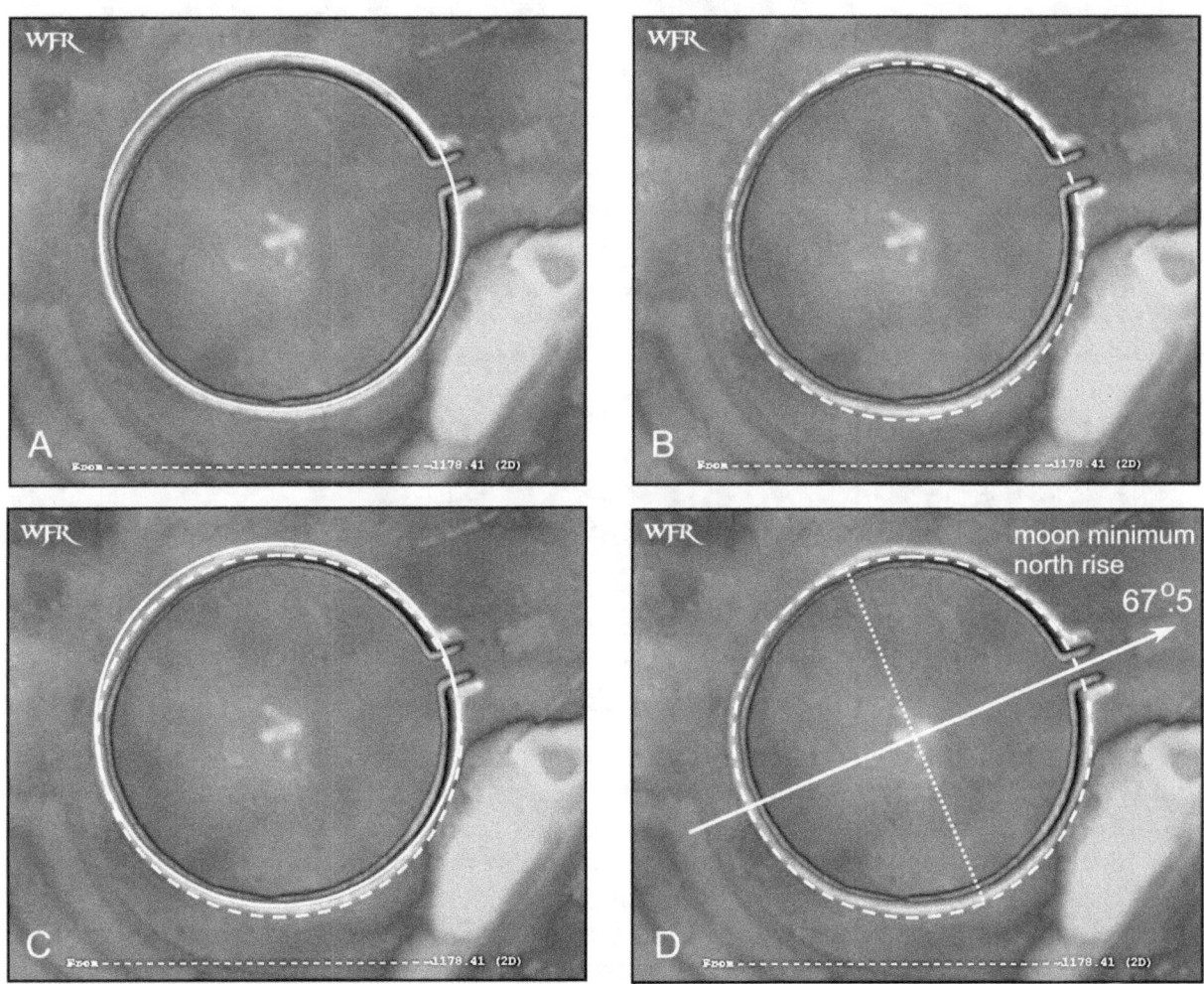

Once the initial work of laying-out the perimeter of the Great Circle was completed, a decision would had to have been made to use either the north or south planning circle for orientation purposes. It appears that the south planning circle (dashed line) was chosen. Using this circle as the guide, the 'center' of the Great Circle was established. From that 'center' the moon's minumum north rise azimuth was used to establish where the entrance to the earthwork would be (figure D).

Figure 3.15. The Newark Earthworks: Great Circle Alignment

The Newark Great Circle is aligned to the moon's minimum north rise as viewed from the center of the Circle through the gateway entrance (where 'center' is established per Figure 3.14).

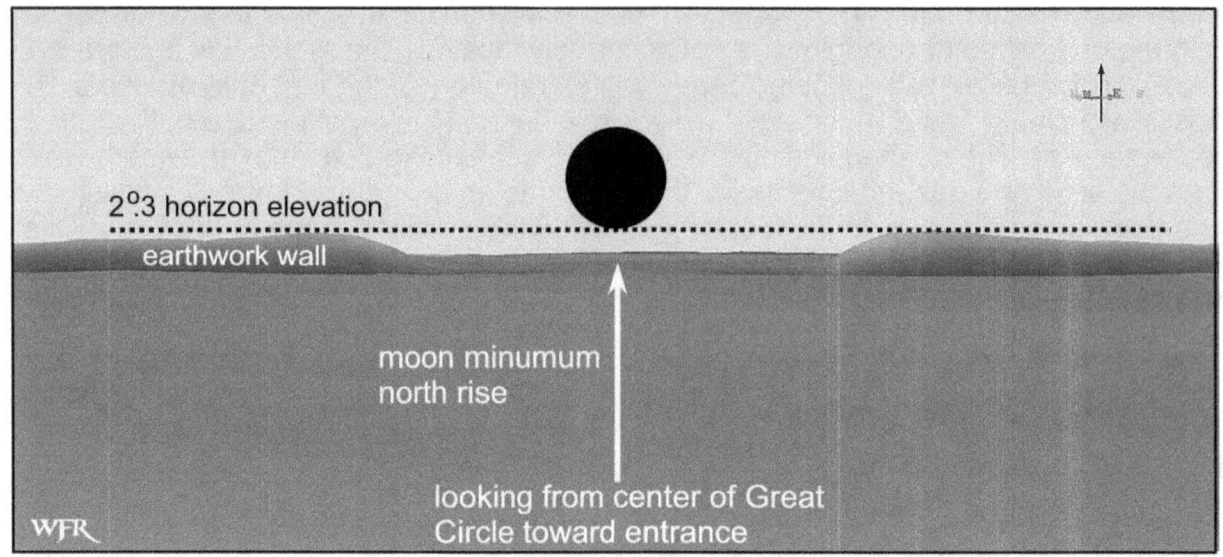

Above: LiDAR image showing simulated moon rise viewed from the center of the Great Circle. The moon will appear centered in the gateway opening, level with the entrance walls (note 9).

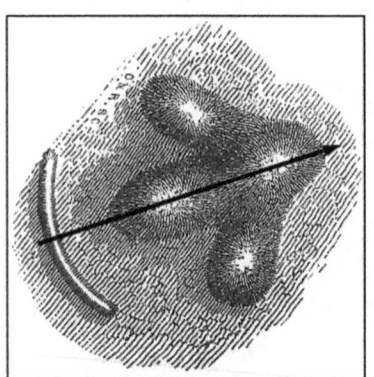 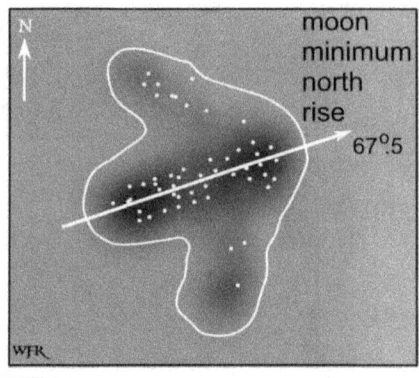 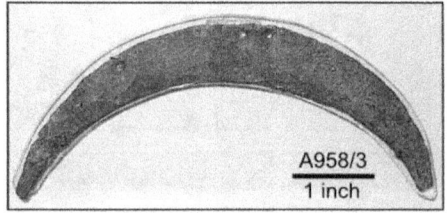

Above left: in addition to its lunar alignment, further support for the association of the Great Circle with the moon is indicated by the existence of a crescent-shaped mound adjacent to the so-called Eagle Mound in the center of the Great Circle (Squier and Davis 1848:fig. 12, arrow added). Note that the moon minumum north rise azimuth bisects the lunar crescent shape of the mound.

Above center: excavation revealed that the so-called Eagle Mound in the center of the Great Circle covered a temple, or Spirit House structure (Greenman 1928). In the figure above, the posthole pattern as found by Greenman is superimposed over a LiDAR image. From this it appears that the Spirit House was aligned along its major axis to the moon's minimum north rise through the entrance gateway.

Above right: among the items fround in the Eagle Mound (Greenman 1928) were a clay basin containing burned bone and a copper piece resembling the moon's crescent shape. Photo by the author.

Figure 3.16. The Newark Earthworks: Ellipse

Except for remnants of a few burial mounds, the Newark Ellipse has been destroyed by urban expansion. Early accounts suggest the enclosure was used as a burial ground. The Salisburys (1862:10-11) reported "eleven mounds of various sizes" within the enclosure and give the dimensions of the Ellipse as 1,800 x 1,500 feet. Below is an enlarged detail from the Salisbury 1862 map, courtesy of the American Antiquarian Society.

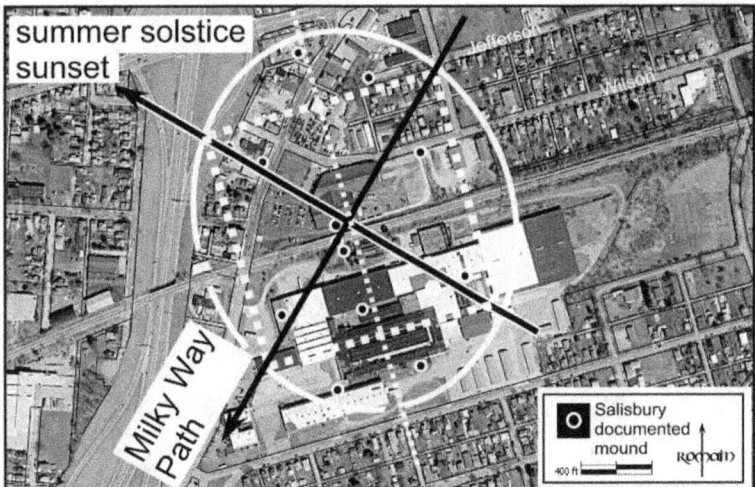

Triangle shows mound containing mica and "numerous skeletons"; X marks where "many skeletons in a good state of preservation were found"; center feature (arrow) was three conjoined mounds with a "tier of skeletons" in the center mound (Salisbury 1862:12-13). (The earlier shown 1860 Wyrck map shows 17 mounds within the Ellipse.)

Below: using the Salisbury (1862) map and other information, the location of the Ellipse can be plotted with reasonable accuracy. Aerial photo from Licking County Integrated Mapping System.

Within limits of accuracy constrained by use of the Salisbury map, the diagonal axis of the Ellipse appears oriented to the summer solstice sunset. Half of the mounds within the Ellipse are situated along this axis. Two others flank the south entrance to the earthwork. The parallel walls leading southwest from the Ellipse are orthogonal to the summer solstice sightline. Further, these parallel walls extend along the trajectory of the Milky Way on the night the summer solstice sunset. In terms of geometry, the width of the Ellipse (1,500 ft.) is nearly equal to the diagonal of a square having sides equal to 1 HMU (1,054 feet) (diagonal of 1 HMU square = 1,489 ft.).

Figure 3.17. The Newark Earthworks: Geller Hill Triangle

Geller Hill is a glacial kame prominently situated in the otherwise flat Newark Valley (note 10). Geller Hill was integral to the design and layout of the Newark Earthwork Complex in several ways.

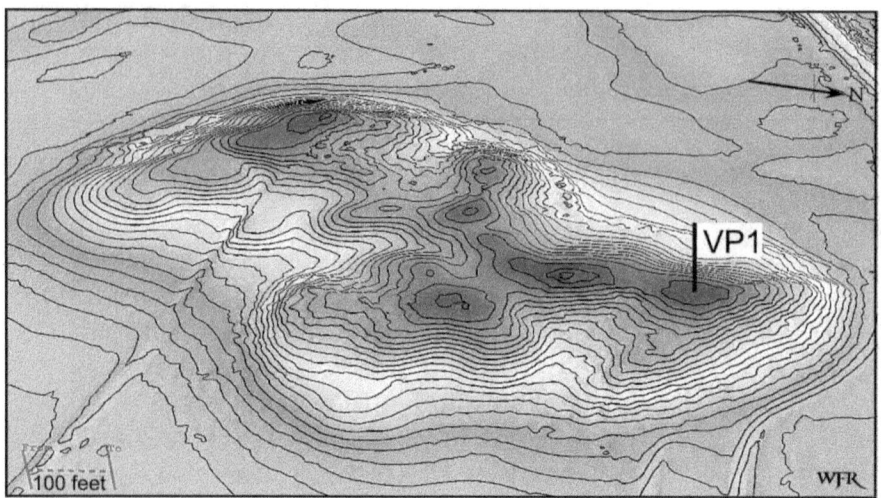

Above: LiDAR image of Geller Hill with contour lines at 2-foot intervals.

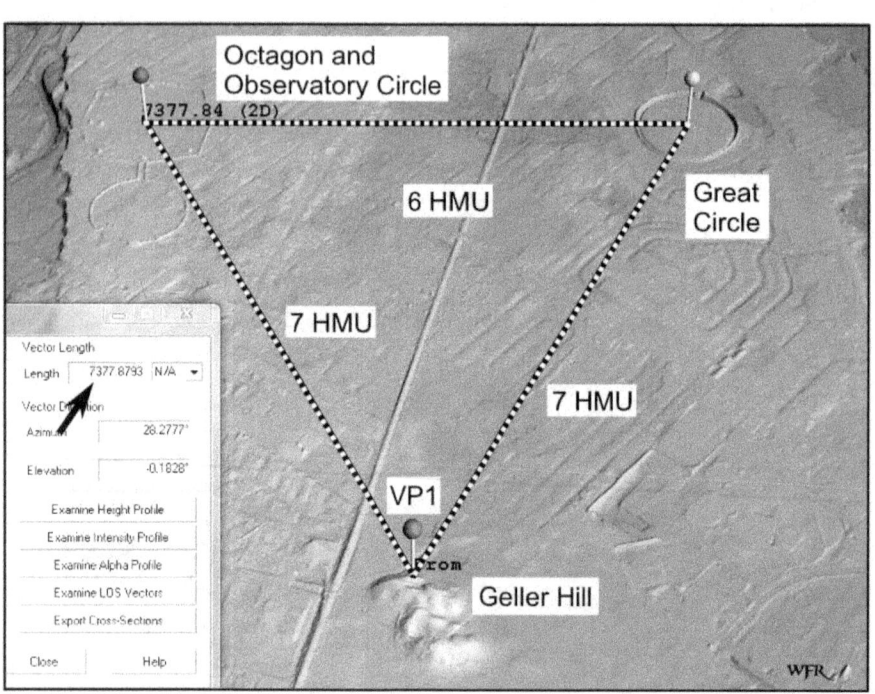

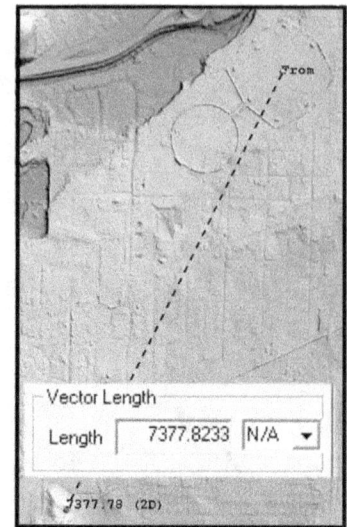

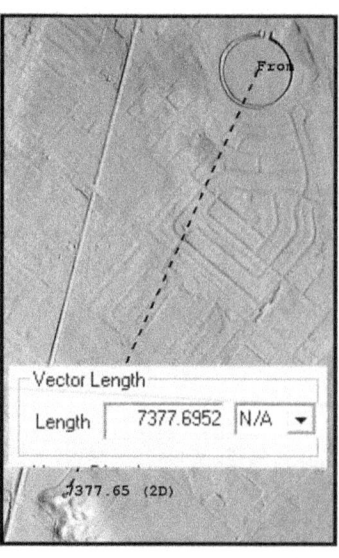

Above left: the distance from vantage point 1 (VP1) on Geller Hill to the center of the Octagon and to the center of the Great Circle is 7,378 feet (or 7 HMU) to within less than one percent. The base leg of the triangle is 6 HMU to within 2 percent. The result is a near-perfect isosceles triangle across distances of more than one mile.
Above right: details of LiDAR measurements.

Figure 3.18. The Newark Earthworks: Moonrise from Geller Hill

Due to its height above the Newark Valley and easy access by foot, Geller Hill provides an ideal location for skywatching (note 11).

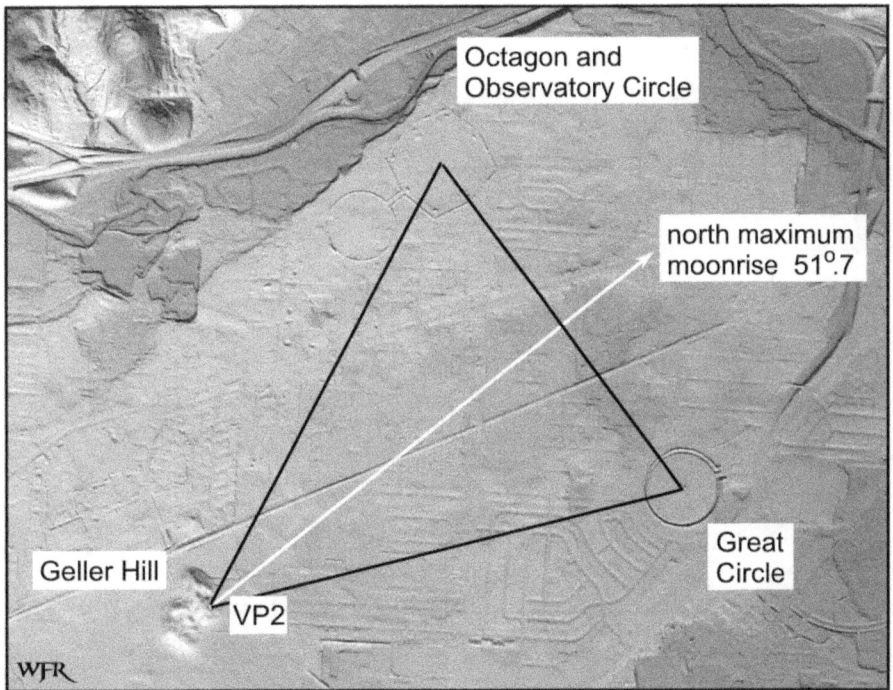

Above: during Hopewell times, an observer standing at vantage point 2 (VP 2) on Geller Hill would have seen the maximum north moonrise over the Newark Valley every 18.6 years.

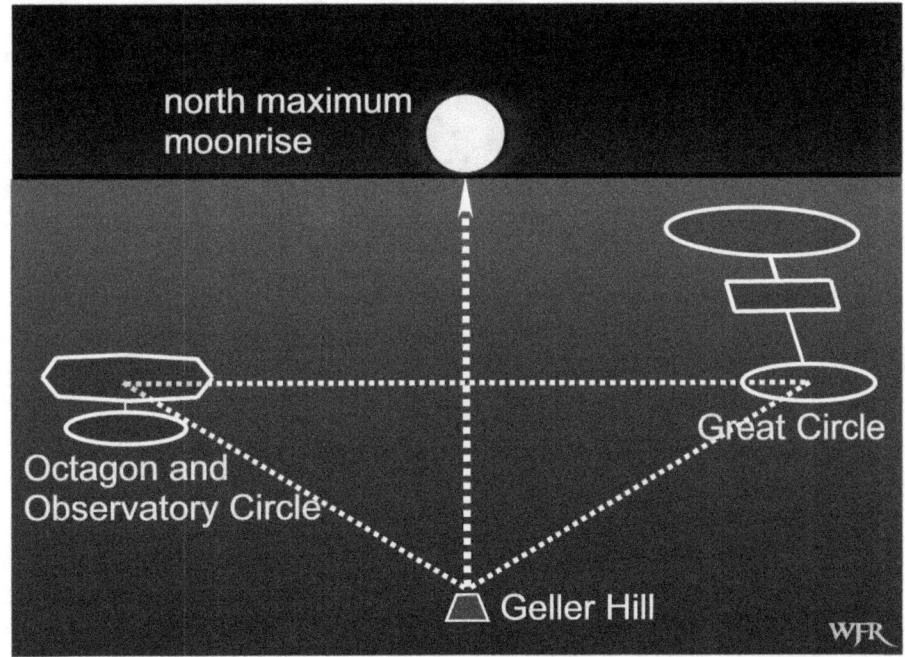

Above: as the moon rose above the distant horizon it would have appeared balanced between the Octagon and Great Circle.

Figure 3.19. The Newark Earthworks: Solstice Alignment

The Newark complex incorporates both solar and lunar alignments in its design. The images below show how central features of the Great Circle and Observatory Mound are aligned to the summer solstice sunset (note 12). Viewed from the inside entrance to the Great Circle, the summer solstice sun will set in alignment with the Observatory Mound. The Observatory Mound is located at the far end of the Observatory Circle, on the major axis of that earthwork.

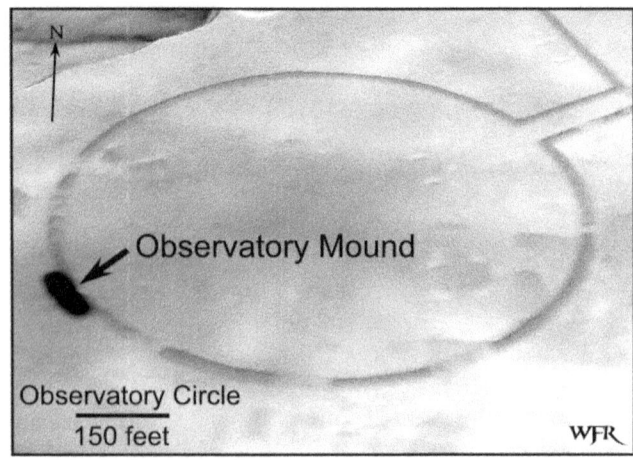

Above: Squier and Davis (1848:fig. 16) sketch of the Observatory Mound.

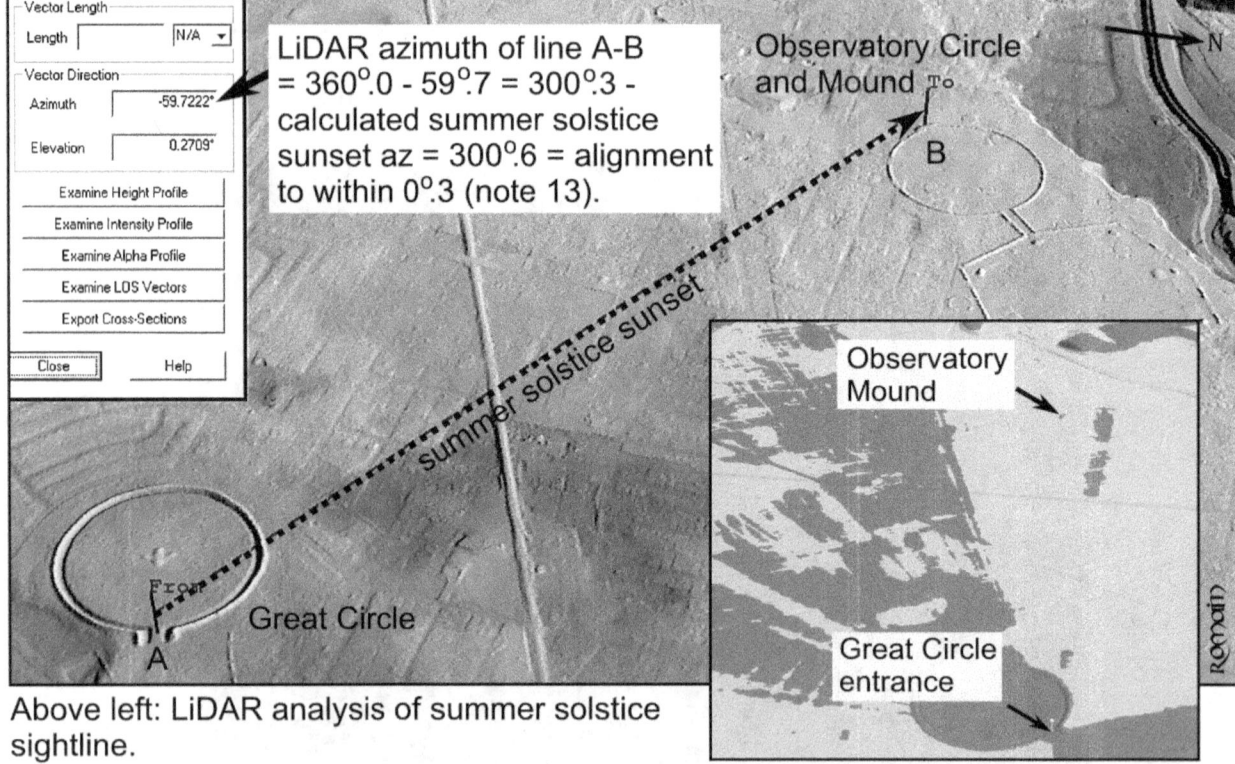

Above left: LiDAR analysis of summer solstice sightline.

Above right: to test if the Observatory Mound would have been visible from the Great Circle entrance along the solstice sightline, LiDAR line-of-sight analysis is used. In the bottom image, all areas in dark gray - including the Observatory Mound, would have been visible from the inside entrance to the Great Circle (assuming current restored wall heights and observer height of 6 feet).

68

Figure 3.20. The Newark Earthworks: Great Hopewell Road

The Great Hopewell Road was a set of low parallel walls, about 150 feet apart, that extended from near the Octagon and Observatory Circle to Ramp Creek - a distance of about two and one-half miles. The Road has been mostly obliterated, except for a small section in a woodlot north of the Newark-Heath airport.

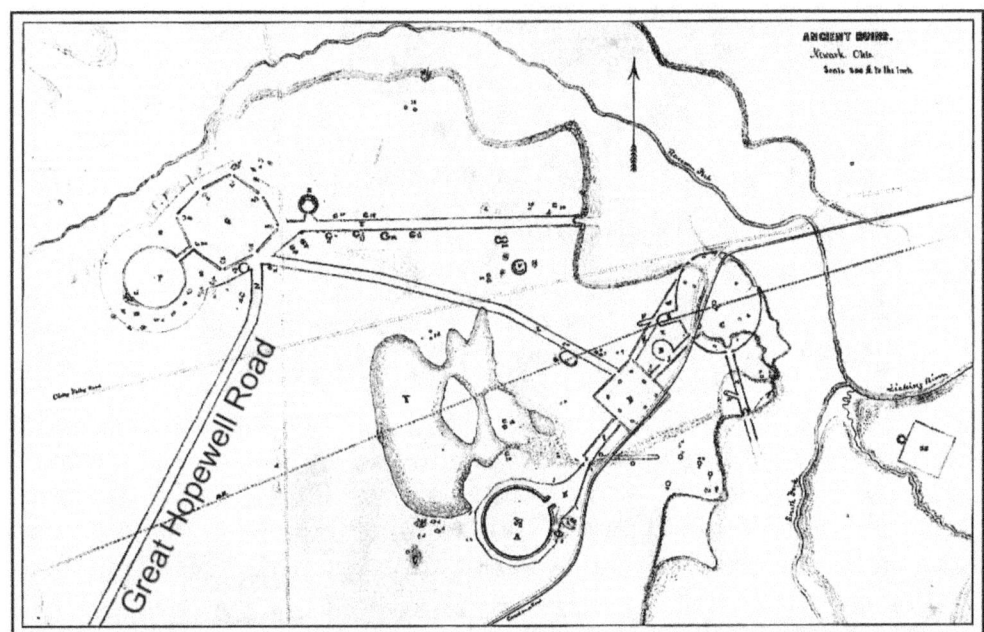

Above: 1862 Salisbury map (annotation added). Courtesy of the American Antiquarian Society.

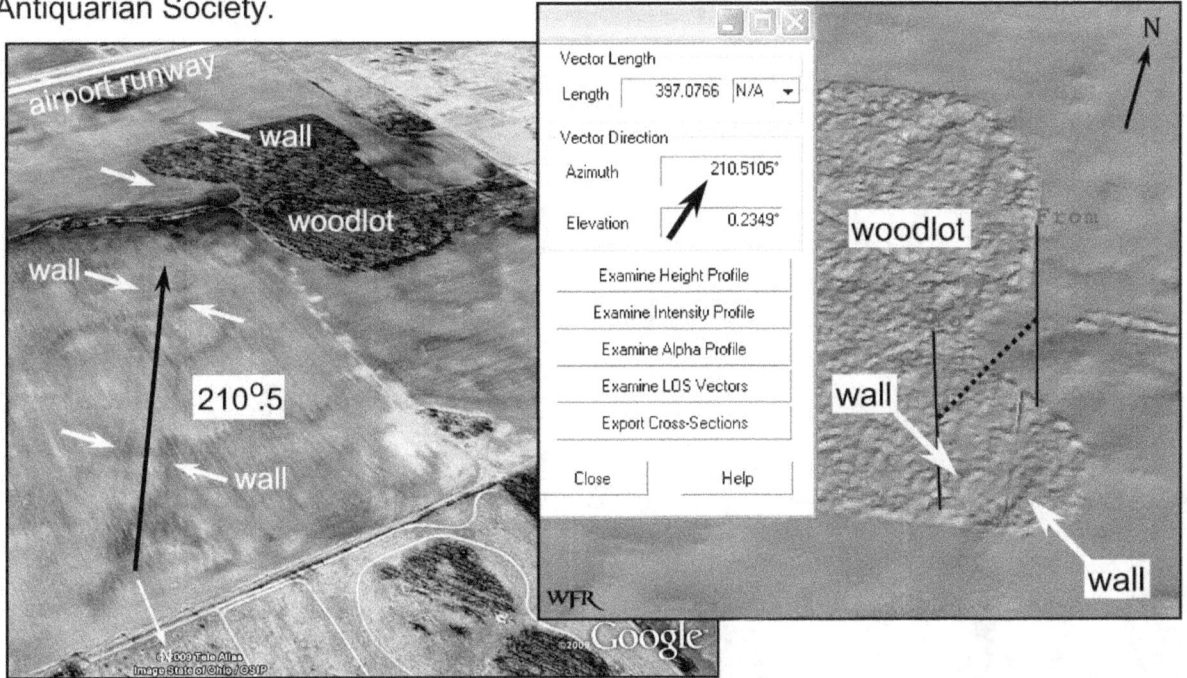

Above left: when conditions are just right, discolorations in the soil show the GHR. Above right: LiDAR analysis of a topographically visible section of the Road in a woodlot north of the Newark-Heath airport provides an azimuth of 210.5 degrees (theta corr of 0.03 is negligible).

Figure 3.21. The Newark Earthworks: Great Hopewell Road in Woodlot

Above: view of extant section of Great Hopewell Road situated in a woodlot north of the Newark-Heath airport. White arrows show the southeast edge of one of the parallel walls. Dashed arrow shows the trajectory of the wall segment. Photo by the author. Below left: aerial photo of Great Hopewell Road north of the airport taken in the 1930s. From the Dache Reeves Collection, National Anthropological Archives, Smithsonian Institution.

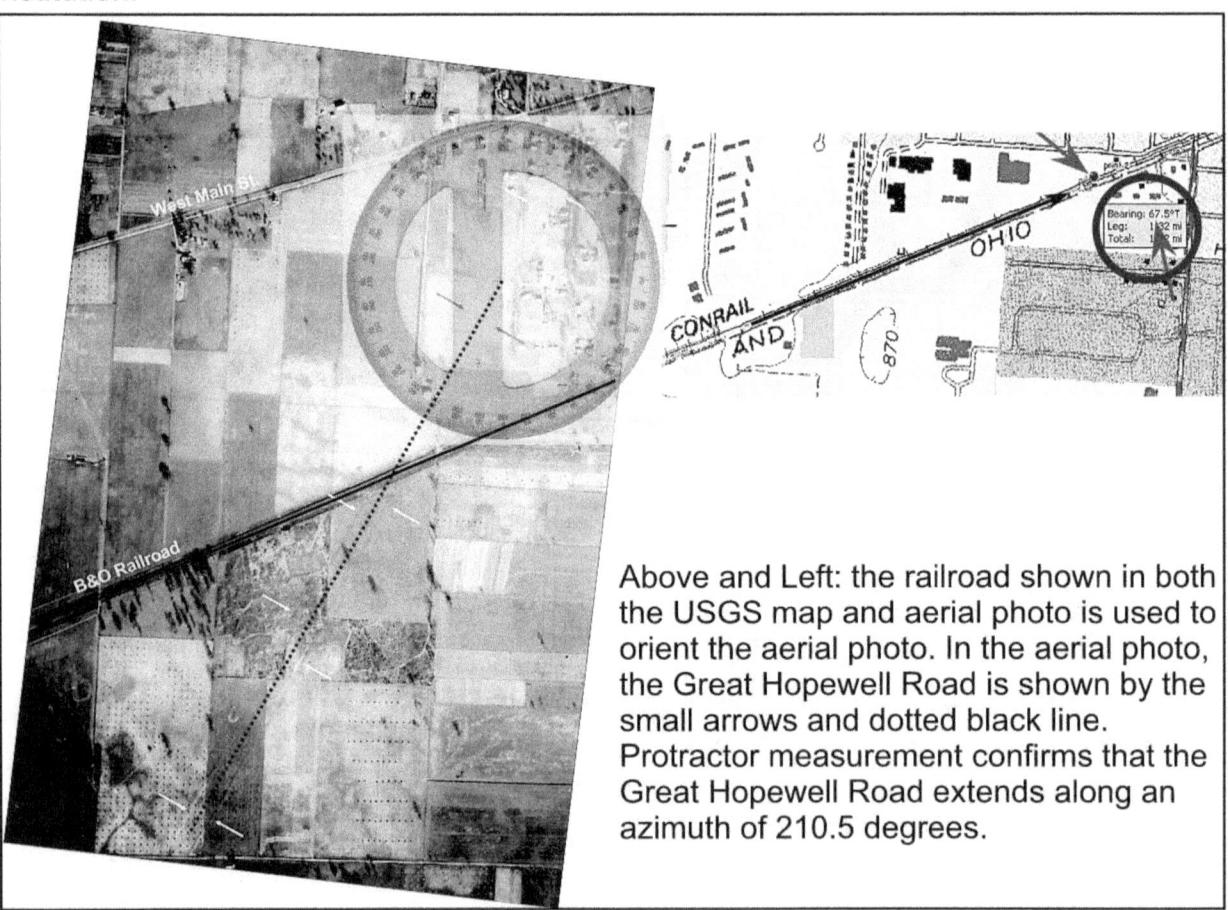

Above and Left: the railroad shown in both the USGS map and aerial photo is used to orient the aerial photo. In the aerial photo, the Great Hopewell Road is shown by the small arrows and dotted black line. Protractor measurement confirms that the Great Hopewell Road extends along an azimuth of 210.5 degrees.

Figure 3.22. The Newark Earthworks: Great Hopewell Road Profile and Origin
The Great Hopewell Road is not a road in the traditional sense. In profile it is concave. This concave shape may have had cosmological significance (note 14).

Above left and right: LiDAR profiles across the Great Hopewell Road. Each dashed line in the image on the left yielded one of the profile lines shown in the image on the right.

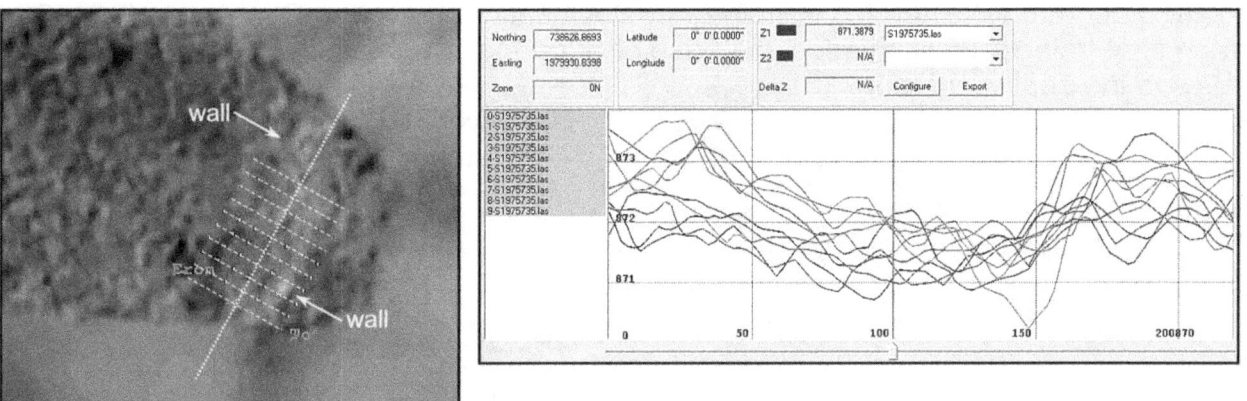

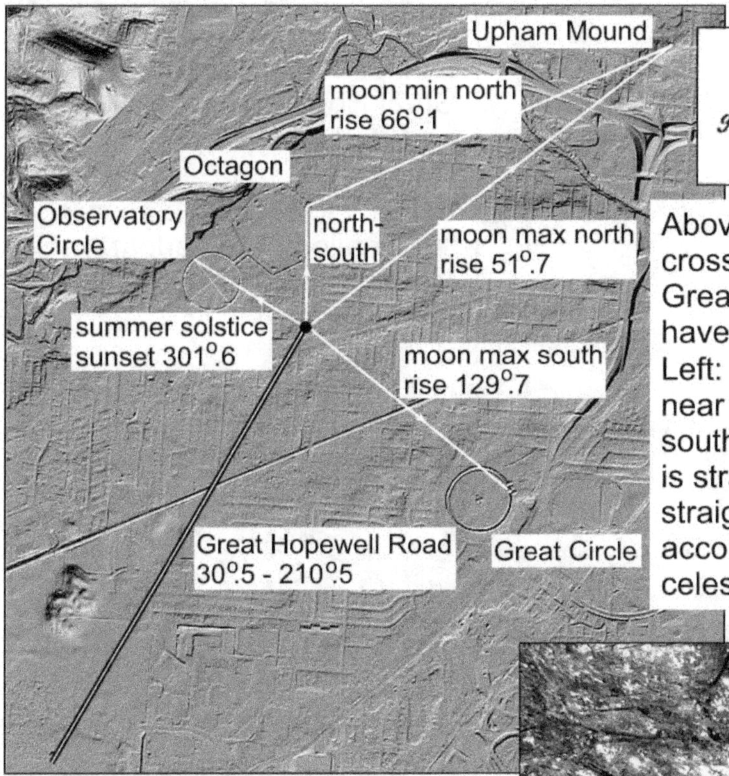

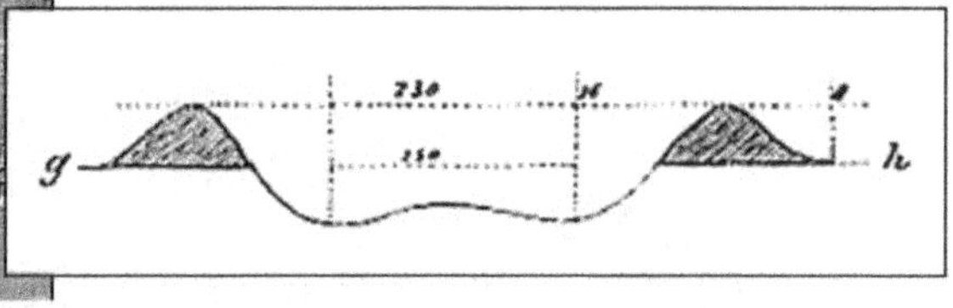

Above: Squier and Davis (1848:Pl. XXVI) cross section of Marietta Sacra Via. The Great Hopewell Road and Sacra Via have similar profiles.
Left: the Great Hopewell Road begins near the Octagon and curves to the south. For the rest of its course, the Road is straight. The point of origin for the long, straight section of the Road can be accounted for by the intersection of celestial azimuths (note 15).

Above: point of origin for straight section of Great Hopewell Road shown by black dot. Trajectory and location for straight section of Road based on aerial photos and woodlot location.
Right: photo of the Upham Mound.

71

Figure 3.23. The Newark Earthworks: Great Hopewell Road Trajectory

LiDAR data do not reveal evidence for the Great Hopewell Road south of Ramp Creek in Newark-Heath. If, however, the LiDAR-determined trajectory of the GHR (az = 210.5 degrees) is extended from the woodlot location where the Road can still be seen, that trajectory will intersect the base of Sugarloaf Mountain in Ross County - 50 miles to the southwest (note 16).

Right: starting point for plotting the trajectory of the Great Hopewell Road. Map provided by MyTopo.

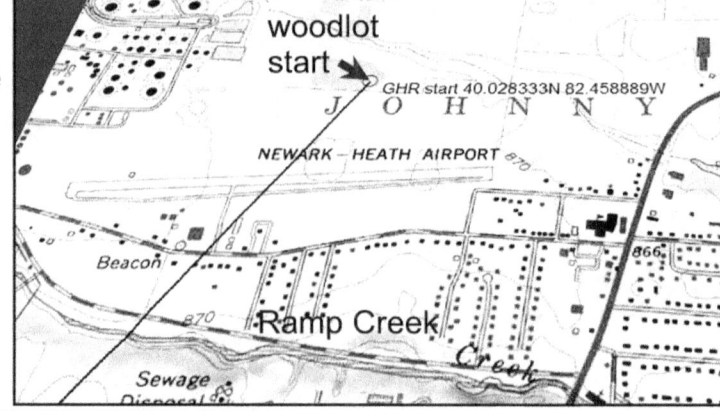

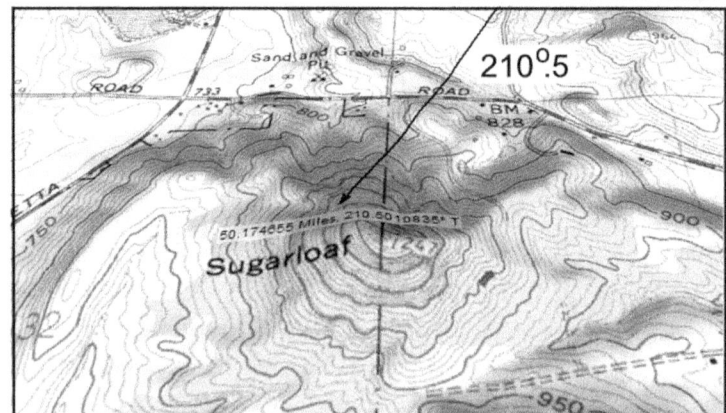

Left: end point for plotting the Great Hopewell Road. Map provided by MyTopo.

Below: LiDAR image showing the location of Sugarloaf Mountain in relation to the Scioto River Valley.

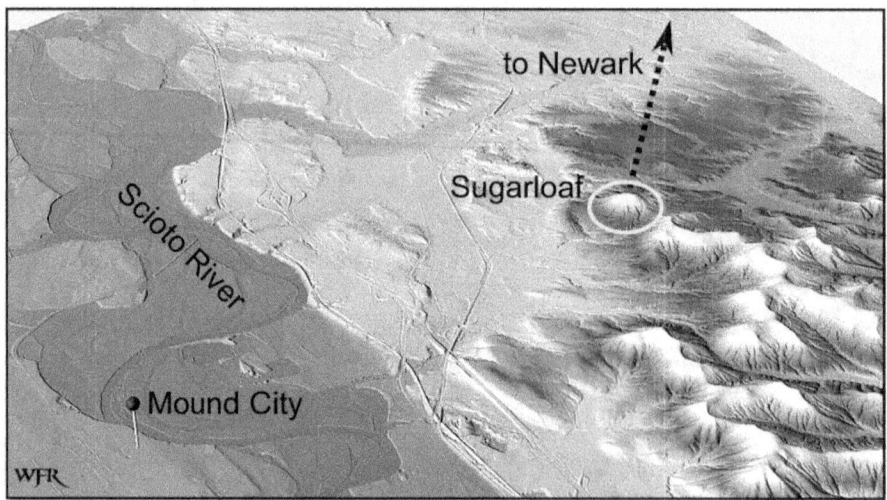

Above: Sugarloaf is a prominent mountain that marks the Chillicothe earthwork area. Because Sugarloaf is a bit of an outlier, it stands out against the horizon and is visible for miles around. As discussed in Chapter 5, Sugarloaf Mountain may have been an important axis mundi and portal to the Otherworld for souls of the dead.

Figure 3.24. The Newark Earthworks: Milky Way Direction

In addition to its terrestrial alignment to Sugarloaf Mountain, the Great Hopewell Road has celestial associations to include an alignment to the Milky Way. During Hopewell times in south-central Ohio, on the date of the summer solstice, just after sunset - at nightfall, the Milky Way would have become visible (note 17). It would have appeared as a band of light arcing across the night sky.

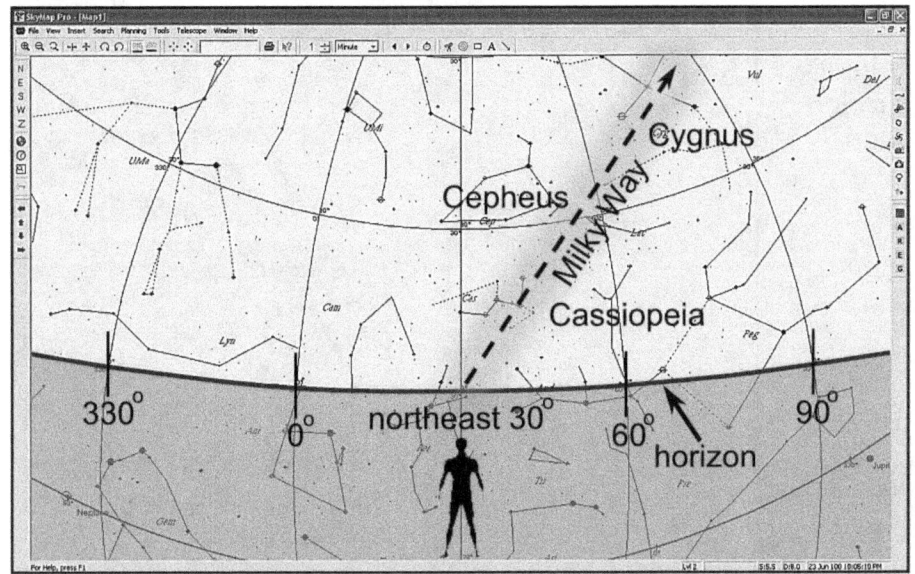

The above view shows a central Ohio Hopewell observer looking northeast at nightfall on the date of the summer solstice any year between 100 B.C. to A.D. 200. Looking in this direction, the Milky Way appears to emerge from the distant horizon. Milky Way constellations in this direction include Cassiopeia, Cepheus, and Cygnus.

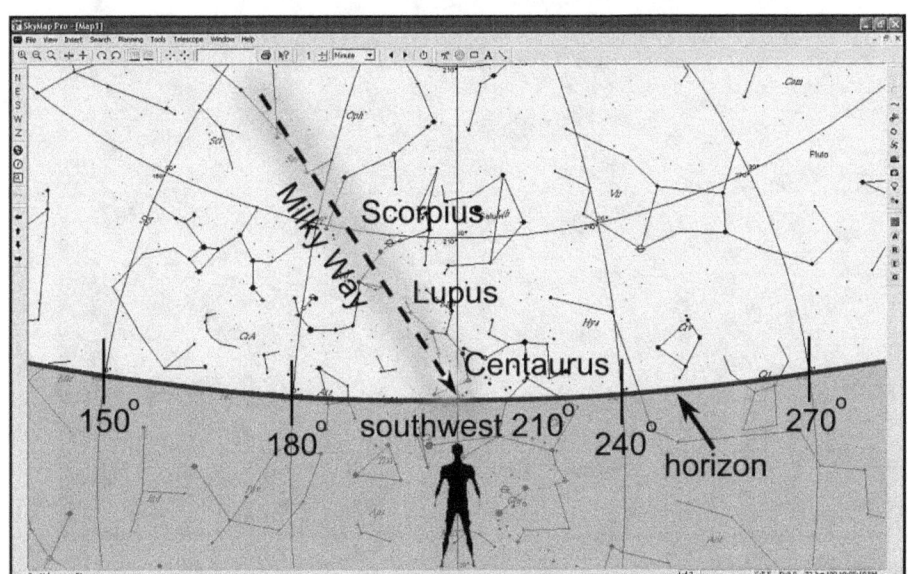

In this view, the observer is facing southwest - where the Milky Way appears to descend into the horizon. Milky Way constellations in this direction include Scorpius, Lupus, and Centaurus. Hopewell observers probably did not draw the constellations the same way we do. However, the stars that make-up the Milky Way were the same. Above maps created using SkyMap Pro, annotation added.

Figure 3.25. The Newark Earthworks: Milky Way - Great Hopewell Road

The Milky Way arcs across the sky. If, however, the ground trajectory of the Milky Way is plotted from its apparent origin in the northeast, through an observer's position on the ground to its southwest terminus, that trajectory will extend along the 30°- 210° azimuth. (Below zenith view star map for south-central Ohio summer solstice at nightfall A.D. 100 ± 200 yrs.)

Right: annotated SkyMap Pro map showing ground trajectory of the Milky Way.

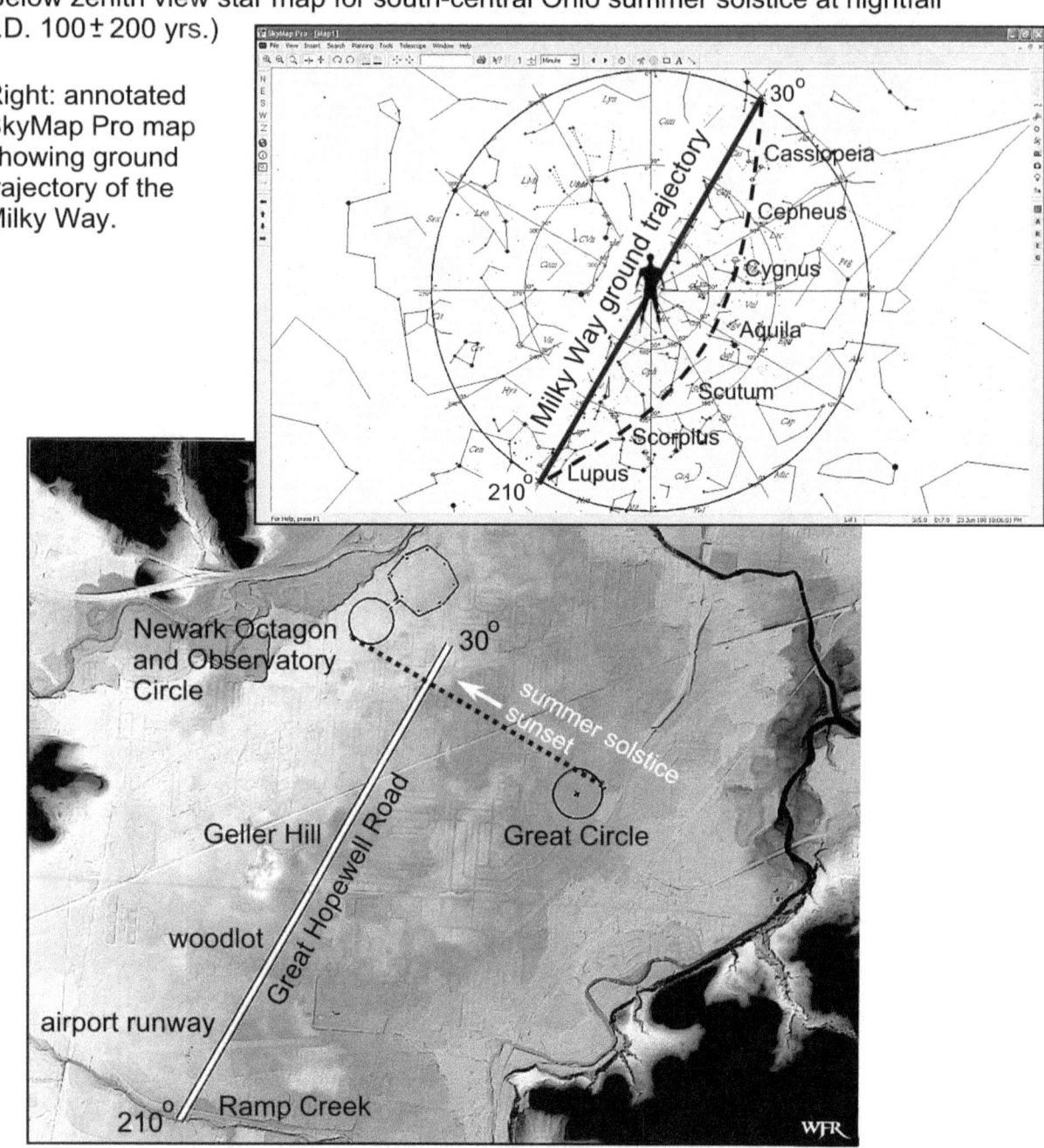

As shown above, the ground trajectory of the Milky Way and Great Hopewell Road extend in the same direction. During Hopewell times, at nightfall on the date of the summer solstice, the Great Hopewell Road mirrored on earth, the trajectory of the Milky Way. Also note, the summer solstice sunset sightline between the Great Circle and Observatory Mound (dotted) extends orthogonally to the Road azimuth (Romain 2005d).

Figure 3.26. The Newark Earthworks: Sun and Star Alignments

As if to emphasize the Milky Way alignment, the sightline from the VP1 overlook on Geller Hill to the center of the Octagon extends parallel to the Great Hopewell Road-Milky Way Path.

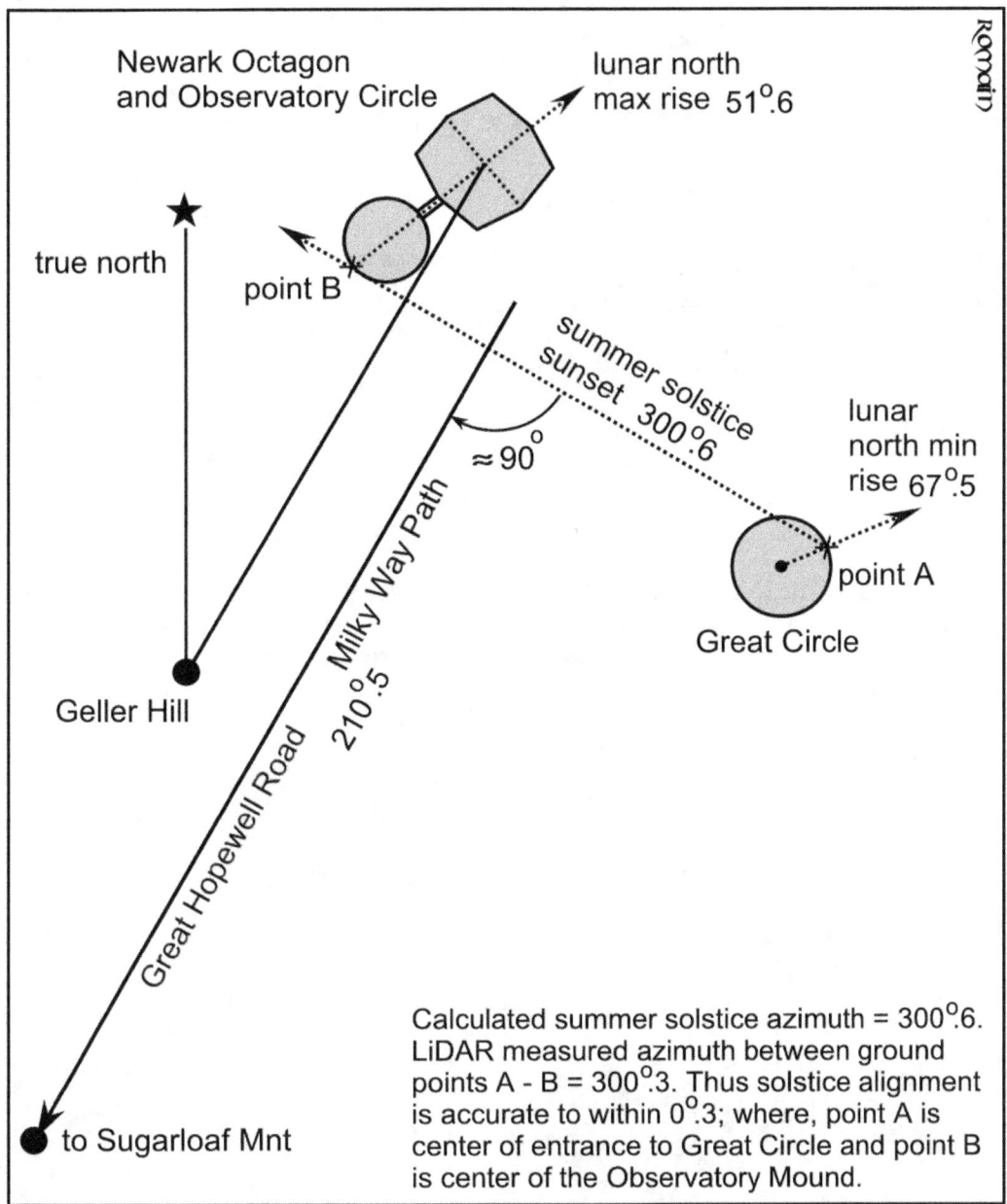

The multiple ways in which solar, lunar, and stellar alignments are linked through the earthworks suggests that the Moundbuilders perceived these things as related - temporally, geometrically, and cosmologically. Together with the living and dead, celestial events - further linked to the landscape, comprised a relational web. We cannot know for certain, but maybe the dead were believed to begin their journey to the Otherworld on the night of the summer solstice, during the years of the moon's maximum or minumum standstills. Maybe it was at this time that souls of the dead made their way from the Ellipse burial ground, along the Great Hopewell Road - Milky Way Path to the Sugarloaf Mountain axis mundi and beyond.

3.27. The Newark Earthworks: Small Circle

Sometimes overshadowed by its larger brethren, the Newark Small Circle has design features that relate it to the Newark Great Circle in several ways.

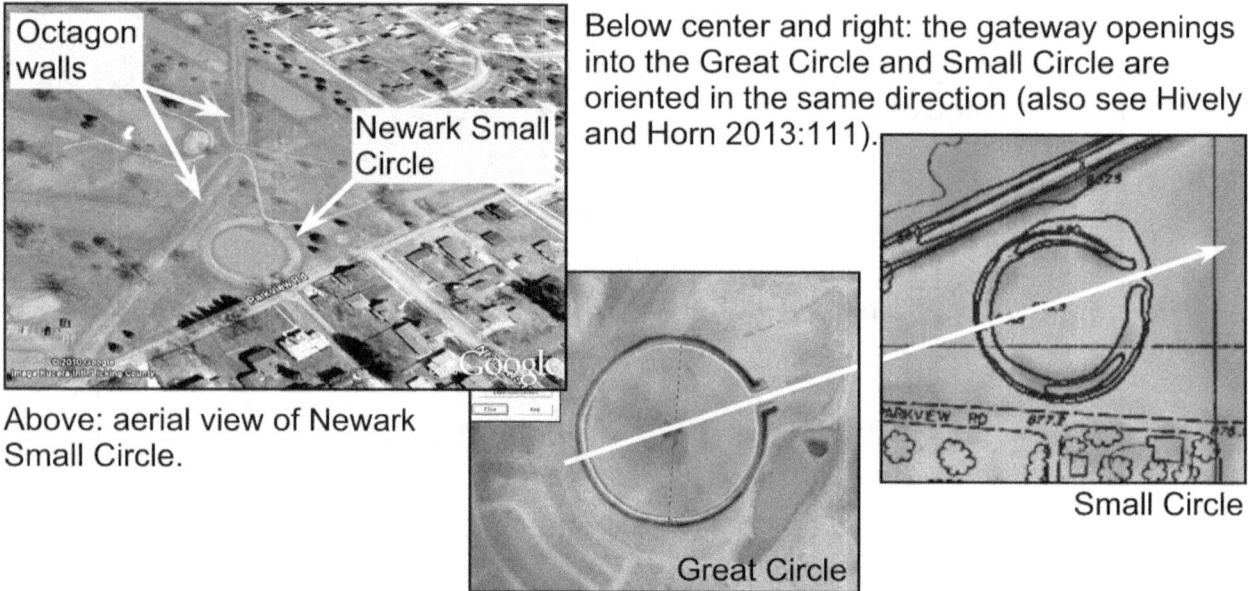

Below center and right: the gateway openings into the Great Circle and Small Circle are oriented in the same direction (also see Hively and Horn 2013:111).

Above: aerial view of Newark Small Circle.

Below center and right: the Great Circle is larger than the Small Circle by an even multiple - i.e., the Great Circle is eight times larger than the Small Circle (147.3 ft. x 8 = 1,178 ft.).

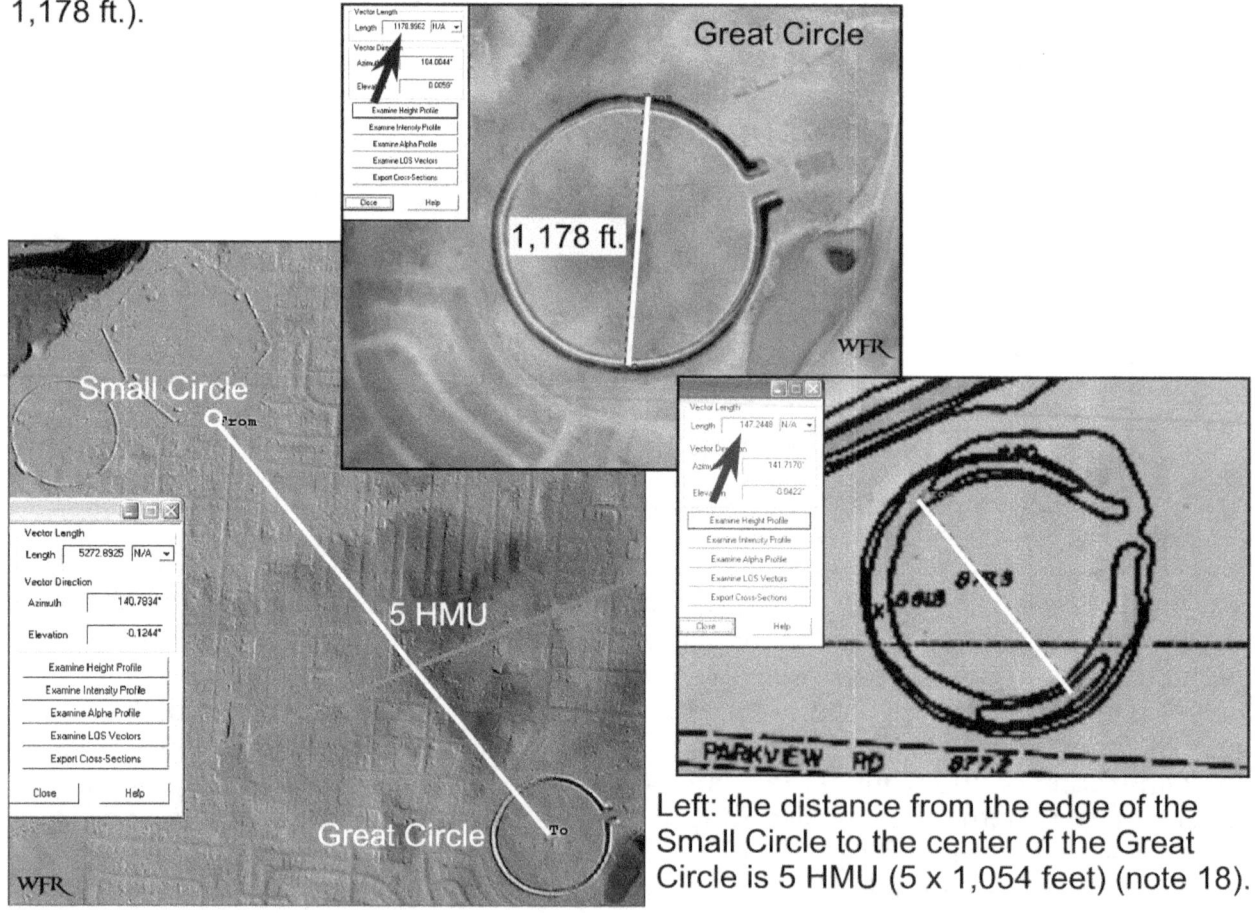

Left: the distance from the edge of the Small Circle to the center of the Great Circle is 5 HMU (5 x 1,054 feet) (note 18).

Figure 3.28. The Newark Earthworks: Small Circles

In addition to the small circle shown in Figure 3.27, a significant number of other small circle earthworks were located in the Newark complex. Many of these circles had crescent-shaped mounds in their interiors - similar to the crescent-shaped mound once located in the Great Circle. Only one crescent-in-circle earthwork still exists. That earthwork - known as Yost, is located 11 miles to the southeast in Perry County.

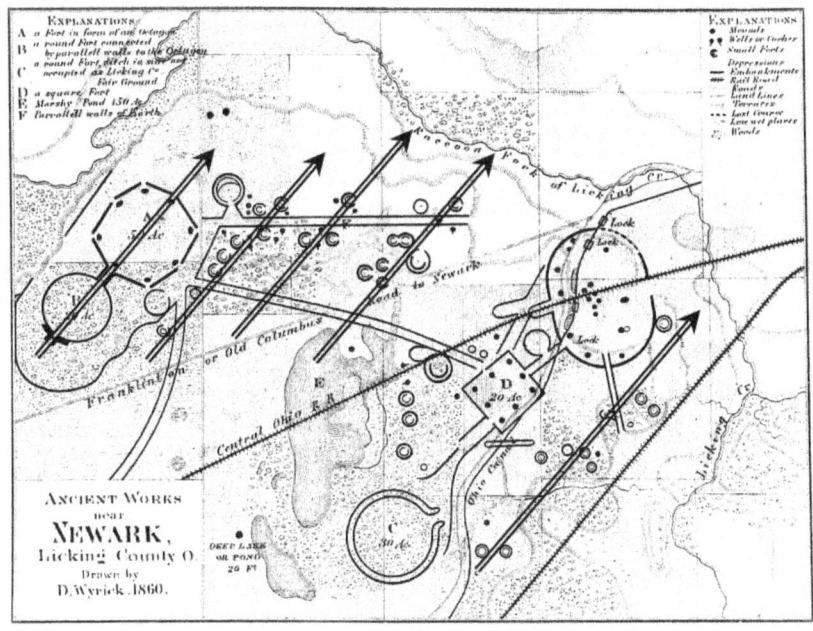
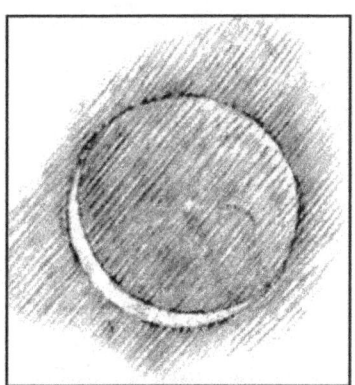

Above left: Wyrick 1860 map of Newark. If the map is generally accurate, then there seems to have been a tendency to situate many of the small circle earthworks on a trajectory parallel to the Octagon's major axis - i.e., to the moon's maximum north rise. Above right: the crescent in circle design may have its origins in a phenomenon known as 'earthshine.' Earthshine occurs during the moon's waxing and waning phases. Drawing by Leonardo daVinci (note 19).

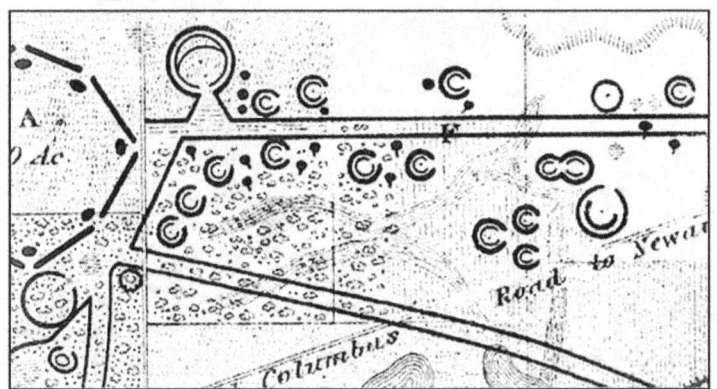
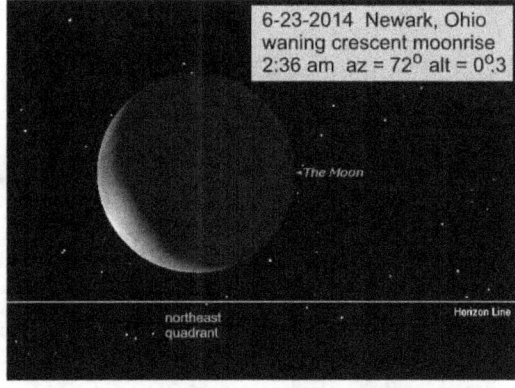

Above left: detail from Wyrick 1860 map. Following the symbolic associations for the Great Circle, the small circles with inner crescents may have been intended as moon symbols. Note how the circles open to the northeast (note 20).
Above right: computer view of waning crescent moon in northeast pre-dawn sky at Newark, June 23, 2014. Annotated image from Stellarium.

Figure 3.29. The Newark Earthworks: Relationships to Earth and Water
In addition to its celestial alignments, there are other special things about Newark.

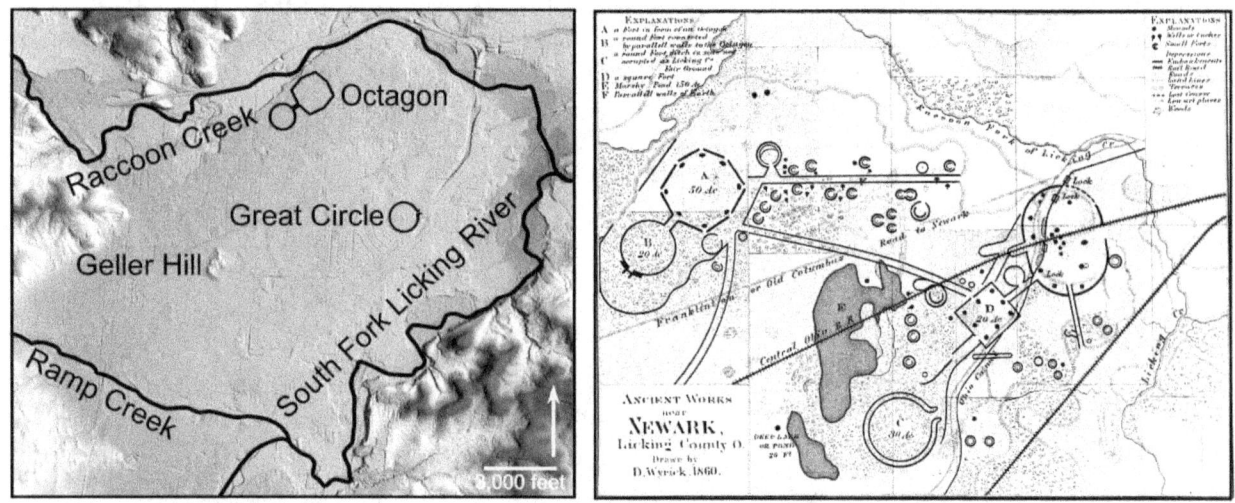

Above left: watercourses surround most of the Newark Earthworks Complex.
Above right: early maps show that before the area was drained, a lake was situated in the approximate center of the complex. The pond to the southwest of the Great Circle was 20 feet deep. Wyrick 1860 map, shading added.

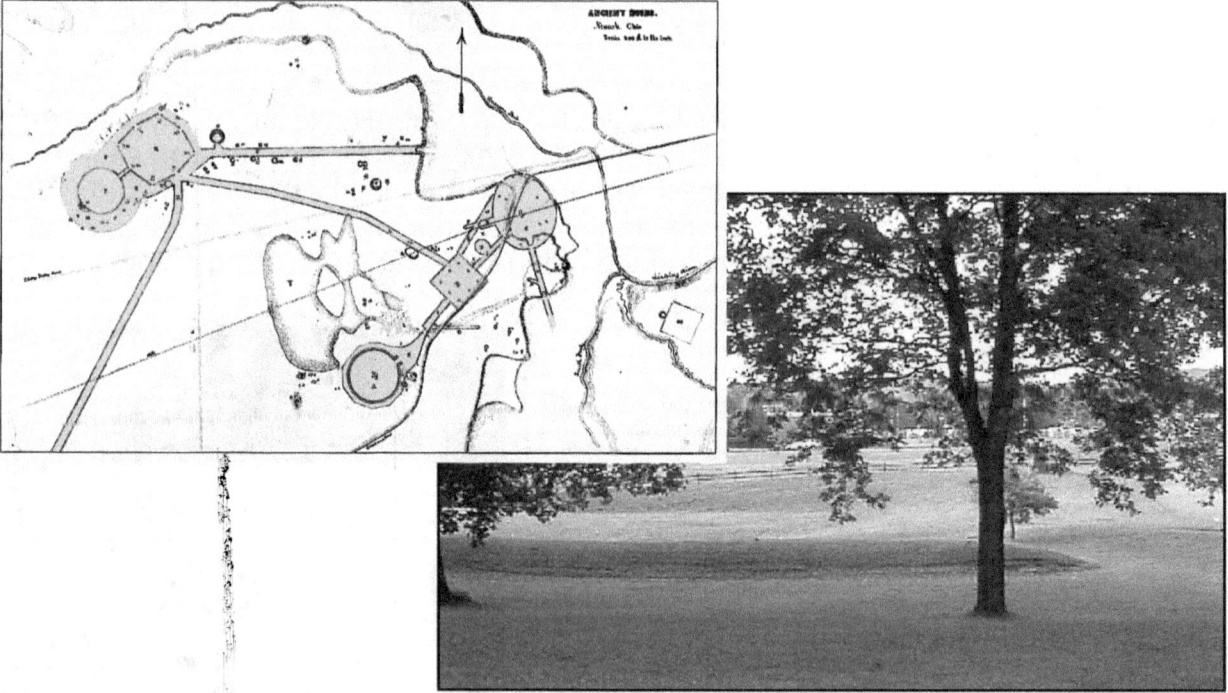

Above left: Salisburys (1862:Pl. 7 and pages 3-4) document the existence of a low wall surrounding the Great Circle and Octagon and Observatory Circle. Together with the parallel walls and walls of the Wright Square and Ellipse, a special area within is delineated. Given its low height, the perimeter wall feature was probably not meant to keep people or animals in or out. Perhaps it was a symbolic spirit barrier - to keep souls of the dead in, or other kinds of spirits, out (note 21).
Above right: restored section of the low perimeter wall that surrounds the Great Circle.

Figure 3.30. Flint Ridge

Flint Ridge extends roughly 10 miles in an east to west direction. Flint Ridge was the main source of Vanport Flint used extensively by the Hopewell. Vanport Flint is a very colorful flint and is the official gem stone for Ohio. In the LiDAR image below, prehistoric flint quarries appear as small pockmarks.

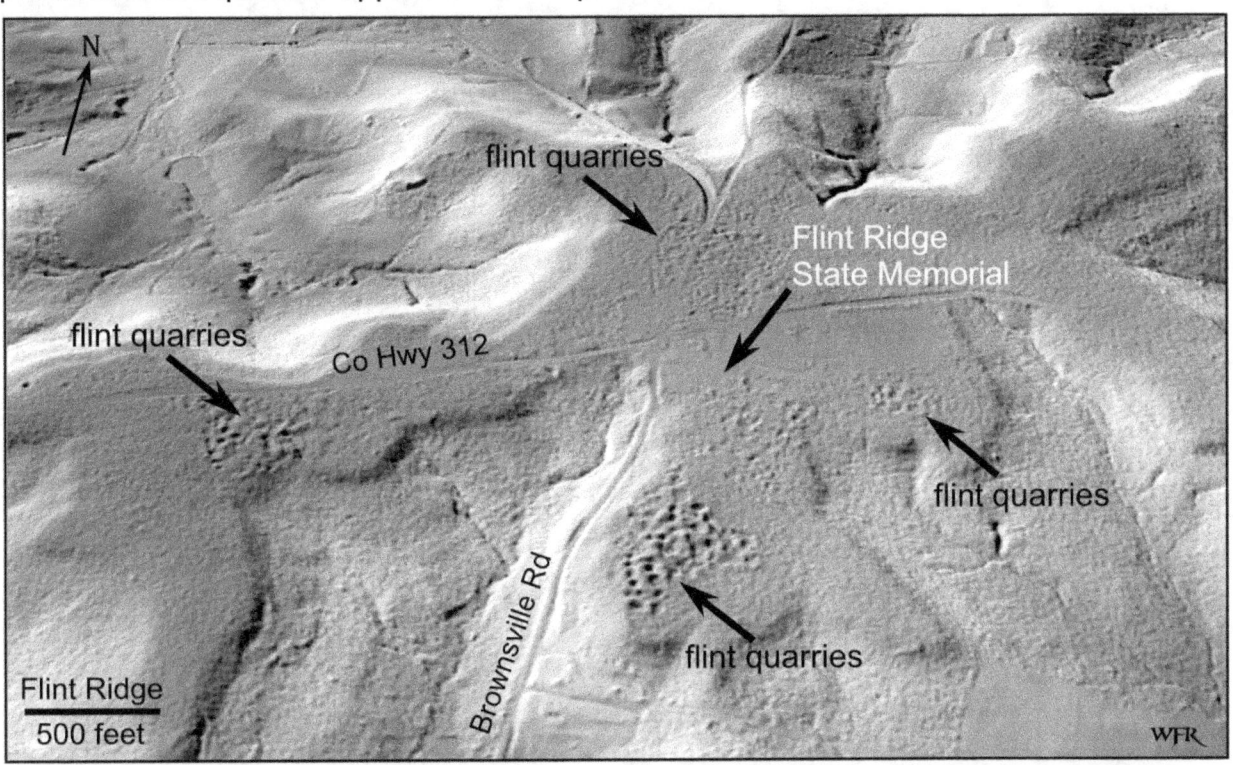

Above left: entrance to Flint Ridge Museum with flint nodules at base of sign.
Above right: water-filled prehistoric quarry at Flint Ridge.

Figure 3.31. Flint Ridge: Claylick Valley and Lunar Alignment

From Newark, the easiest way to Flint Ridge would have been to follow the valley formed by Swamp Run and Claylick Creek upstream for a distance of 10 - 12 miles.

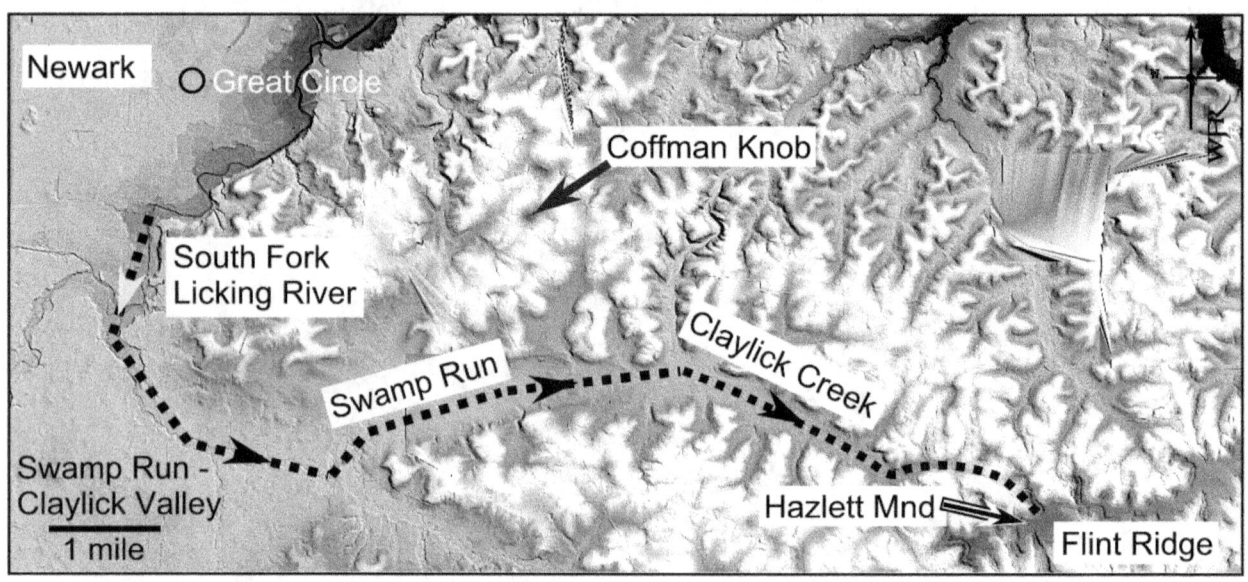

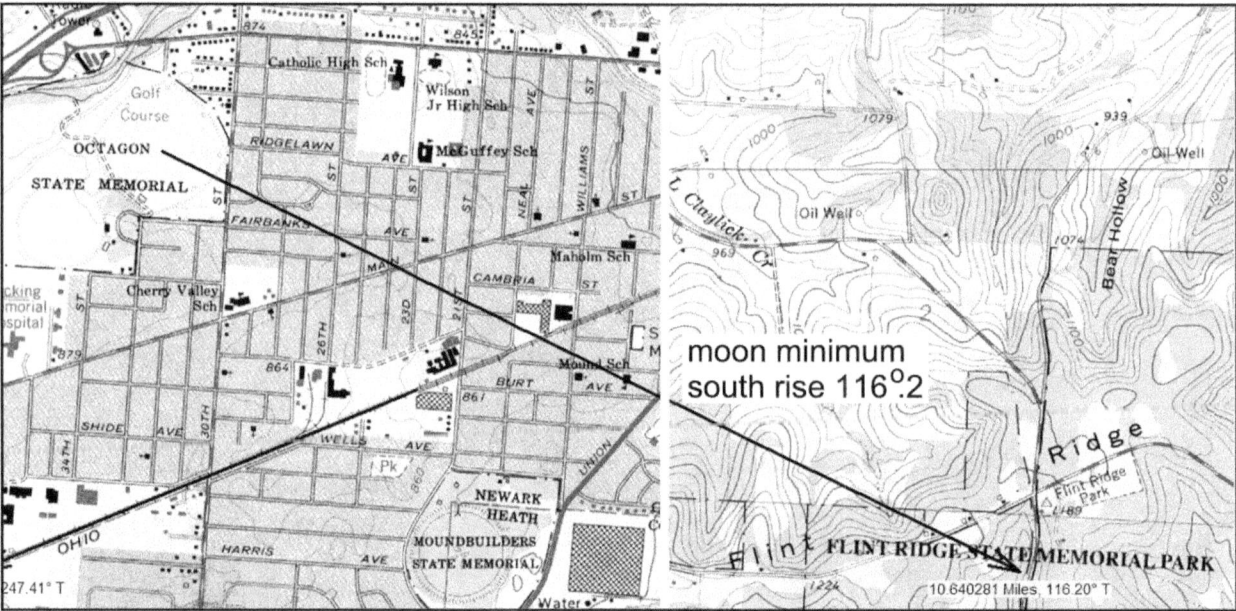

Above: Flint Ridge is located on the moon's south minimum rise azimuth as plotted from the center of the Newark Octagon (A.D. 100, corr h = 1.8). Map by MyTopo, annotation added.

Figure 3.32. Flint Ridge: Hazlett Mound

There are more than half-a-dozen prehistoric mounds and structures on the western aspect of Flint Ridge. Of these, the Hazlett Works were the most intricate. The Hazlett Works consisted of a low perimeter wall of flint blocks with several mounds within. The largest mound contained an elaborate Hopewell burial in a tomb made of flint boulders (Mills 1921).

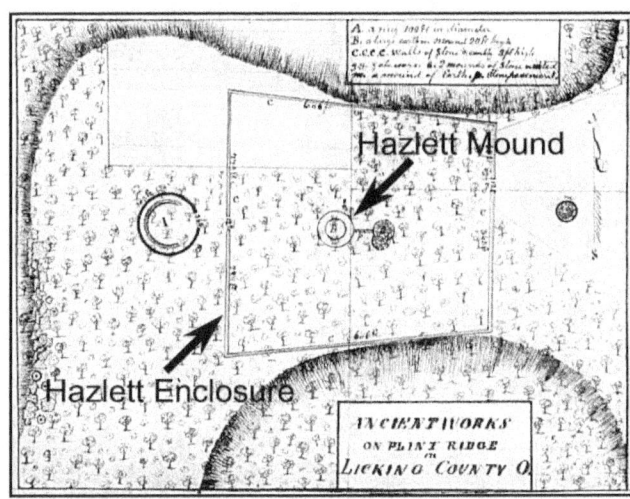
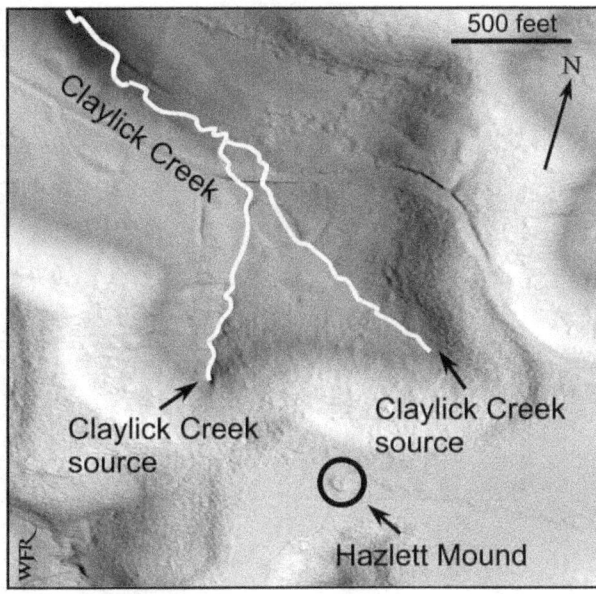

Above left: map of Hazlett Earthworks from mid-1800s by J. Unzicker. By permission of the Western Reserve Historical Society.
Above right: LiDAR view of the Hazlett Mound. The Hazlett Works overlook the source of Claylick Creek.

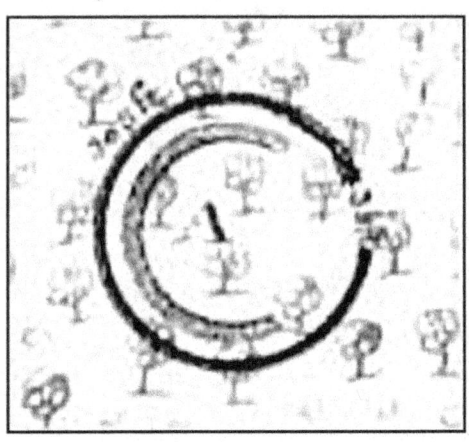

Above left: detail from Unzicker map of Hazlett showing crescent-shaped wall inside circle embankment.
Above right: photo showing what remains of the Hazlett Mound.

Figure 3.33. Flint Ridge: Hazlett Mound Solstice Alignment

Coffman Knob is located 2 - 3 miles southeast of Newark. It is the highest mountain in the Newark area.

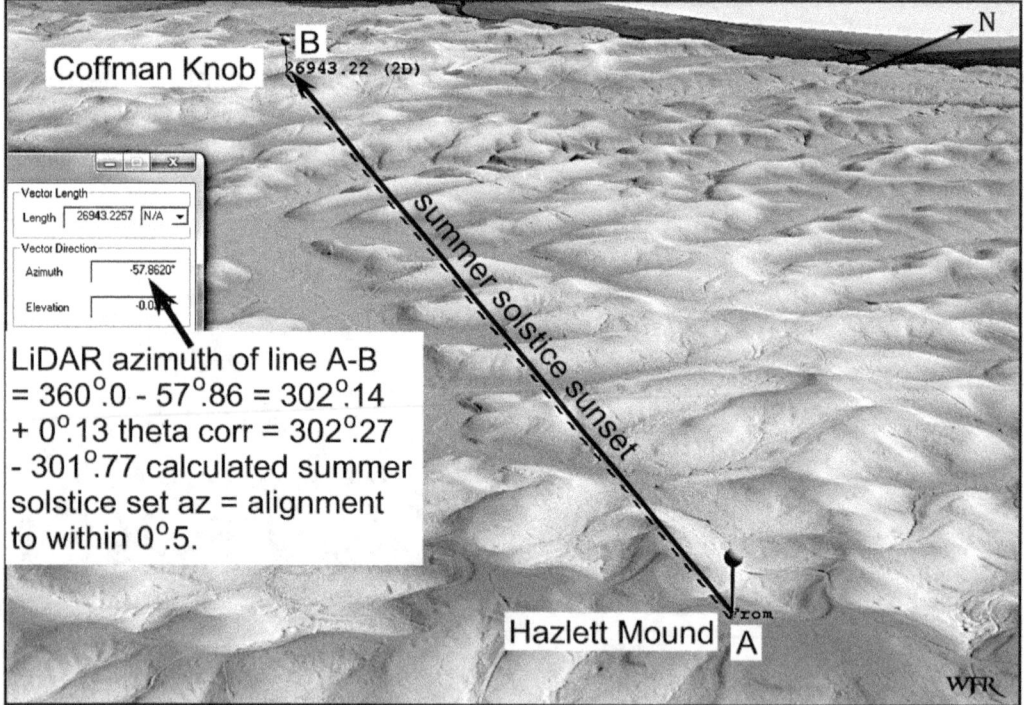

LiDAR azimuth of line A-B
= 360°.0 - 57°.86 = 302°.14
+ 0°.13 theta corr = 302°.27
- 301°.77 calculated summer solstice set az = alignment to within 0°.5.

Above: viewed from the Hazlett Mound, the summer solstice sun will set in alignment with Coffman Knob to within one-half of one degree (note 22). Coffman Knob is about 5 miles northwest of the Hazlett Mound.

Figure 3.34. Glenford Earthworks

The Glenford-Yost-Roberts earthworks are located in Perry County, 12 - 14 miles southeast of Newark, and 6 - 8 miles southwest of Flint Ridge. Each earthwork overlooks a branch of the Jonathan Creek - Valley Run confluence.

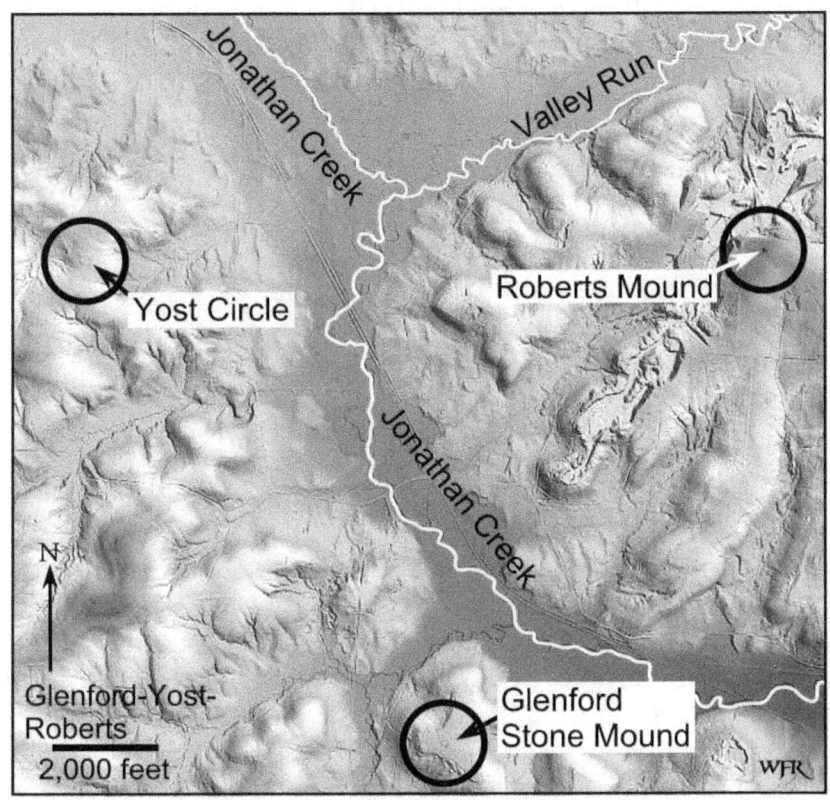

Below: LiDAR line-of-sight analysis shows that the three earthworks would have been intervisible. Intervisible areas are in very light gray.

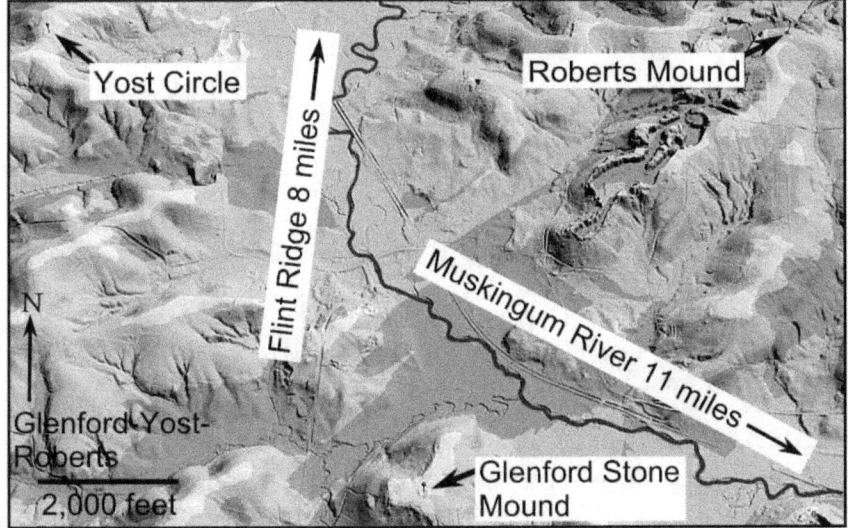

One of the things that makes the Glenford-Yost-Roberts area special is that the earthworks overlook the southern approach to Flint Ridge - from the Muskingum River, north along the Jonathan Creek Valley (note 23). The Muskingum River in turn, is a major waterway that flows directly into the Ohio River.

Figure 3.35. Glenford: Glenford Hilltop Enclosure

The Glenford site is a hilltop enclosure. A large stone mound is located within the enclosure. Early reports state the mound was 100 feet in diameter and 12 feet high (Thomas 1894:471). Radiocarbon dating of material found in the mound (270 B.C.) (Dutcher 1988) suggests an Adena affiliation (also see Moats 2011). A large pit is visible at the center of the mound where it has been dug into.

One of the things that makes the Glenford unique is that the hill itself, perimeter enclosure walls, and stone mound are comprised of a high quality sandstone. Nearby beds of this sandstone are commercially quarried for their sand. Hopewell burial mounds often incorporate layers of sand thus suggesting sand was in some way special.

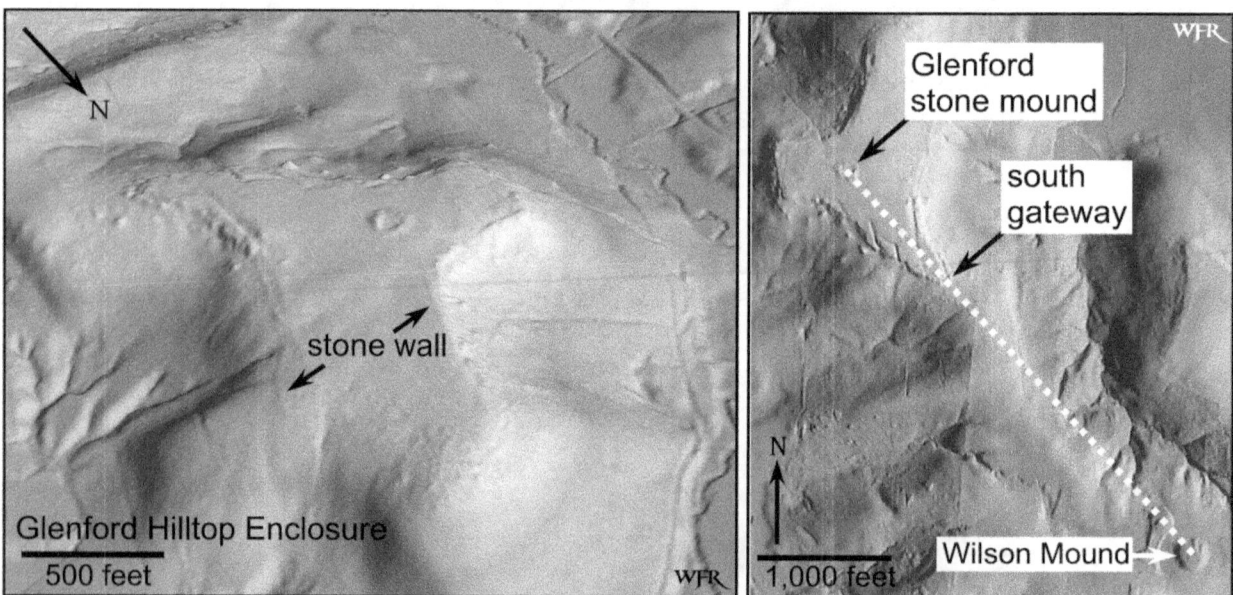

Above left: LiDAR reveals visible remnants of Glenford's stone walls.
Above right: LiDAR image. Of interest is that the Wilson Mound is located on a straight line trajectory established by the Glenford Stone Mound and the southeast 'gateway' into the Glenford enclosure. A low ridge forms a natural walkway between the Wilson Mound and Glenford Enclosure. The Wilson Mound is an 18-foot high mound that covers nearly an acre (Moorehead 1899:142).

Figure 3.36. Glenford: Yost Earthworks

The Yost Circle is one of the few surviving examples of a circle earthwork with an interior crescent feature. In the 1800s the Yost Circle was 4 feet in height (Moorehead 1897:174). A ditch separates the crescent feature from the circle embankment.

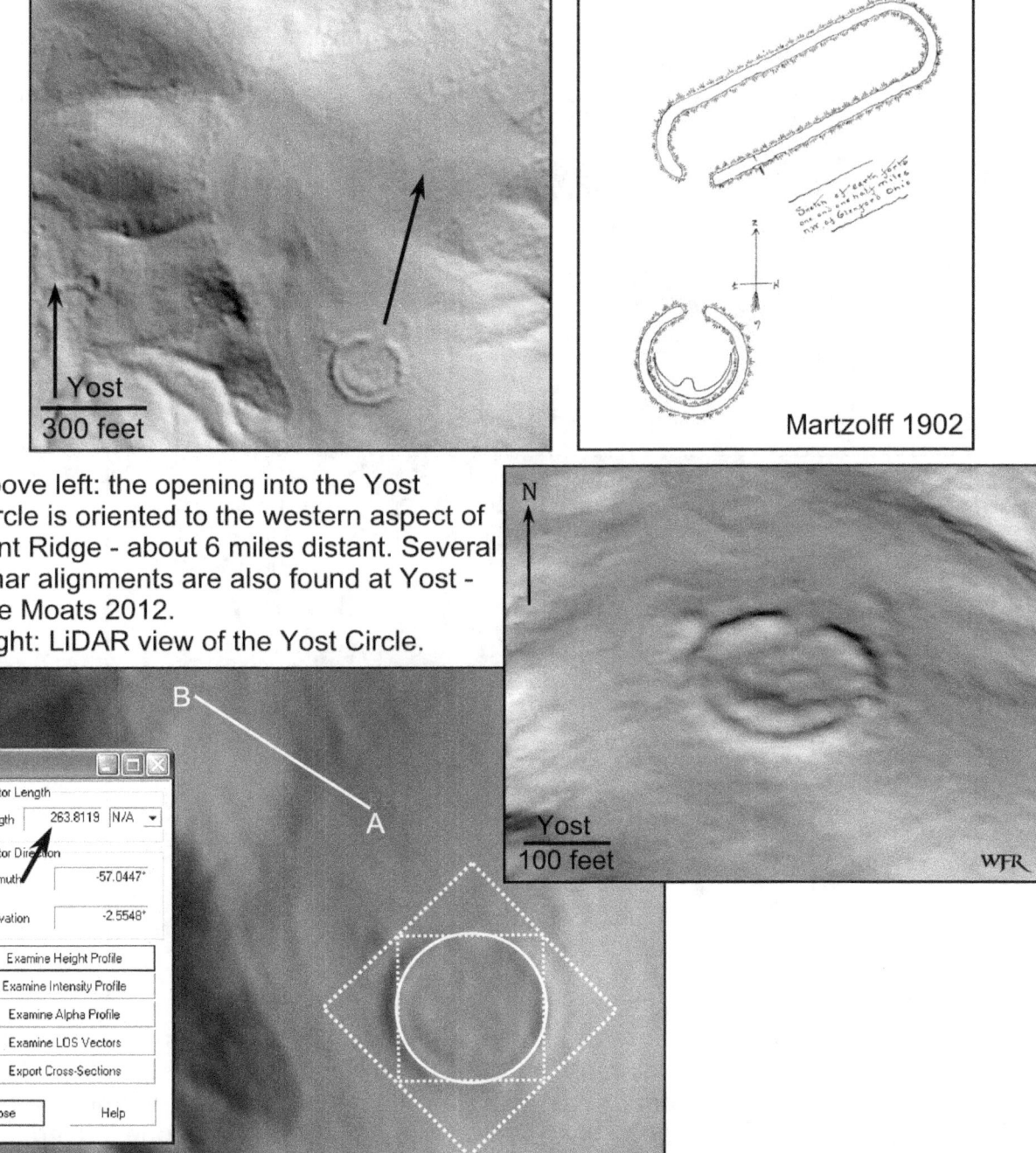

Above left: the opening into the Yost Circle is oriented to the western aspect of Flint Ridge - about 6 miles distant. Several lunar alignments are also found at Yost - see Moats 2012.
Right: LiDAR view of the Yost Circle.

The Yost Circle uses the 1/4 HMU length (263.5 feet) in its design. In the above left figure, the LiDAR program has generated a 263.8-foot line for scale purposes (A-B). That line is used to make a 263.5-foot square and nested figures. As shown, the 1/4 HMU-derived drawn circle matches the circumference of the earthwork (note 24).

Figure 3.37. Glenford: Intersite Relationships

Moorehead (1899:138) describes the Roberts Mound as "one of the highest tumuli east of the Scioto." He gives the dimensions of the mound as 120 feet in diameter and 27 feet high. Excavation in 1897 found a partially cremated skeleton near the center.

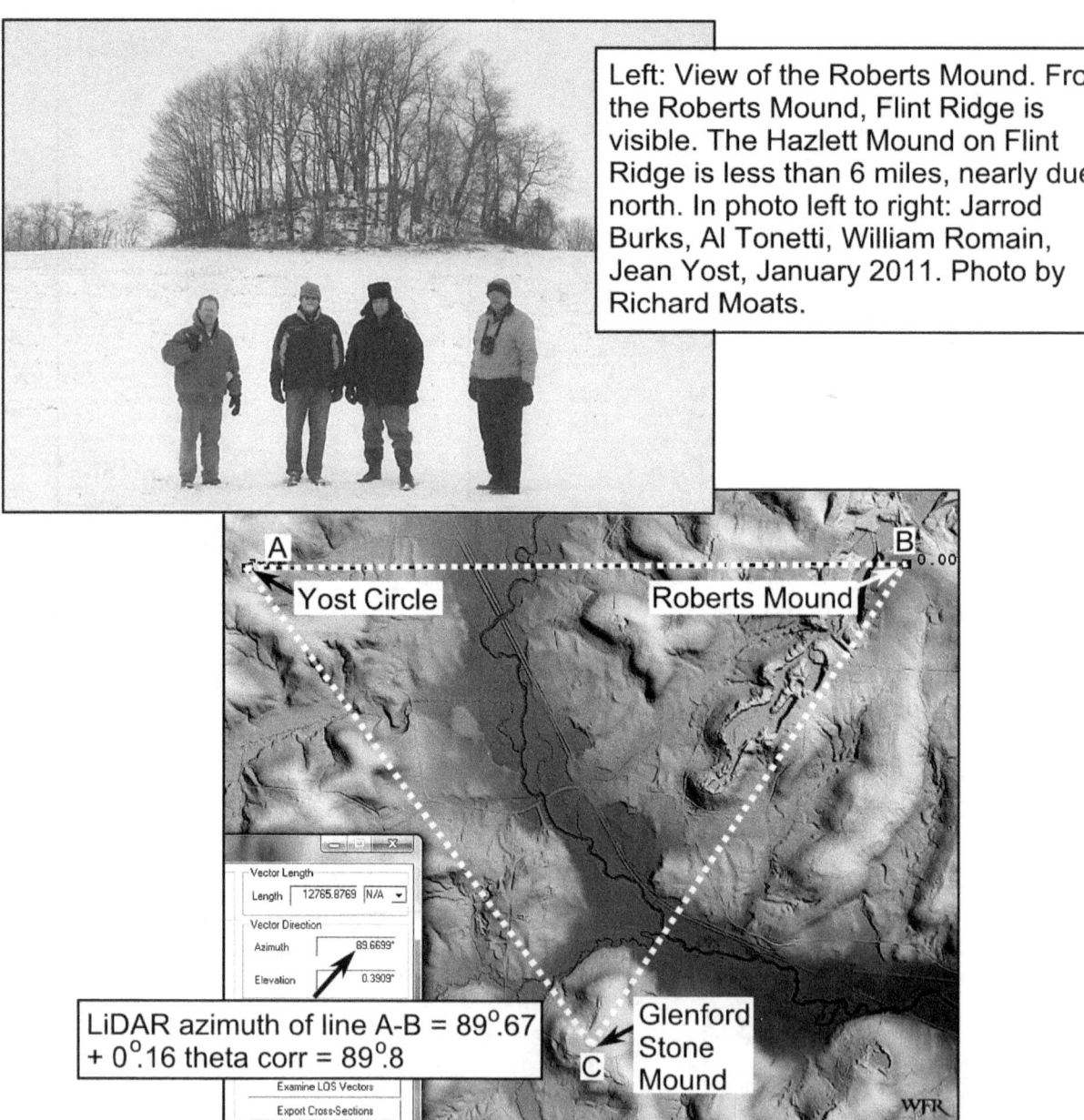

Left: View of the Roberts Mound. From the Roberts Mound, Flint Ridge is visible. The Hazlett Mound on Flint Ridge is less than 6 miles, nearly due north. In photo left to right: Jarrod Burks, Al Tonetti, William Romain, Jean Yost, January 2011. Photo by Richard Moats.

LiDAR azimuth of line A-B = 89°.67 + 0°.16 theta corr = 89°.8

Above: LiDAR view of Glenford area. The sightline between the entrance to the Yost Circle and Roberts Mound extends east-west to within one-quarter of one degree (note 25). If a triangle is drawn between Yost, Roberts, and the Glenford Stone Mound, sides A-C and B-C will be equal to each other to within 300 feet - across a distance of more than 2 miles (note 26). These earthworks were associated with Flint Ridge in terms of intervisibility, access routes, and orientation.

Figure 3.38. The Marietta Earthworks

The Marietta Works are situated at the confluence of the Muskingum and Ohio Rivers. Although most of the complex has been mostly covered over by urban growth, several features are more or less intact to include the Conus Mound, Sacra Via, Capitolium and Quadranaou mounds.

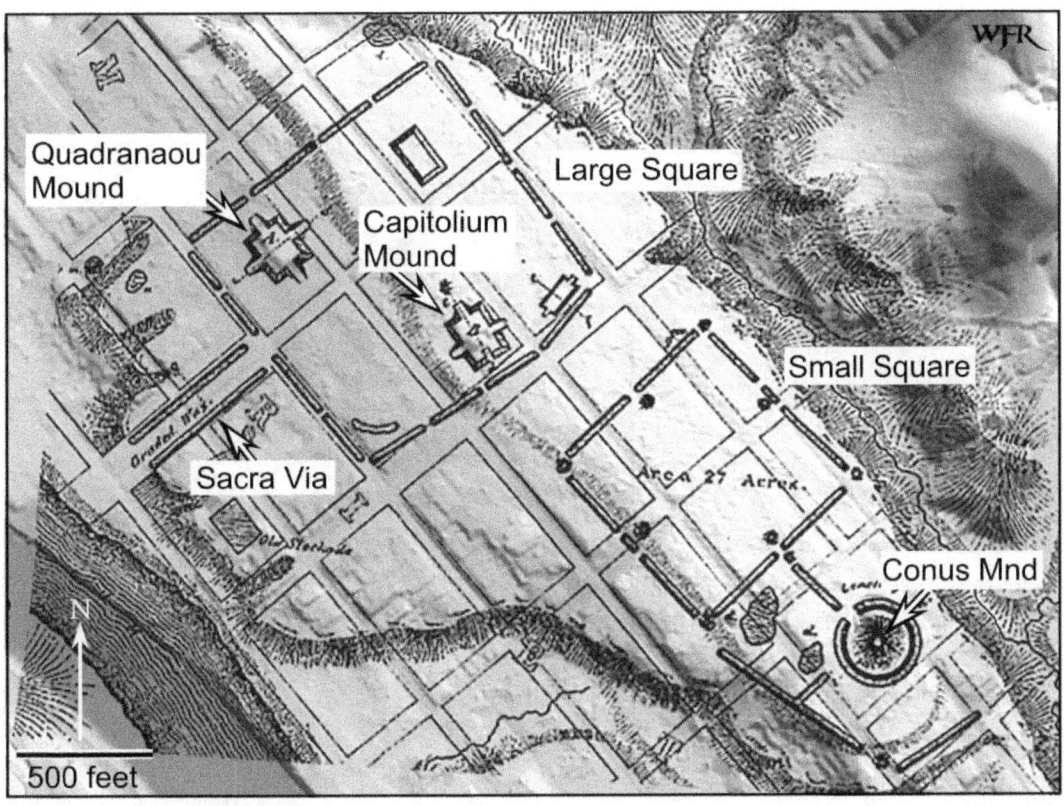

Above: Squier and Davis (1842:Pl. XXVI) map of Marietta superimposed over LiDAR image, annotation added.

Above left: aerial view of Marietta showing Muskingum and Ohio rivers.
Above right: the major axis of the Marietta Earthworks extends parallel to the lay of the land - established in this case by the river terrace to the southwest and the narrow ravine to the northeast - both shown by dashed lines.

Figure 3.39. The Marietta Earthworks: Conus Mound

The Conus Mound is a 30-foot high structure surrounded by a low circle embankment and interior ditch. It is a 'classic' Adena mound. Shallow excavation into the top of the mound in the early 1800s found the "bones of two or three skeletons" (Hildreth 1842:340).

Above left: View of the Conus Mound.
Above right: Squier and Davis (1848:facing p.139) sketch of the Conus Mound.
Right: Photo showing interior ditch and wall surrounding the Conus Mound.

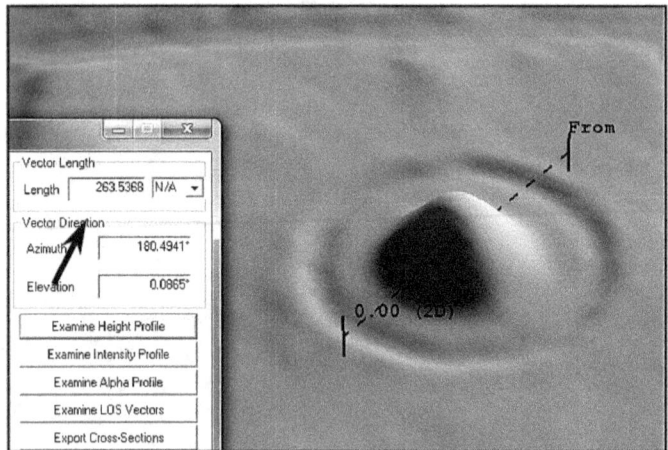

Above: LiDAR analysis shows that the outside diameter of the Conus Mound embankment is equal to 1/4 HMU (1,054 ft. / 4 = 263.5 ft.).

Below: LiDAR profile of the Conus Mound, surrounding ditch and embankment.

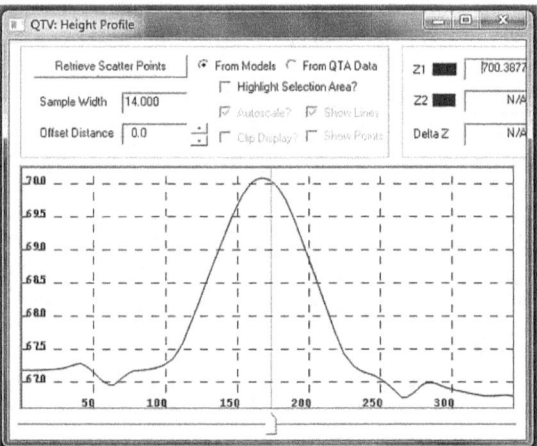

Figure 3.40. The Marietta Earthworks: Conus Mound Solstice Sunrise
The Conus Mound overlooks a bifurcated valley to the east. Significant hills, 150 - 170 feet in height form the north and south sides of this valley.

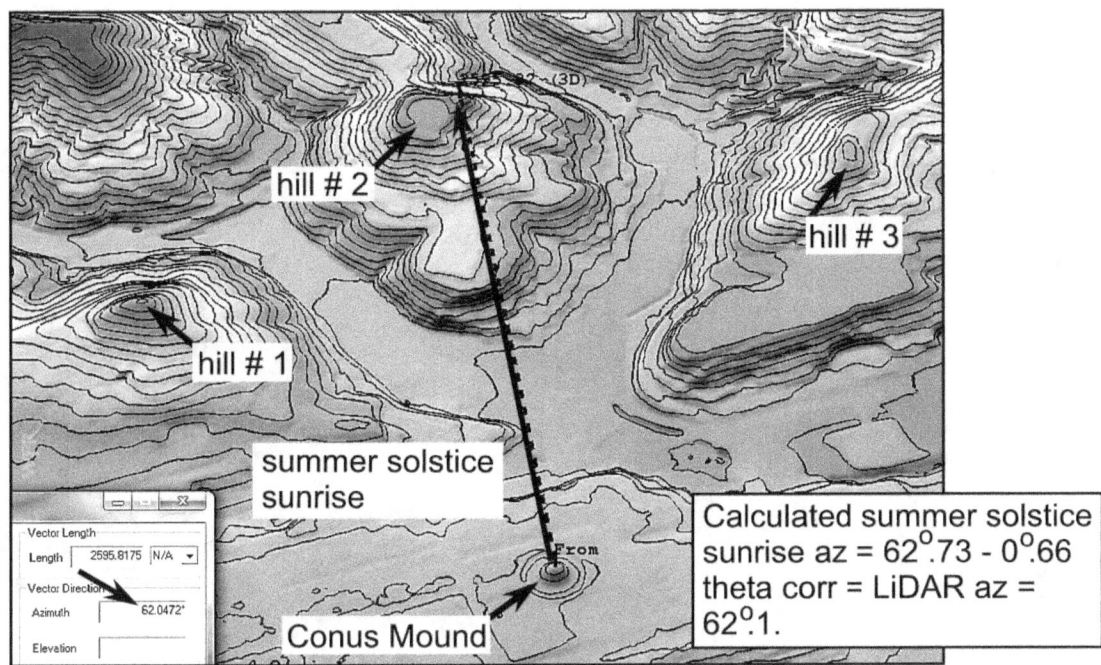

Above: LiDAR contour map. Viewed from the Conus Mound, the summer solstice sun will rise behind hill # 2, seemingly balanced over the valley (note 27).

Above right: on this 1924 USGS map, the original water tank is shown as a small dot.

Above left: View of hill # 2 from the top of Conus Mound. A large water tank is situated on hill # 2 - complicating horizon elevation calculations. The present tank replaced a smaller tank shown on the earliest available USGS 7.5-minute series map for the area - from 1924 (based on and updated from a 1902 survey) (above right). Since the footprint for the new, two-million gallon water tank involved significant grading of the summit of hill # 2, the 1924 USGS map is used to determine the horizon profile and summit elevation. Unknown, however, is if solstice observations were made from the top or bottom of Conus Mound, for a difference of 30 feet in elevation. Given these uncertainties, the posited solstice azimuth could differ from that given here by plus or minus one degree.

Figure 3.41. The Marietta Earthworks: Conus Mound Solstice Sunset

Viewed from the base of the Conus Mound, the summer solstice sunset occurs in alignment with a small knoll located on hill # 4, about 2.8 miles northwest of the Conus Mound (note 28).

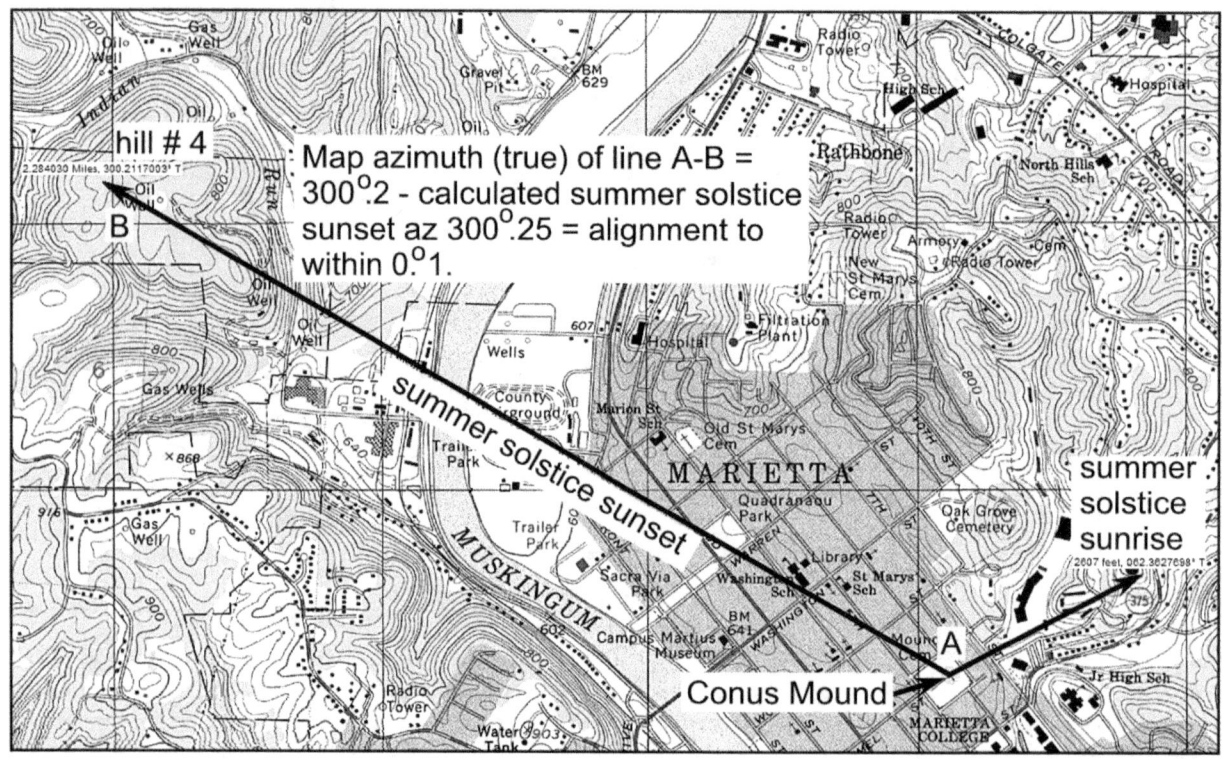

Above: the location of the Conus Mound seems to have been established by locating the point on the river terrace from which both the summer solstice sunrise and sunset would be visible over selected features - in this case knolls on nearby hilltops. USGS 7.5 series topo map provided by MyTopo, annotation added.

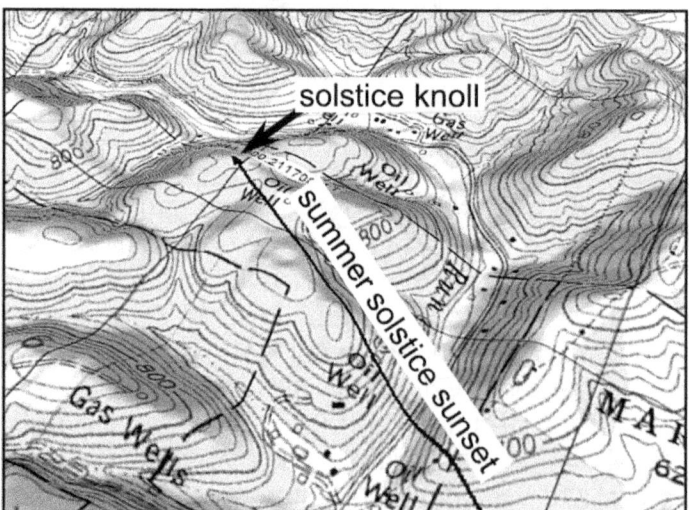

Above: Three-dimensional map view of hill # 4 and knoll that serve as the foresight for the summer solstice sunset alignment from the Conus Mound. USGS 7.5 series topo map provided by My Topo, annotation added.

Figure 3.42. The Marietta Earthworks: Quadranaou Mound Geometry
The Quadranaou Mound is a large platform mound, about 10 feet in height.

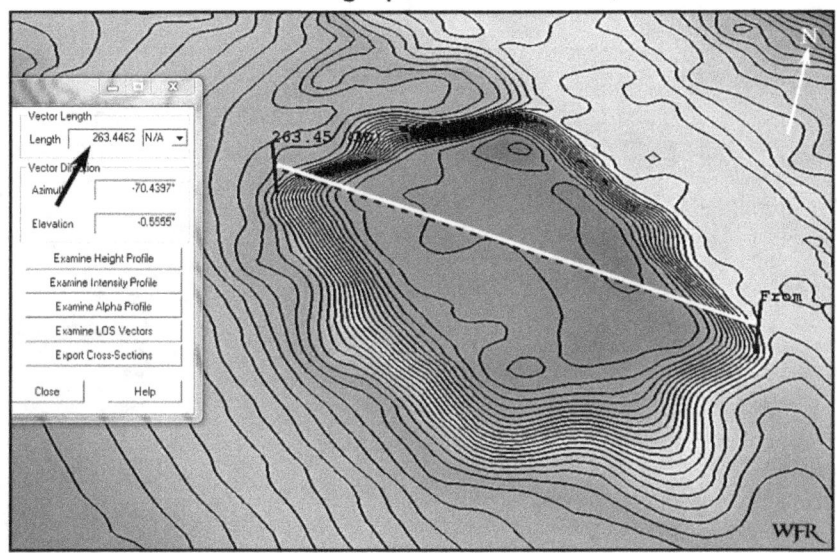

Above: the Quadranaou Mound uses the 263.5-foot or 1/4 HMU length in its design. LiDAR analysis shows the diagonal of the earthwork is equal to about 263.5 feet, or 1/4 HMU.
Below: the dimensions of the Quadranaou Mound suggest that it was designed using the concept of the 30-60-90 degree triangle.

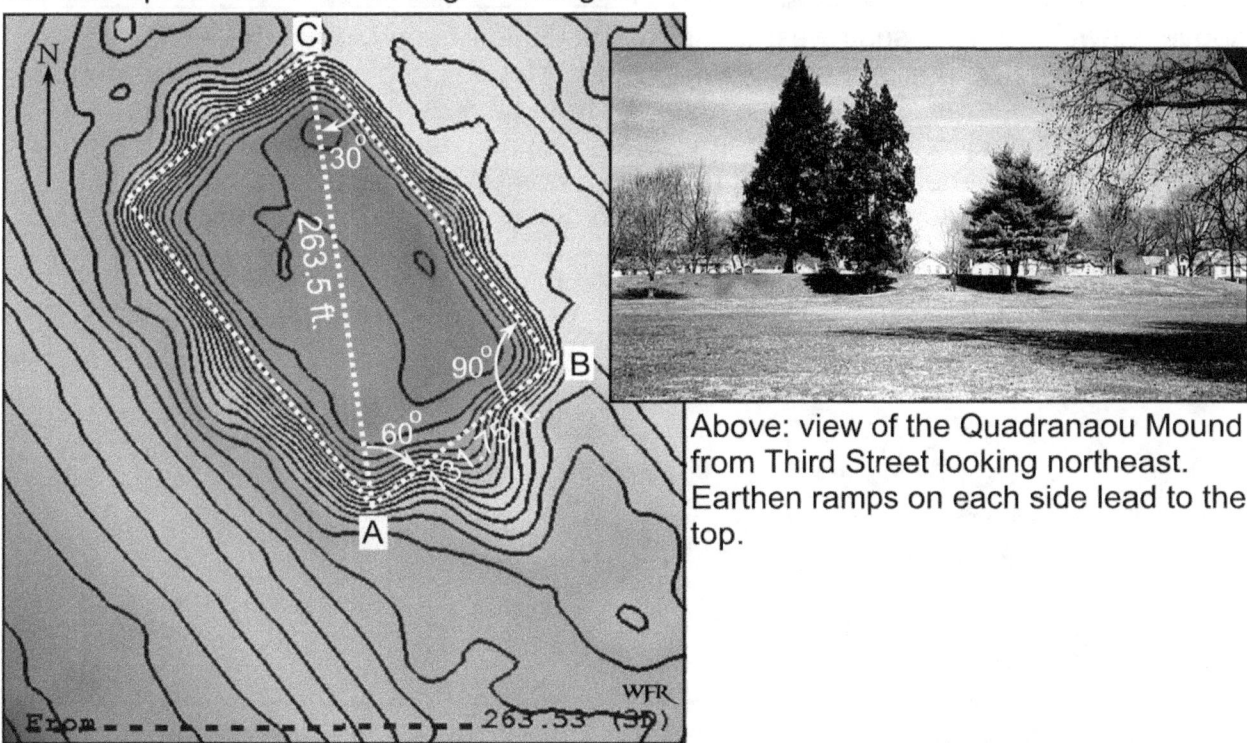

Above: view of the Quadranaou Mound from Third Street looking northeast. Earthen ramps on each side lead to the top.

Above: in a 30-60-90 degree triangle, the base leg (A-B) is one-half the length of the hypotenuse (A-C). In the above image, two 30-60-90 degree triangles are placed back-to-back. If the hypotenuse of these triangles (line A-C) is 263.5 feet (i.e., 1/4 HMU) then, the two short sides of the Quadranaou Mound will be 131.75 feet - or 1/8 HMU.

Figure 3.43. The Marietta Earthworks: Quadranaou Mound Solstice Sunrise
The Quadranaou Mound is aligned to both the winter solstice sunrise and winter solstice sunset (note 29).

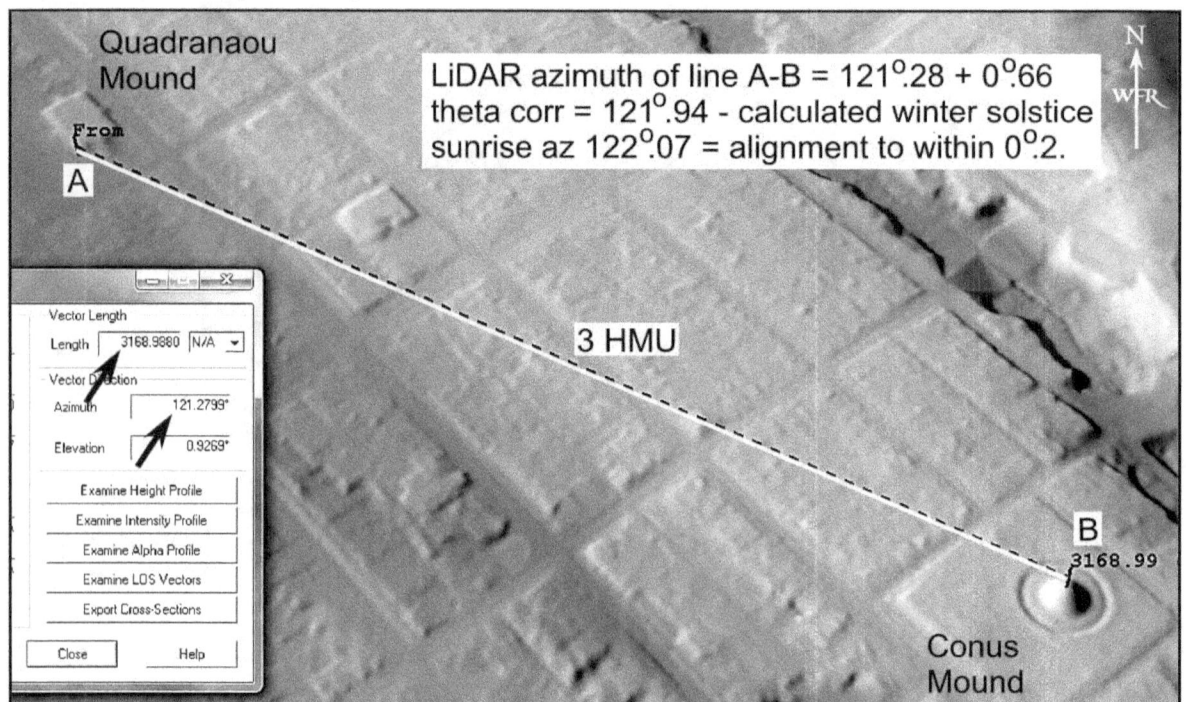

Above: viewed from the southwest corner of the Quadranaou Mound the winter solstice sun will rise in alignment over the Conus Mound. Two observations support the intentionality of this alignment: 1) the distance from backsight point A at the corner of Quadranaou Mound to the center of the Conus Mound is equal to 3 HMU (3,162 ft.) to within less than one-half of one percent; 2) from point A the Conus Mound appears sillouhetted and balanced against the bowl-like distant horizon - see below.

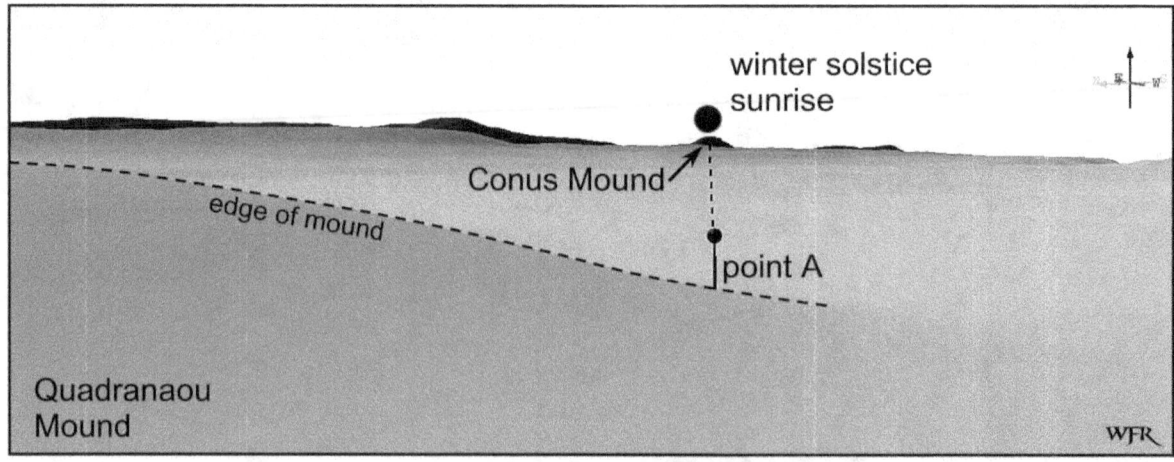

LiDAR view from Quadranaou Mound point A looking toward the Conus Mound. No vertical exaggeration applied. Simulated winter solstice sun. Viewed from point A, the Conus Mound and solstice sunrise appear balanced between the peaks on either side.

Figure 3.44. The Marietta Earthworks: Quadranaou Mound Solstice Sunset
In addition to its relationship to the winter solstice sunrise, the Quadranaou Mound is aligned to the winter solstice sunset along its minor axis.

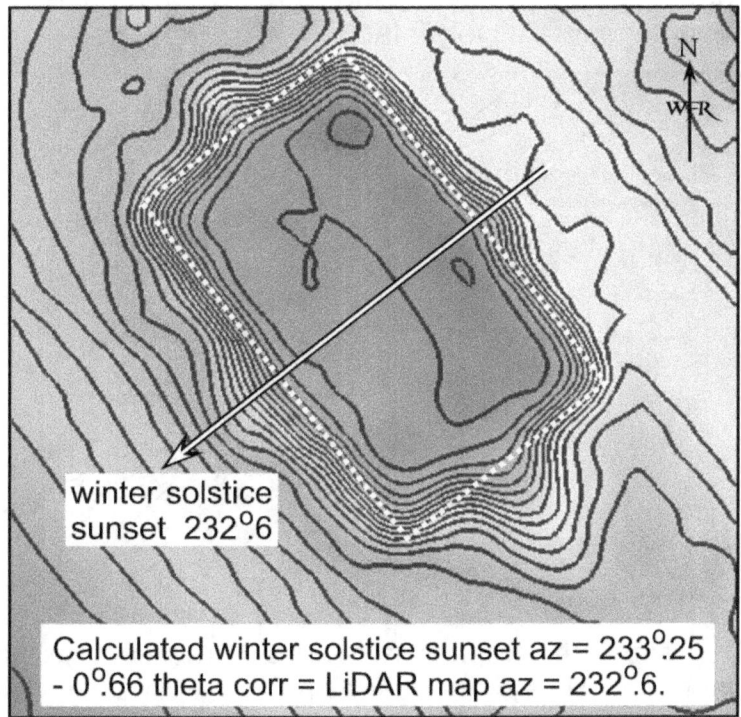

In the above image, the calculated winter solstice sunset azimuth (theta corrected) is superimposed over a LiDAR contour map of the Quadranaou Mound (note 30).

Above left: view of the northeast side of the Quadranaou Mound. Large tree is nearly centered on the minor axis.
Above right: winter solstice sunset along the minor axis of the Quadranaou Mound, December 21, 2001. View is from the northeast center ramp area looking southwest across the top of the Quadranaou Mound.

Figure 3.45. The Marietta Earthworks: Capitolium Mound

The Capitolium Mound is a platforrm mound about 6 feet in height. Modifications to the mound were made in the early 1900s to accomodate construction of the library building now situated on its summit (note 31). For the most part, however, the size, shape, and orientation of the mound appear relatively intact.

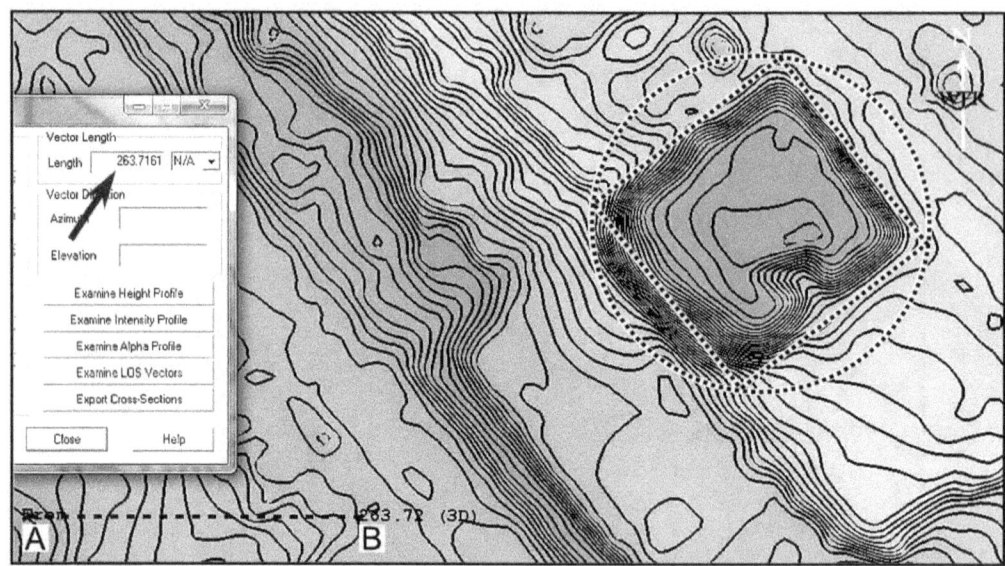

Above: LiDAR contour map of the Capitolium Mound (6 inch contour interval). The dimensions of the Capitolium Mound are based on the 1/4 HMU length (263.5 ft.). In the above image, a circle having a diameter equal to 1/4 HMU is constructed using line A-B for scale. A square is inscribed within the circle. The constructed square closely corresponds to the dimensions of the mound on three sides.

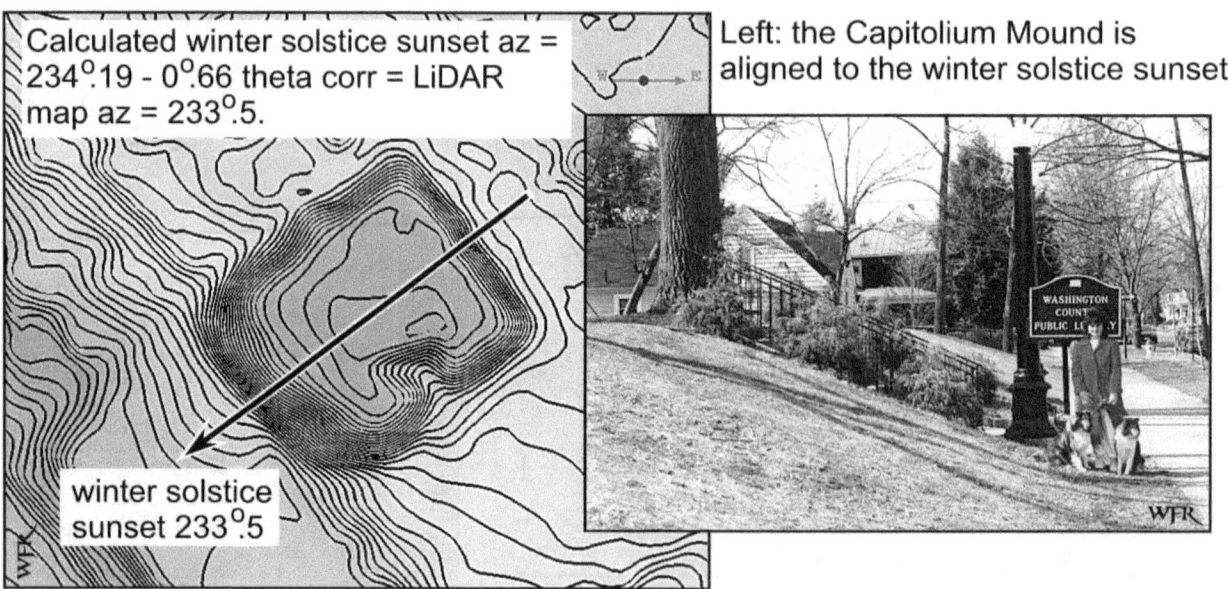

Above left: calculated winter solstice sunset azimuth (theta corrected) superimposed over LiDAR contour map of the Capitolium Mound (note 32).
Above right: view of the Capitolium Mound along Fifth Street.

Figure 3.46. The Marietta Earthworks: Sacra Via

The Sacra Via was a parallel-walled structure that led from the Large Square downslope to the Muskingum River. Squier and Davis (1848:74) describe it thusly: "It is six hundred and eighty feet long by one hundred and fifty wide between the banks, and consists of an excavated passage....the earth, in part at least, is thrown outward upon either side, forming embankments from eight to ten feet in height."

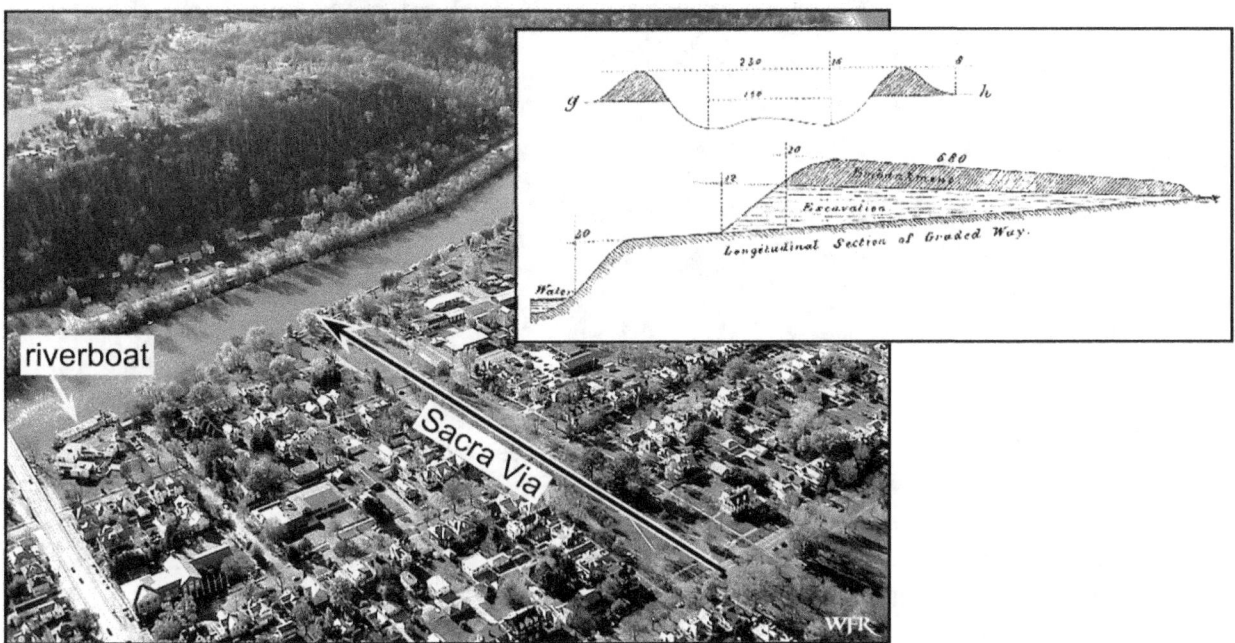

Above left: aerial view of the Sacra Via. In this photo, the Valley Gem sternwheeler is visible and gives a sense of scale for this earthwork.
Above right: enlarged detail of the Sacra Via, from Squier and Davis 1848:Pl. XXVI.

Above: the embankment walls of the Sacra Via are no longer visible; and two narrow cobblestone streets now run along its length. However, the center of the Sacra Via is preserved as a city park. In the distance is a high ridge that forms the horizon.

Figure 3.47. The Marietta Earthworks: Sacra Via Winter Solstice Sunset
The Sacra Via is aligned to the winter solstice sunset.

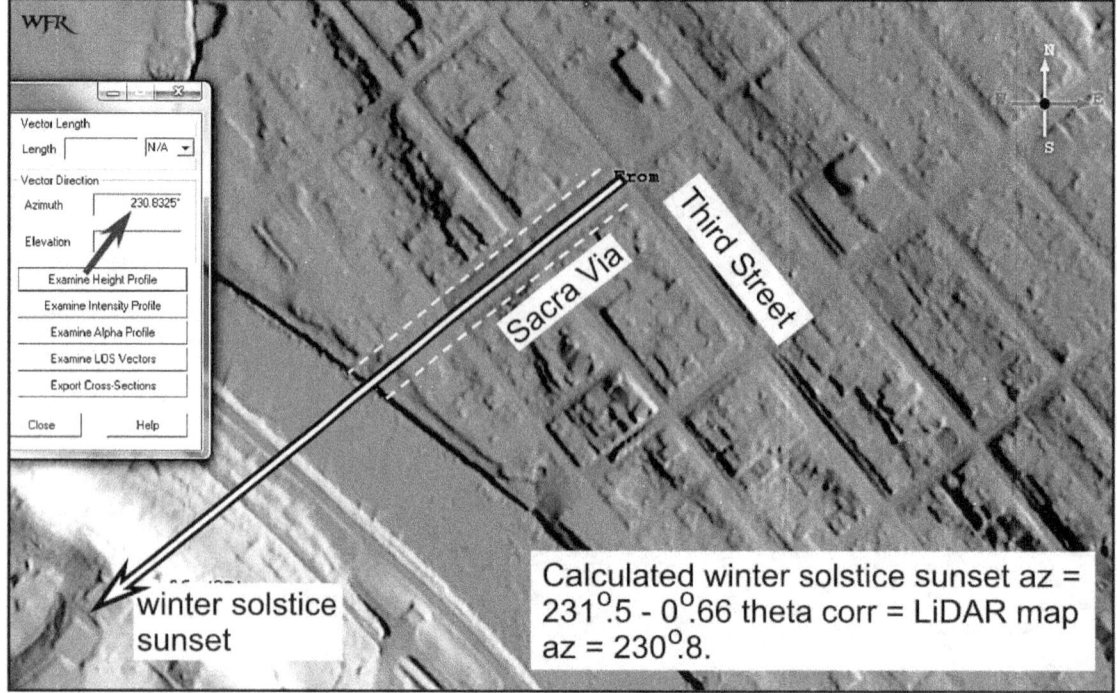

Above: calculated winter solstice sunset azimuth (theta corrected) superimposed over LiDAR image (note 33). Solstice azimuth calculated for horizon elevation as viewed from the intersection of Sacra Via and Third Street - which is the entrance into the Sacra Via nearest to the gateway opening into the Large Square.

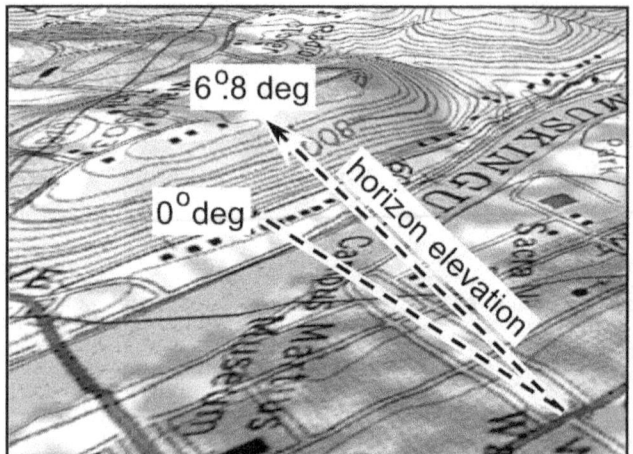

Above: three-dimensional map showing how the ridge across from the Sacra Via affects the horizon elevation. Map by MyTopo, annotation added.
Right: winter solstice sunset alignment along Sacra Via, December 21, 1996.

CHAPTER 4
HOCKING RIVER WATERSHED

Proceeding west from the Muskingum River, the next region where earthworks are found is in the Hocking River Watershed. The Hocking River is a tributary of the Ohio River, drains about 1,197 square miles, and is about 102 miles long. Although a significant number of mounds are found in the region, there are few geometric enclosures.

Rock Mill Earthworks

The most northern earthworks considered in this chapter are the Rock Mill Works in Fairfield County. This earthwork is situated near Lancaster, Ohio, among low, rolling hills that were once covered by successive glaciations. The cultural affiliation of the earthwork is not known with certainty. As shown by the accompanying figures, however, a combination of factors made this location special.

First, the Rock Mill Works are located on a hill that affords an impressive view of the surrounding landscape, to include the source of the Hocking River, just a couple of miles to the north. Similar instances of earthworks overlooking water sources are found elsewhere (e.g., Hazlett Mound) thus suggesting that in some cases, this was intentionally selected for.

Second, the earthwork is less than one-half mile from a unique waterfall and plunge pool. In a normally quiet forested environment, the sound of this waterfall may have been the loudest sound that most people would ever hear. In its unusual nature, this experience may have been considered an auditory manifestation of the sacred.

Perhaps further enhancing the special nature of the locale was the associated plunge pool. Like all plunge pools, this one results from the force of falling water carving-out the bedrock below the falls. The plunge pool is about 15 feet deep. What makes the plunge pool of potential interest is that, as a result of the action of the waterfall, the water in the plunge pool becomes highly oxygenated. During summer months, when water levels are low and slow, plunge pools are a preferred place for fish to gather, to take advantage of the oxygenated water. For the Adena-Hopewell, the gathering of fish at Rock Mill may have been recognized as unusual and like the auditory aspect of the site, perhaps a sensual manifestation of the sacred – hence, an auspicious place to build an earthwork.

Wolf Plains Group

The largest and best-known earthwork group in the Hocking River Watershed is the Wolf Plains Group. This group of earthworks is the largest Adena complex known for Ohio and surrounding states (Murphy 1989:195). The group is comprised of 22 mounds and 9 circle earthworks. The complex is located in the un-glaciated Appalachian Plateau area, at the confluence of Sunfish Creek and the Hocking River. The complex is situated on one of the few, relatively flat terrace areas found in the lower Hocking River Valley. Surrounding the Wolf Plains Group are rugged hills and steep terrain.

One of the most important resources found in the Hocking Valley is iron ore – specifically, hematite. The Wolf Plains Group is located only 6 miles from Bessemer, Ohio – a commercially utilized source for iron ore. As archaeologist James Murphy (1989:40) notes, nodular hematite "more than a foot in diameter" is found in abundance in eastern Athens County.

Hematite was of special significance for the Adena and Hopewell. According to Murphy (1989:40), celts and other implements made of hematite are often found in local Athens County collections. And, "pieces of rubbed and un-worked hematite are common in the fill of burial mounds, as are large boulders of burned hematite" Murphy (1989:40). Mostly though, hematite was crushed into powder and heated to make red-colored pigments. For Native Americans, the color red held special significance and red pigments were often applied to designate the sacred status of people and things. Given this, it may be that the largest of the Adena earthwork complexes – i.e., the Wolf Plains Group, was situated where it is in order to make use of the nearby hematite. Indeed, as Murphy (1989:40) notes, hematite was really the only "exportable resource" that

Valley residents had. Another possibility is that the prevalence of red earth in the area might have been considered as an indicator of the areas' overall special and sacred nature, thus making the location appropriate for one of the largest Adena complexes known.

Figure 4.1. Map of sites in the Hocking River Watershed

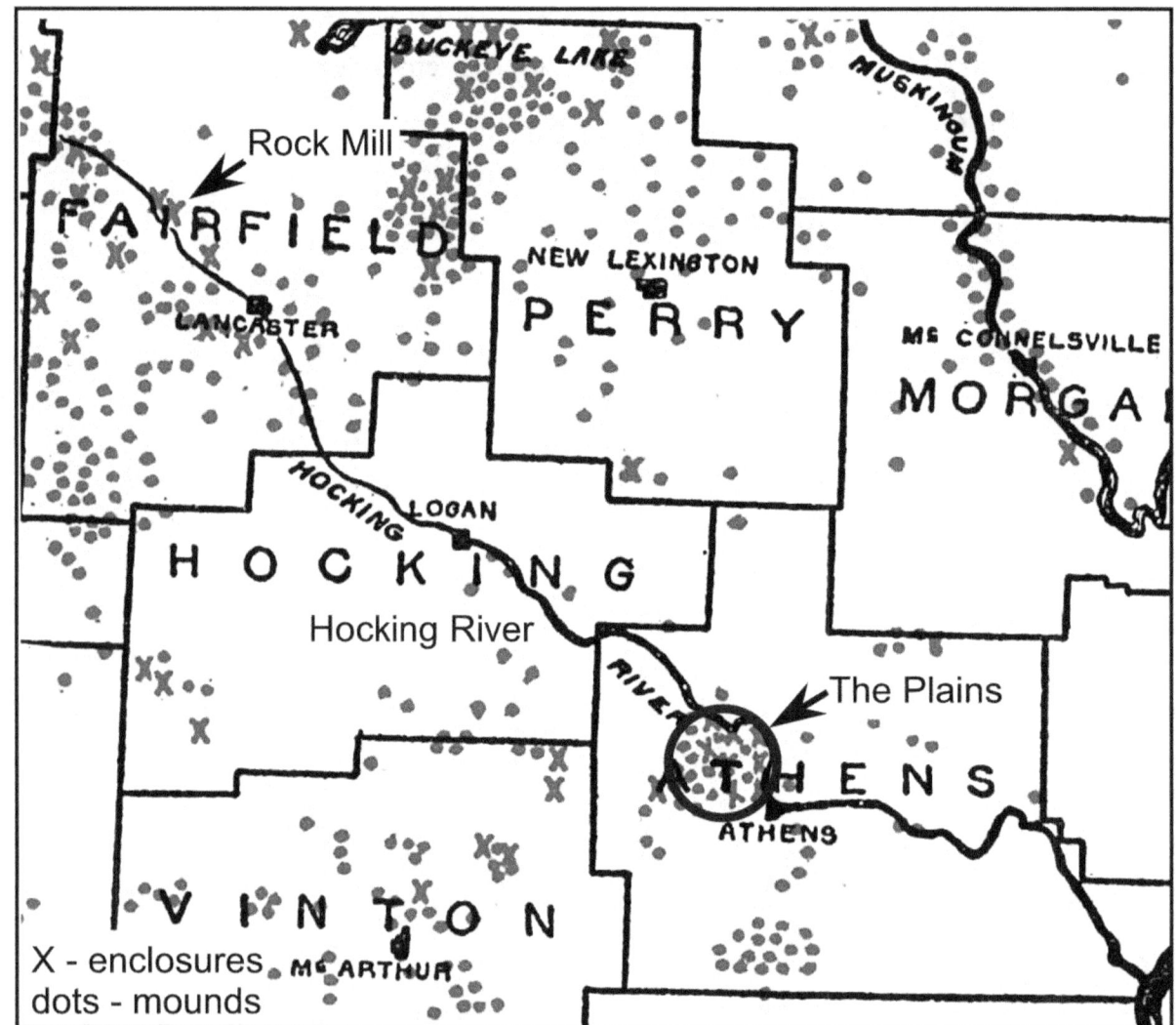

Above: detail from Mills (1914:Pl. XI) showing mounds and earthworks in the Hocking River Watershed. Annotation added.

Figure 4.2. Lancaster: Rock Mill Works

The Rock Mill Works is an interesting earthwork that seems to combine Adena circles with a Hopewell square. The earthwork has been mostly obliterated by a housing development. One of the interesting things about the site, however, is its location. The site is situated on top of a hill that overlooks the source of the Hocking River. The site is also less than one-half mile from a visually impressive waterfall and plunge pool.

Left: map of Rock Mill site by Squier and Davis 1848: Pl. XXXVI, no. 3.

Above left: view of the Hocking River at the base of the hill that the Rock Mill Works are located on. Earthworks are at the top of the hill to the upper right.
Above right: waterfall and plunge pool near Rock Mill Works. A covered bridge crosses the river chasm. In the 1800s a grist mill was built at this location. It still stands.

Figure 4.3. Lancaster: Rock Mill Works: Alignments

Although little remains of the Rock Mill site, by combining information from several images, a fair assessment can be made of potential alignments - presuming that the only map we have of the site - made by Squier and Davis, is accurate at least in terms of the shape of the square.

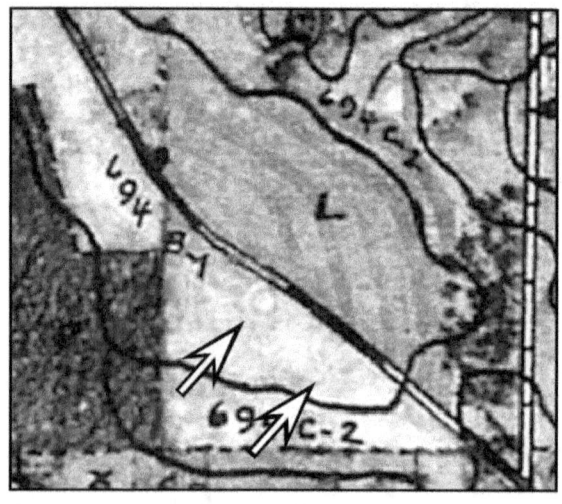
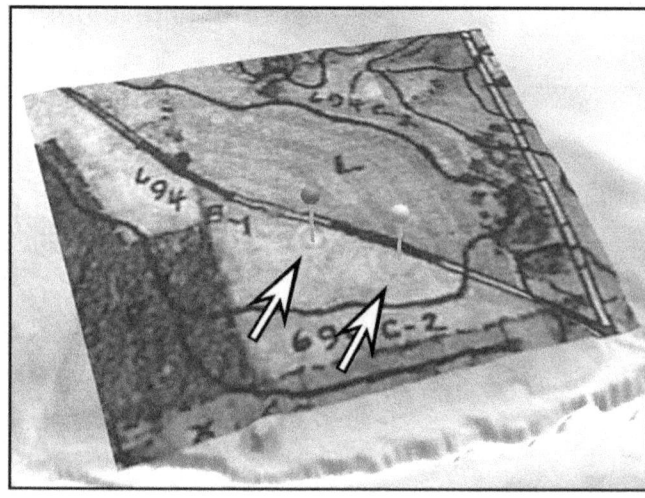

Above left: step 1: detail from 1938 US Department of Agriculture aerial. In this image the two circle earthwork features are still visible.

Above right: step 2: the 1938 aerial photo is georeferenced and superimposed over a LiDAR image of the area. Marker pins are placed at the centers of the two circles.

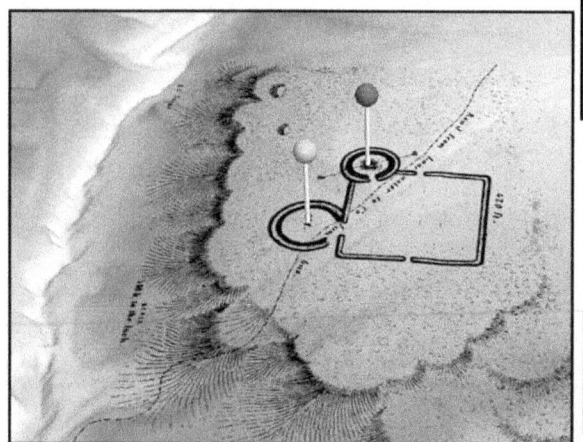
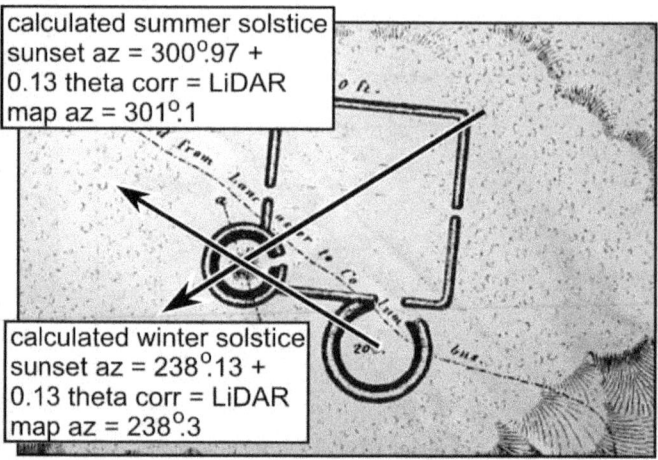

calculated summer solstice sunset az = $300°.97$ + 0.13 theta corr = LiDAR map az = $301°.1$

calculated winter solstice sunset az = $238°.13$ + 0.13 theta corr = LiDAR map az = $238°.3$

Above left: step 3: with the marker pins left in place, the 1938 photo is replaced by the Squier and Davis map - which is oriented by placing it so that the marker pins are at the centers of the two circles.

Above right: step 4: with the Squier and Davis map georeferenced, potential alignments can be assessed. As indicated, winter and summer solstice sunset alignments may be incorporated into the earthwork design (note 1). Ground-truthing of the square is needed for confirmation. The posited summer solstice sunset alignment is interesting, since the azimuth between the circle features can still be assessed using the 1938 aerial photo.

Figure 4.4. The Wolf Plains Group

The Wolf Plains Group is a cluster of mounds and circle earthworks located in the Hocking River Valley, within the city named, The Plains. Twenty-two mounds and nine circle earthworks have been identified. Most are damaged or entirely destroyed. Below: LiDAR view of the Wolf Plains Group (numbering convention after Blazier, Freter, and Abrams 2005).

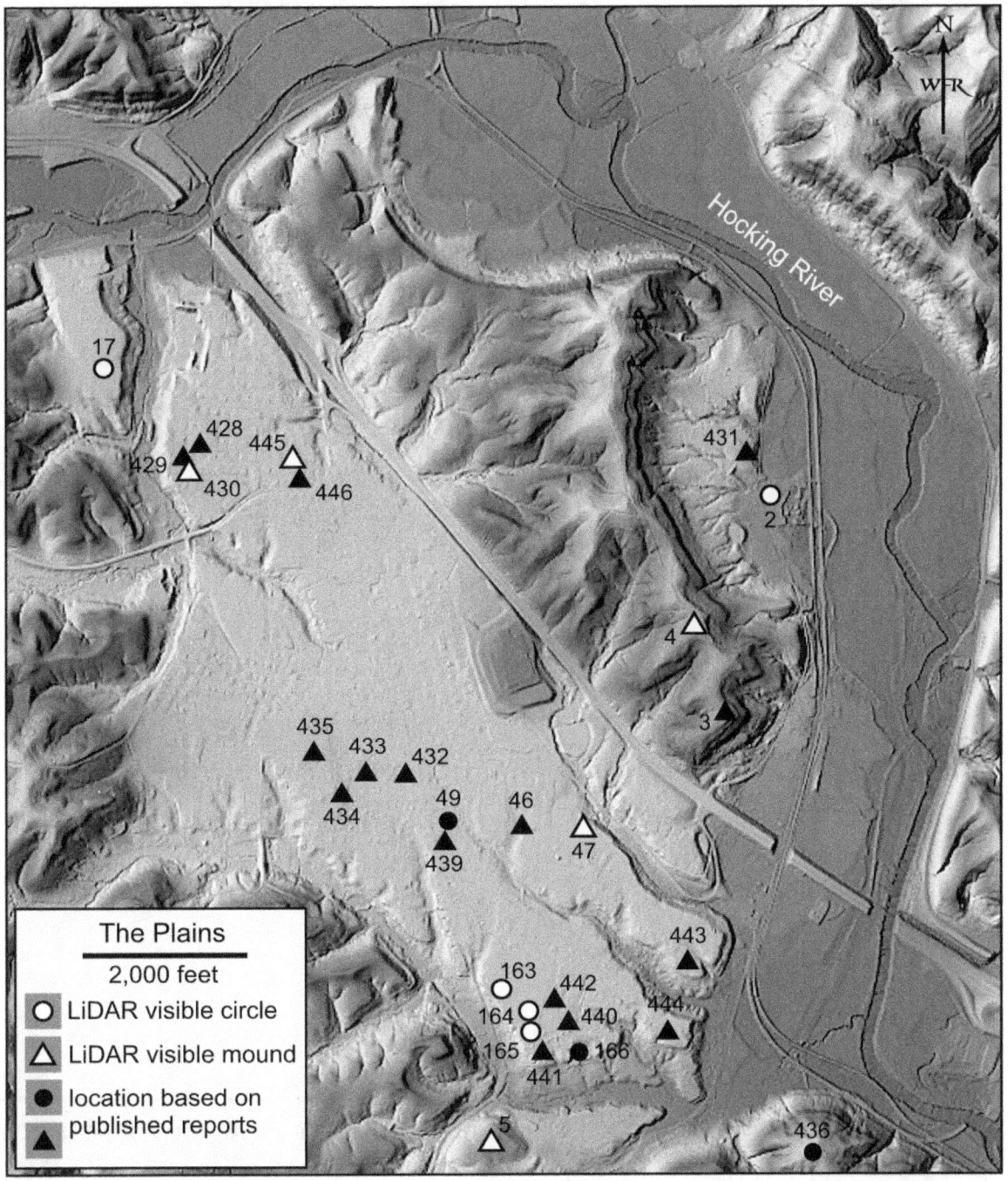

Radiocarbon-dating indicates that "the earthworks in The Plains were built from ca. 50 B.C to A.D. 250, making this center coeval with the larger Hopewell complexes found elsewhere in Ohio" (Blazier, Freter, and Abrams 2005:111).

Figure 4.5. The Wolf Plains: Mounds

Human remains have been recovered from most of the mounds at The Plains. Records, however, are quite sketchy (e.g., Andrews 1877; but also see Skinner and Norris 1984).

Left: Connett Mound 5 (33AT430). This mound is about 15 feet high and 90 feet in diameter. It overlooks a deep ravine 700 feet to the west.

Right: Hartman Mound (33AT445). This is the largest of The Plains Group mounds.

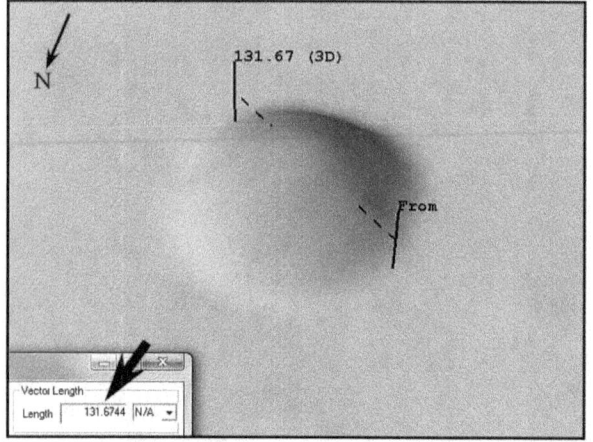

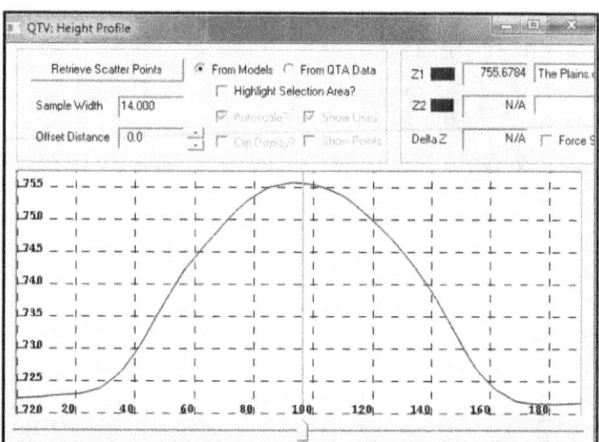

Above left: LiDAR analysis shows that the the diameter of the Hartman Mound is equal to 1/8 HMU (1,054 ft. / 8 = 131.75 feet).
Above right: LiDAR profile through the Hartman Mound. The profile shows this mound to be more than 30 feet in height.

Figure 4.6. The Wolf Plains: Circle 33AT2

Five circle earthworks can be detected at The Plains using LiDAR. All five incorporate a unit of length equal to 186.3 feet in their diameters (note 2). The 186.3-foot unit is related to the HMU in the following way.

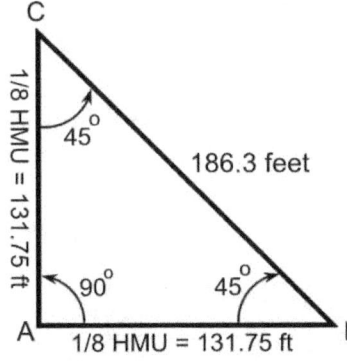

Left: a 45-45-90 degree right triangle is drawn. If sides A-B and A-C are 1/8 HMU (or 131.75 feet) in length then, side B-C will be 186.3 feet.

Given that The Plains circles were contemporaneous with Hopewell, it is not surprising to find sophisticated geometric concepts in their design.

Below left: LiDAR image of circle 33AT2. This is the best-preserved of The Plains circles.

Below center: 186-foot circle superimposed over 33AT2.

Below right: oblique view showing 186-foot length.

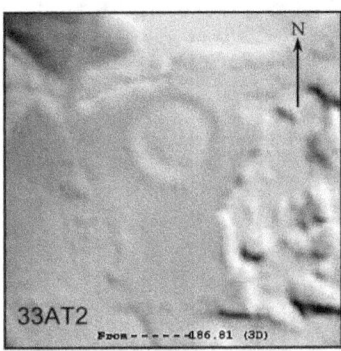 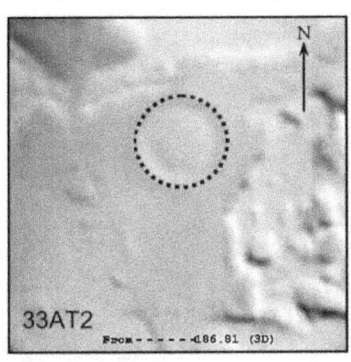 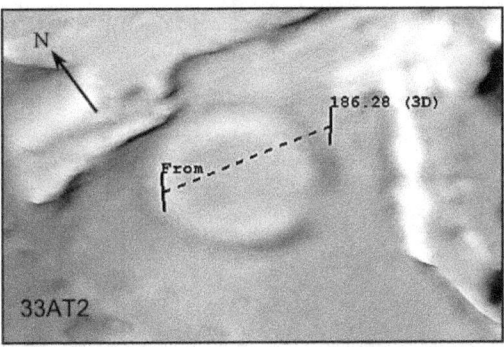

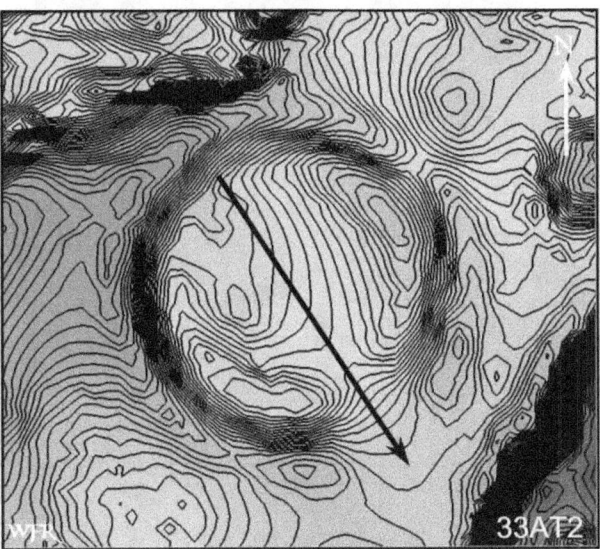 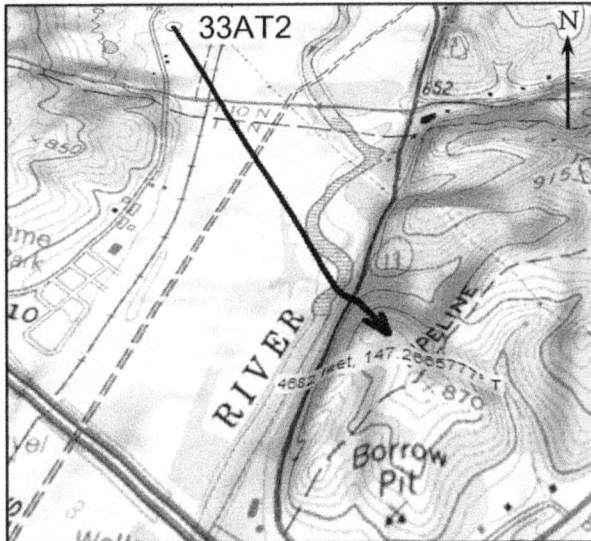

Above left: LiDAR contour map of 33AT2 (approximate 2 inch contour interval).
Above right: like several other Adena earthworks, the axis of 33AT2 is oriented to a landscape feature - in this case, to a prominent mountain peak about 1 mile to the southeast. Map provided by MyTopo.

Figure 4.7. The Wolf Plains: Circles 17 and 163
Circles 33AT17 and 33AT163 also incorporate the 186-foot length.

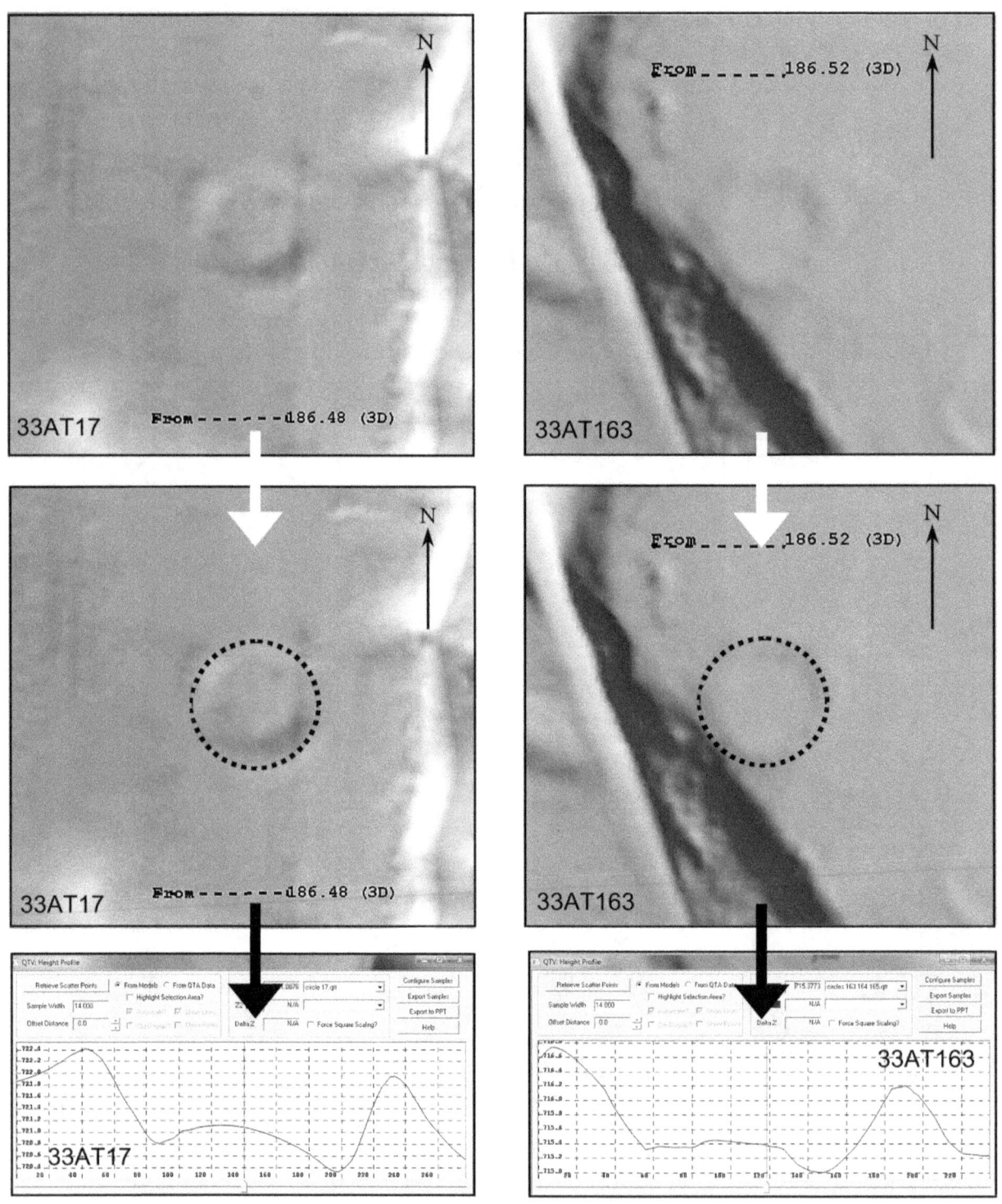

Upper left: LiDAR image of 33AT17. Middle left: 186-foot circle superimposed over 33AT17. Lower left: LiDAR profile of 33AT17.
Upper right: LiDAR image of 33AT163. Middle right: 186-foot circle superimposed over 33AT163. Lower right: LiDAR profile of 33AT163.

Figure 4.8. The Wolf Plains: Circles 164 and 165

Circles 33AT164 and 33AT165 also incorporate the 186-foot length.

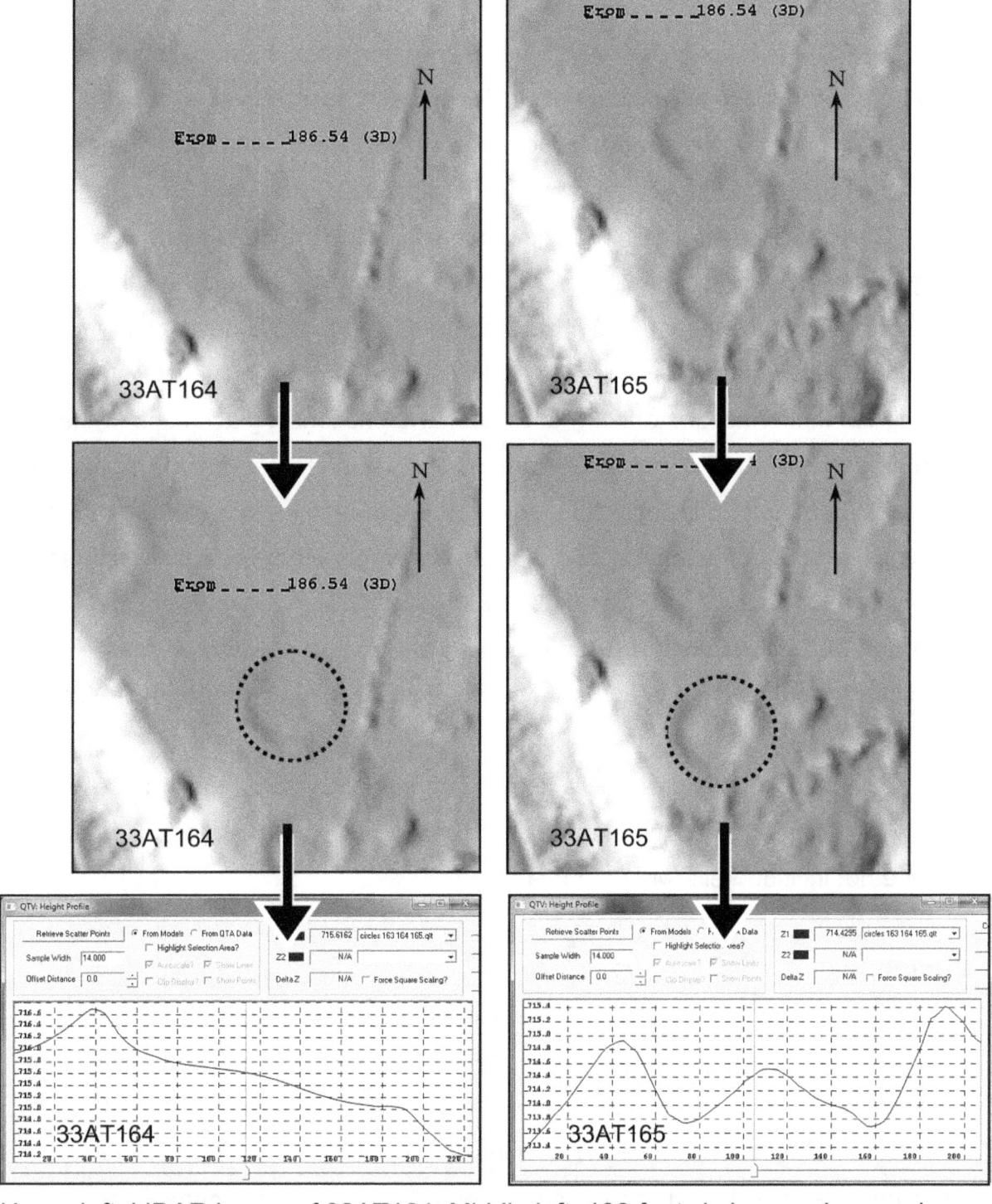

Upper left: LiDAR image of 33AT164. Middle left: 186-foot circle superimposed over 33AT164. Lower left: LiDAR profile of 33AT164.
Upper right: LiDAR image of 33AT165. Middle right: 186-foot circle superimposed over 33AT165. Lower right: LiDAR profile of 33AT165.

CHAPTER 5
SCIOTO RIVER WATERSHED

The Scioto River reaches 230 miles, from north of Columbus, south to the Ohio River. About two miles south of Chillicothe, the Scioto River is joined by Paint Creek from the west. More Hopewell earthworks are located within a ten-mile radius of this confluence than anywhere else in Ohio.

Chillicothe is situated in a unique location, which may explain why so many earthworks are located there. What makes the location unique is that it is at the intersection of several major physiographic and biotic zones including: 1) the border between the Mississippian and Devonian bedrock systems; 2) near the edge of the farthest extent of the Illinoian and Wisconsin glaciations; 3) near the boundaries of the Till Plains, Glaciated Appalachian Plateau, and Unglaciated Appalachian Plateau physiographic regions; 4) near the juncture of three major soil types; and 5) at the intersection of three major forest types (Maslowski and Seeman 1992; Shane 1971; Romain 2000:15-16; 1993). As Maslowski and Seeman (1992:11) explain, "…the result is what was probably the greatest diversity of microenvironments for an area of comparable size in the mid-Ohio Valley."

Space limitations preclude a detailed commentary about each and every earthwork. There are, however, several observations I would like to make here that are not included elsewhere in this chapter.

Mound City

Mound City is also known as the 'City of the Dead.' The title is appropriate enough, considering that at least two dozen burial mounds, each containing multiple burials, have been identified within the perimeter walls of the earthwork enclosure. As such, Mound City is a liminal location where the dead exist in a place between This World and the Otherworld. Mound City is essentially a gateway to the Otherworld.

Contributing to the experience of Mound City as a liminal place is that the site is in a special location relative to the mountains to the east. As shown by Figure 5.11, a powerful visual heirophany positions Mound City in a balanced relationship with the sun and moon as they rise over the mountains to the east. On the dates of its minimum north and minimum south rise, the moon will rise over Sugarloaf Mountain and Mount Logan, respectively. On the dates of the summer and winter solstices, the sun will rise along the edges of the mountain range, basically framing the mountains (Hively and Horn 2010). On the date of the equinox, the sun will rise directly across from Mound City, over Bunker Hill. If Mound City were located north or south of its present position, these relationships would not be visible.

There are two points to be noted with reference to this heirophany. First, the relationship between Mound City and the celestial azimuths just noted is geometric and symmetrical. Geometrically, Mound City is situated at the apex of a triangle formed by the azimuths of the sun and moon. What this means for a Mound City-based observer is that the sun and moon will appear to move from north to south and back north again, along the eastern horizon, with the observer at the apparent center of this movement. The relationship is therefore observer-centric with the sun and moon.

Second, there is a temporal aspect to this. Not only is the Mound City observer at the geometric center relative to the north and south limits of the sun and moon, the observer is also at a center place in time. That is, the passage of time as referenced to the cyclic movements of the sun and moon seems relative to the individual. In this, the observer is at the center of it all and not only enters into the sacred, but further, becomes a focal point for time and space. Like the Newark Earthworks, Mound City is a cosmic center.

Lastly, there is another cosmological interpretation would offer. I propose that Mound City was not just a burial ground; but that it was also intended as a terrestrial and metaphorical equivalent of the Land of the Dead. Consider the following. In many Native American religions, the Land of the Dead is the reverse of This World (e.g., Hall 1997:132-139; Hudson 1976:133). What is up in This

World is down in the Land of the Dead; when it is daytime in This World, it is night time in the Land of the Dead, and so on. According to this logic, in This World, Mound City is a place for the dead. In the reversed Land of the Dead, the Otherworld analogue of Mound City becomes a place for continued existence – i.e., life after death in a different realm.

Further, as shown by Figure 5.8, Mound City is oriented to the summer solstice sunset through its diagonal axis. In the cycle of seasons, the time from summer solstice to winter solstice is a time of maturity, decline, and death. Thus Mound City is oriented in the direction of death. In the reversed Land of the Dead, however, the spirit counterpart to Mound City is oriented in the opposite direction – toward life.

Next consider that each of the more than two dozen burial mounds at Mound City were found to have within them, postmold remains of house structures. In the older archaeological literature these structures were called 'charnel houses' because it is presumed that the Hopewell dead were processed in such structures before mound burial or cremation. My preferred term for these structures is Spirit House. In any case, multiple burials were made within each of these house structures before they were dismantled, burned, and mounded-over with earth. If the cosmological reversal suggested here holds true, then, the burned house structures of Mound City become functioning lodges in the Land of the Dead; and the walled Mound City perimeter that encloses the dead in This World becomes a community of sentient people in the Otherworld. In this view, Mound City was an earth analogue for the Land of the Dead.

Portsmouth Earthworks

Portsmouth is located at the confluence of the Scioto and Ohio rivers, about 42 miles south of Chillicothe. Today, the Ohio River marks the boundary between Ohio and Kentucky. Major earthworks, however, are found on both sides of the Ohio River, in the immediate vicinity of the confluence.

Portsmouth is located in the Unglaciated Allegheny Plateau region. The landscape is one of steep hills and high ridges. The Scioto River Valley at Portsmouth is 1 ½ to 2 miles wide.

The Portsmouth Earthworks was one of the major complexes of the Adena-Hopewell. Indeed, Squier and Davis (1848:78) calculated that, including enclosure and parallel walls, there were "upwards of 20 miles" of embankment walls at Portsmouth.

Among the most intriguing of the Portsmouth earthworks are the U-shaped mounds situated at the heart of the Group B Complex. Originally there were two U-shaped mounds. Today, only one remains. The question is: 'Why U-shaped mounds?' Nowhere else in Adena or Hopewell do we find this earthwork shape.

One answer may be that the U–shaped mounds were intended to mimic the shape of the valley immediately to the north (see Figure 5.44). The resemblance is striking. The length to width proportions between the U-shaped earthworks and the actual valley are equivalent. The U-shaped earthworks are oriented north, as is the real valley. Further, the openings of the U-shaped earthworks correspond to the south-facing opening of the valley.

The imitation of nature in humanly-created designs is a form of mimesis. To emphasize the idea in design applications, natural forms are often stylized and exaggerated, even if miniaturized.

The notion that the Portsmouth U-shaped earthworks mimic a landscape configuration has precedent. Conical-shaped Adena mounds – such as the Conus Mound and Miamisburg Mound, for example, clearly resemble the distinctive conical hills of southern Ohio (Romain 2000). So too, Hopewell loaf-shaped mounds – such as the Seip-Pricer Mound, and several mounds at Mound City resemble nearby loaf-shaped mountains (Romain 2000). Parallel-walled features - i.e., so-called walkways or roads, especially in cases where their cross-sectional profiles are concave (e.g., Newark Great Hopewell Road - see Romain and Burks 2008c), resemble stream beds and appropriately enough, most of the parallel walls lead downslope or downstream to water. A case can even be made that large Hopewell circle enclosures recall the distinctive circular old river channels made, for example, by the Scioto River and Paint Creek near High Bank and other places.

In any case, it is interesting to note that, as Figure 5.44 shows, the Portsmouth Group B Circle containing the U-shaped earthworks is situated a bit off-center from the major axis of the actual valley. This deviation is accounted-for by the observation that for the moon's maximum north rise and set to be seen over their respective mountain peaks (as also shown in Figure 5.44), the observer's location had to be exactly where the Group B Circle is located. Thus the moon controlled the location for the Group B Circle Earthwork, even if the internal geometry of the earthwork was inspired by the local topography.

An interesting question that occurs is: 'Why, two U-shaped earthworks instead of one?' Unfortunately, I do not have a good answer for that. What we can surmise, however, is that through whatever rituals that were performed within the Circle, and through their proxy model of the U-shaped valley, as well as the deliberate action of situating the circle at intersecting lunar sightlines, the Moundbuilders were able to link their rituals and sensory experiences to the landscape - meaning earth and sky.

Figure 5.1. Map of sites in the Scioto River Watershed

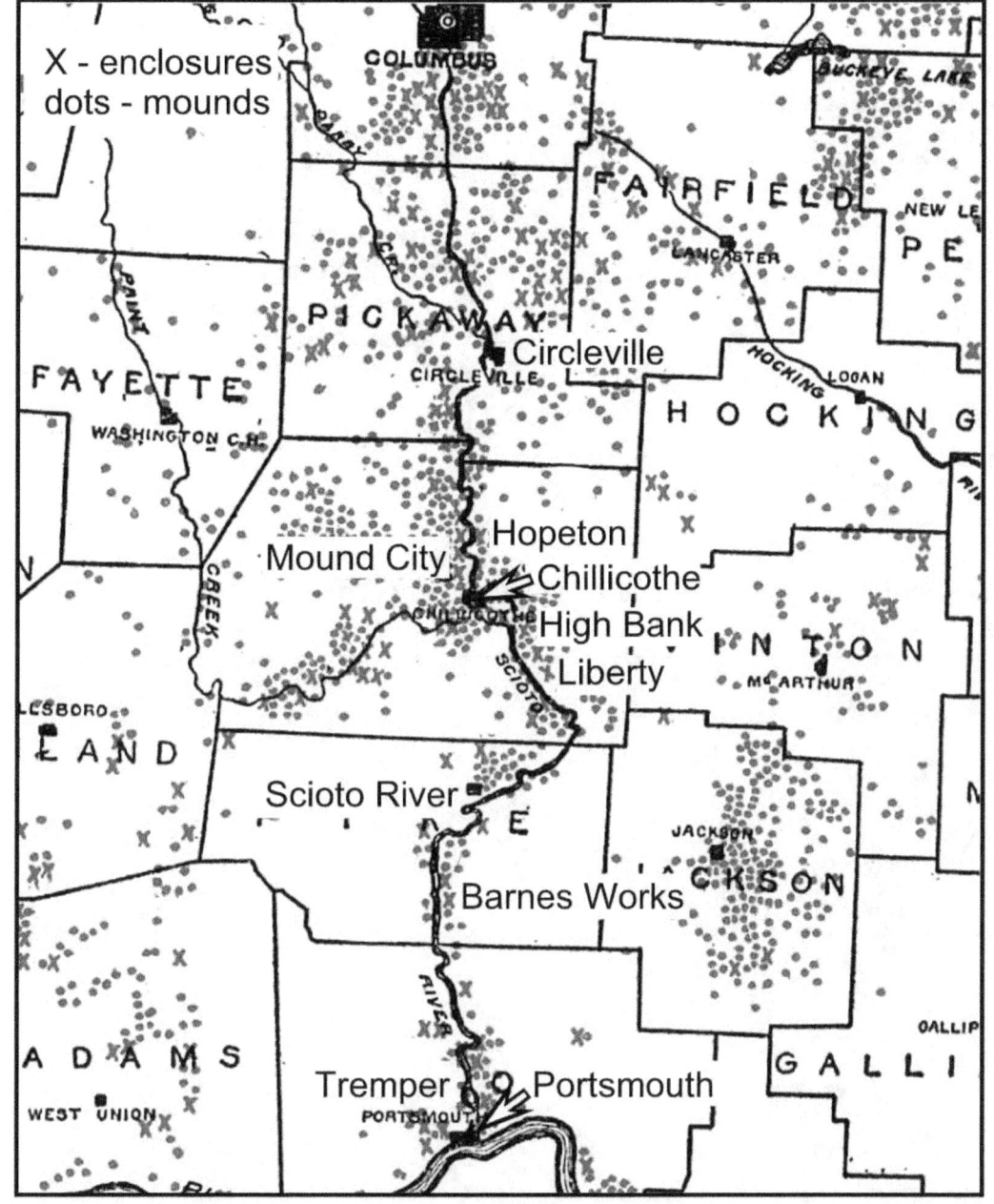

Above: detail of Mills (1914:Pl. XI) map showing earthworks and mounds in the Scioto River Watershed (annotation added).

Figure 5.2. Circleville Earthworks

Located about 17 miles north of Chillicothe, the Circleville Works were the most northern of the complex geometric sites along the Scioto River. The earthworks have been completely destroyed by urban development.

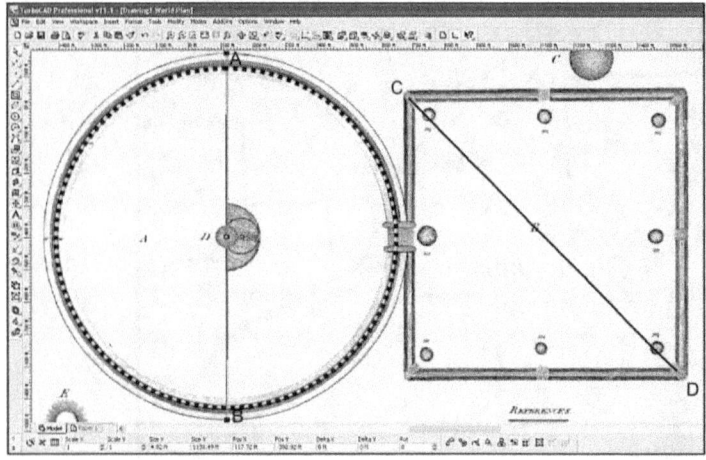

Left: Atwater's (1820:Pl. V) map of Circleville imported into TurboCad and scaled per dimensions given by Atwater (1820:178). The resulting diagonal of the Square is 1,178 feet - a distance related to the HMU as explained in Chapter 3.

Above: if Atwater's data are correct, then the outside diameter of the Large Circle would have been the same size as the Newark Great Circle and Shriver Circle to within about 40 feet.

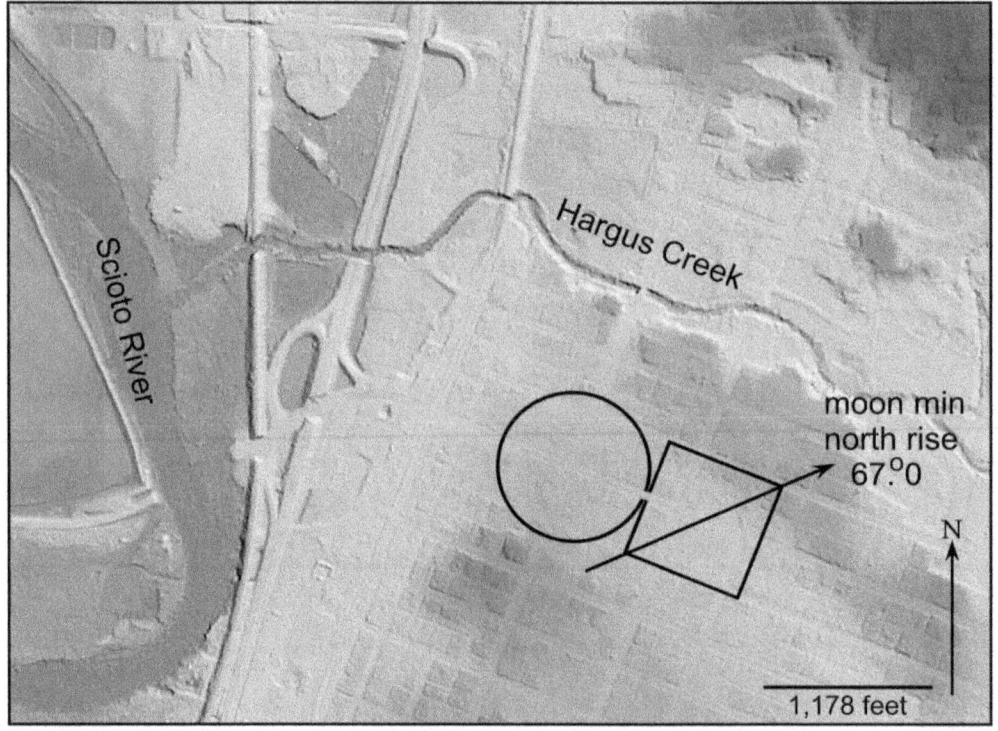

Above: schematic plan of Circleville superimposed on LiDAR image. Circleville was located at the confluence of Hargus Creek and the Scioto River. Its main axis was more or less parallel to the river terrace to the north and Hargus Creek. If the above estimated position and orientation are correct, the site may have been lunar aligned through the diagonal of the Square. (Calculations for A.D. 100, corrected horizon elevation of 1.6 degrees.)

Figure 5.3. The Chillicothe Earthworks

The Chillicothe area is traditionally considered the heartland of Ohio Hopewell. More than 100 mounds and dozens of earthwork enclosures of various shapes and sizes have been recorded for the area. Hills on both sides of the Scioto River form roughly a 2 1/2 mile wide valley offering a rich variety of resources.

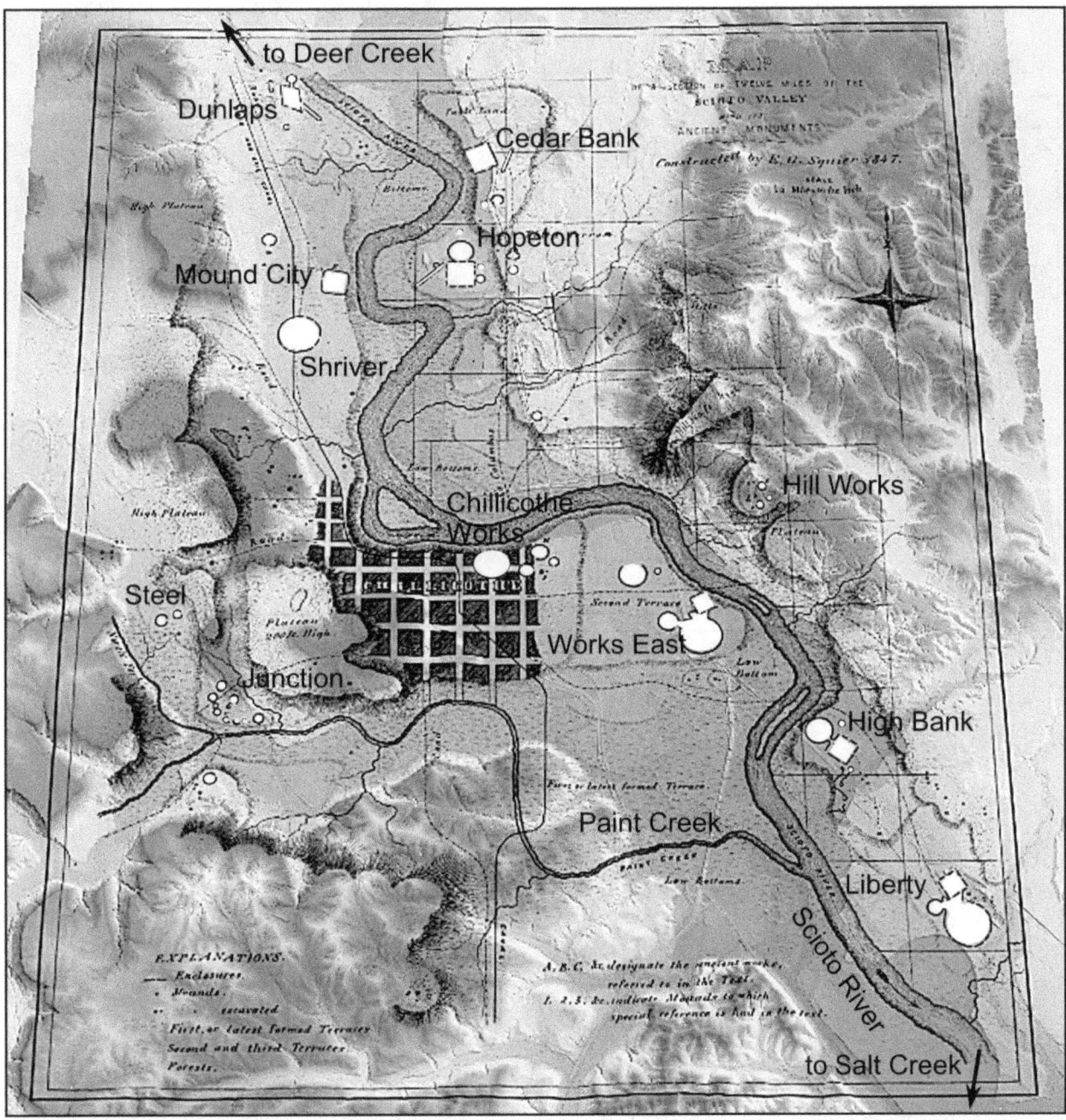

Above: Squier and Davis's (1848:Pl. II) map of the Chillicothe Scioto River Valley superimposed over LiDAR image. Significant enclosures highlighted. Annotation added.

Figure 5.4. Central Scioto River Valley Archaeological Zone

The central Scioto River Valley archaeological zone extends a few miles further north and south than the borders of the Squier and Davis map allow. On the north, the zone is marked by the beginning of hills that form the west side of the Scioto River Valley, the confluence of Deer Creek and the Scioto River, and several sites - to include the Stitt Mound. The south terminus of the zone is marked by the confluence of Salt Creek and the Scioto River, Chimney Rock Mountain, and the Johnson Works. Map provided by MyTopo, annotation added.

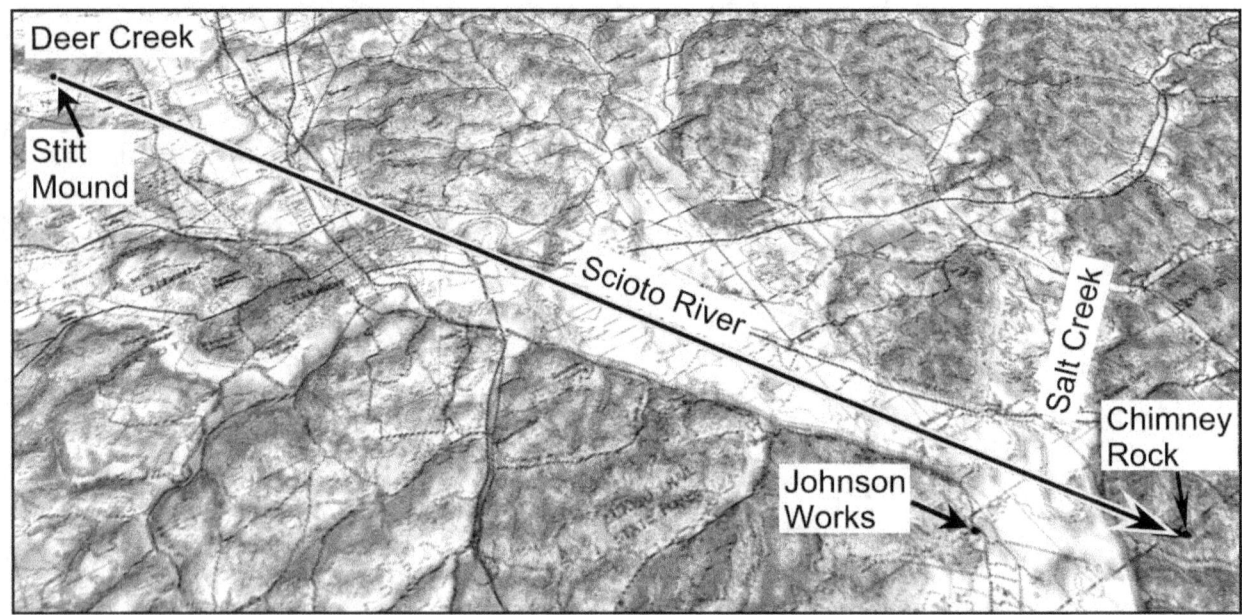

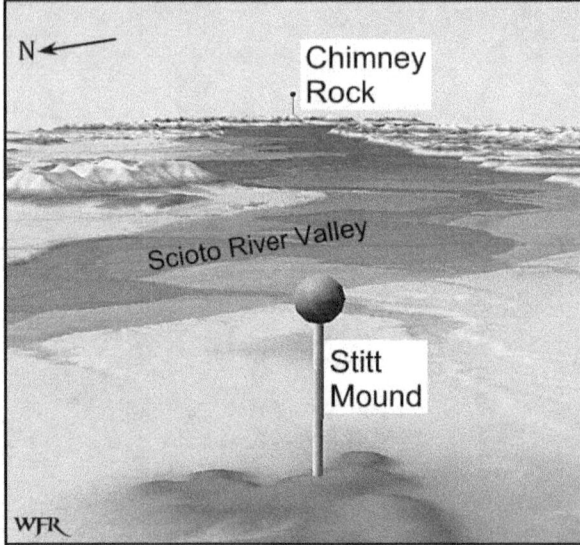

Above left: LiDAR view of the 22-mile length of the central Scioto River Valley. Chimney Rock is at the far end of the valley (note 1).
Above right: postcard photo from the early 1900s showing the rock formation that gave rise to the name, Chimney Rock Mountain - on which the formation is located. Image credit: www.waverlyinfo.com and Jim Henry.

Figure 5.5. Chillicothe Earthworks: Stitt Mound

Stitt Mound is located on the brow of a ridge overlooking the confluence of Deer Creek and the Scioto River. The mound is about 17 feet high and 120 feet in diameter. In the 1890s the mound was tunneled into; but no reports describe what was found.

Above: LiDAR contour map of the Stitt Mound (1.5-foot contour interval.) In addition to providing a view of the Scioto River Valley, Stitt Mound is situated so that the winter solstice sun will appear to rise over Sugarloaf Mountain - about 6 miles distant. Sugarloaf Mountain is the most visually dominant mountain in the area (note 2). Below: winter solstice azimuth from Stitt Mound. Map provided by MyTopo, annotation added.

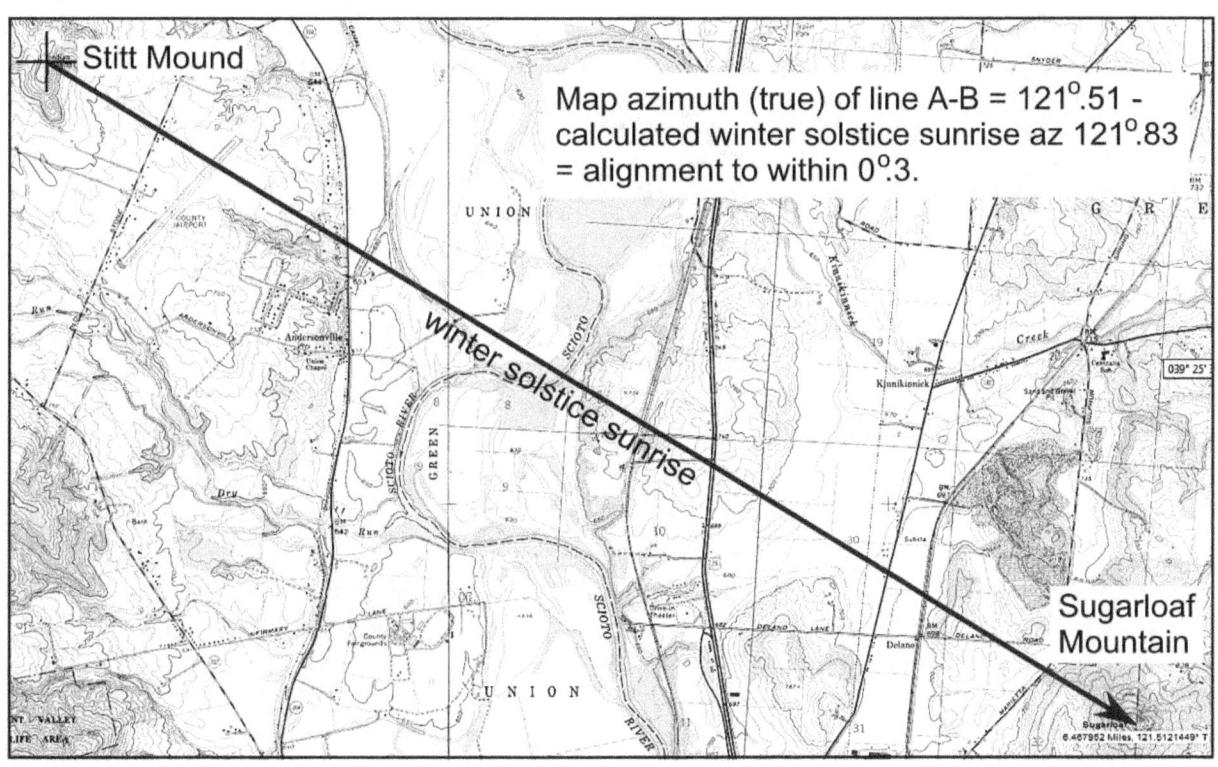

Map azimuth (true) of line A-B = $121°.51$ - calculated winter solstice sunrise az $121°.83$ = alignment to within $0°.3$.

Figure 5.6. North Scioto River Valley Earthworks

In addition to Stitt Mound, several significant earthworks are located at the northern boundary of the central Scioto Valley archaeological zone. Below map provided by MyTopo, annotation added.

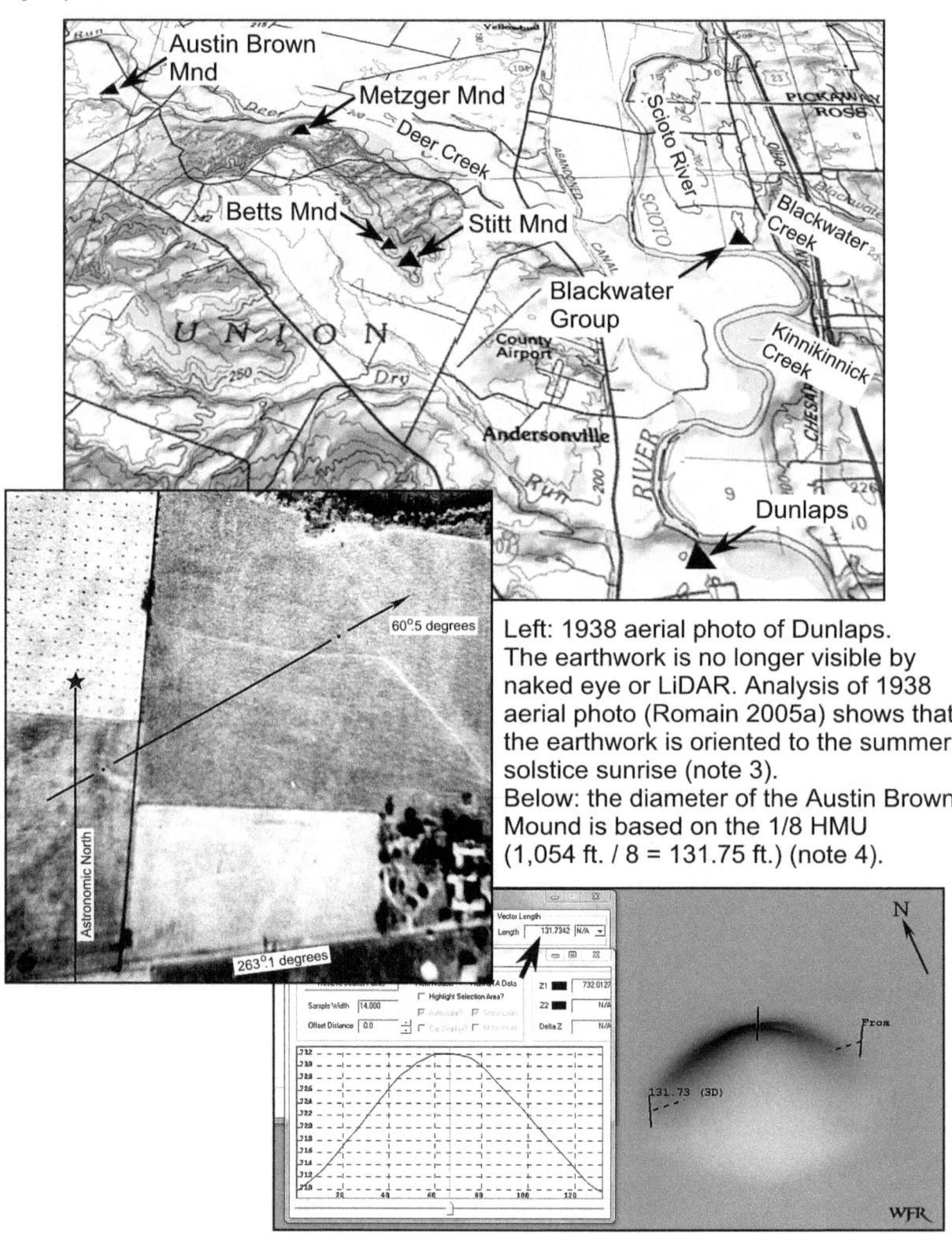

Left: 1938 aerial photo of Dunlaps. The earthwork is no longer visible by naked eye or LiDAR. Analysis of 1938 aerial photo (Romain 2005a) shows that the earthwork is oriented to the summer solstice sunrise (note 3).

Below: the diameter of the Austin Brown Mound is based on the 1/8 HMU (1,054 ft. / 8 = 131.75 ft.) (note 4).

Figure 5.7. Chillicothe Earthworks: Mound City

Mound City is situated on the west bank of the Scioto River, about three miles north of downtown Chillicothe. At least 24 burial mounds were originally located within the enclosure (note 5). Many of the burials at Mound City were accompanied by unique objects and precious materials to include effigy smoking pipes, copper plates, ocean shell, obsidian, shark teeth, galena, and mica.

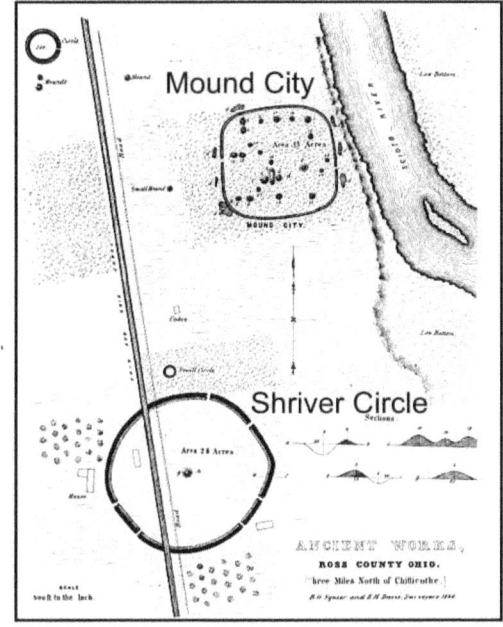
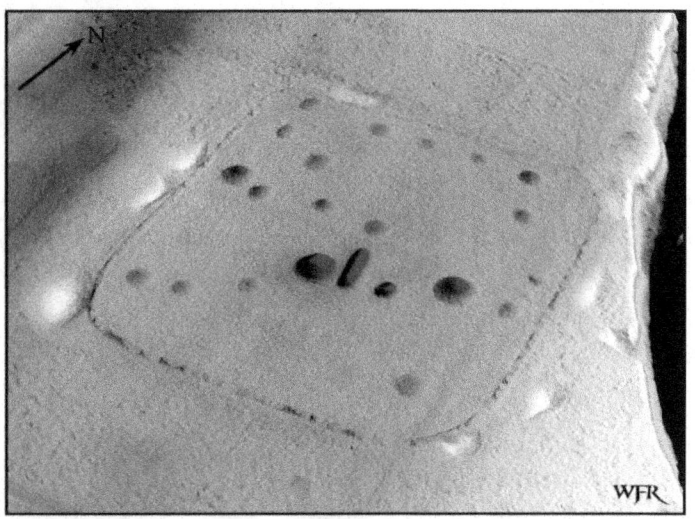

Above left: Squier and Davis (1848:Pl. XIX) map of Mound City and Shriver Circle (annotation added).
Above right: LiDAR image of Mound City (note the dug pits around the perimeter).

Above: View of Mound City looking from outside of the west entrance, across the enclosure to the northeast.

Figure 5.8. Chillicothe Earthworks: Mound City Alignments

Mound City is aligned to the summer solstice sunset along its diagonal axis in one direction and to the trajectory of the Milky Way along its opposite axis (note 6).

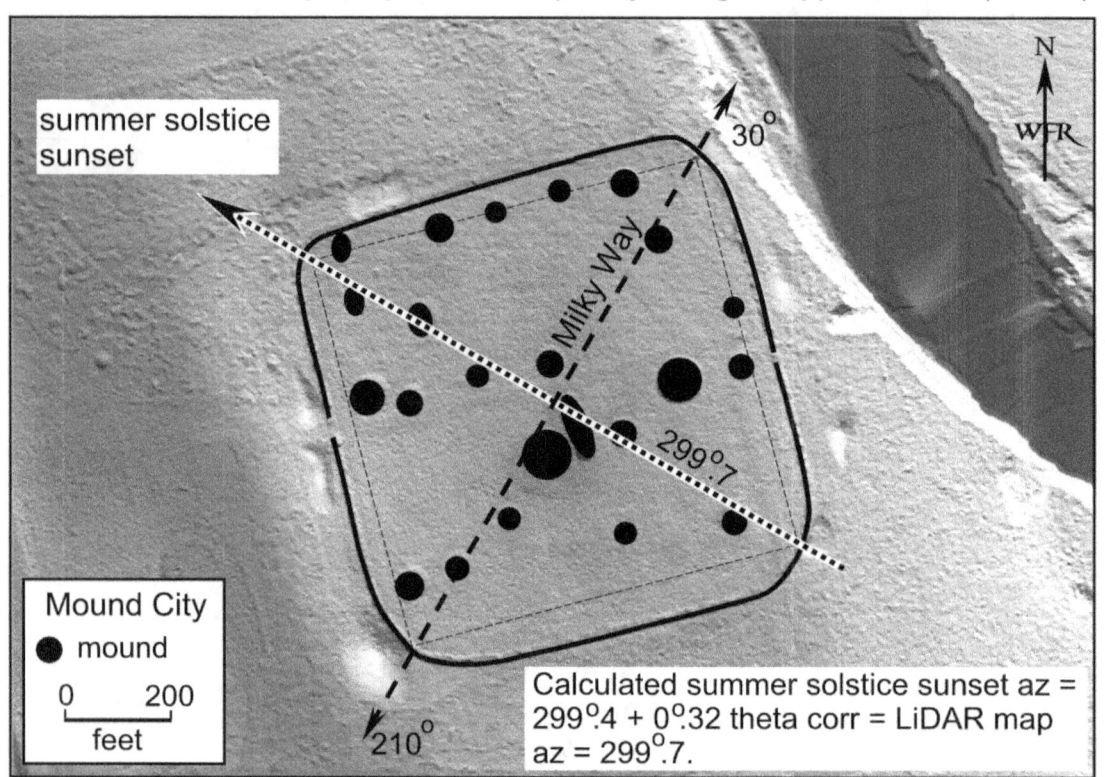

Above: line drawing superimposed on LiDAR image showing solstice and Milky Way alignments (note 7). Many Native American tribes believed that the Milky Way was a Spirit, or Ghost Road traveled by souls of the dead to reach the Land of the Dead.

Above left: aerial view of Mound City showing the summer solstice sunset sightline.
Right: Summer solstice sunset along the diagonal axis of Mound City.

Figure 5.9. Chillicothe Earthworks: Mound City and Cygnus
There is another interesting relationship between Mound City and the Milky Way.

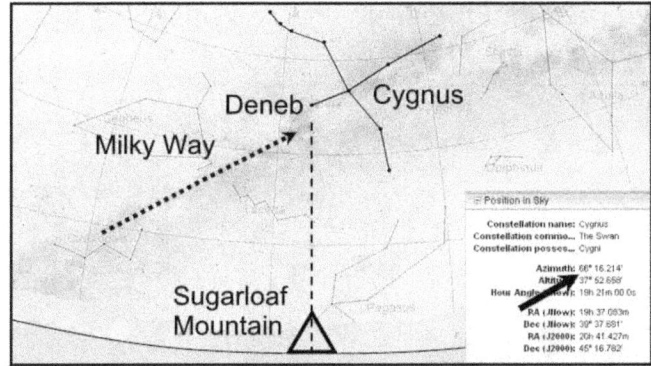

Left: at sunset on the night of the summer solstice in A.D. 100, as viewed from Mound City, the constellation Cygnus and its alpha star, Deneb would have been seen over the Sugarloaf Mountain axis mundi (note 8). Deneb marks the beginning of the Great Rift - the place where the Milky Way splits into two paths.

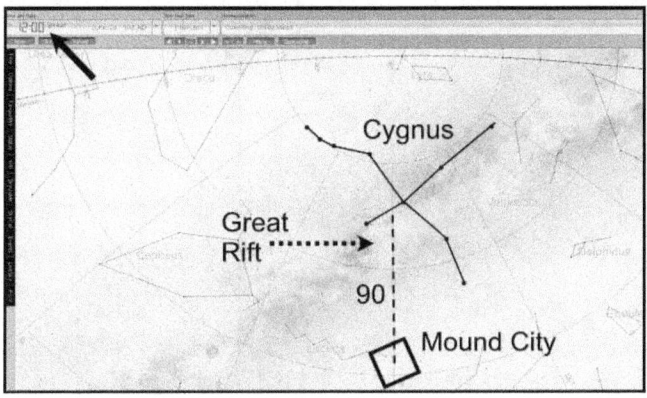

Left: at midnight on the night of the summer solstice, Cygnus and Deneb were located east of Mound City. Many tribes believed that an antagonist-judge (eagle, dog, old man, or old woman) confronted the soul on the Milky Way Path. If the encounter went well, the soul continued along one trail to the Land of the Dead. If not, the other trail led to oblivion (Lankford 2011, 2007c). Maps by Starry Night, annotation added.

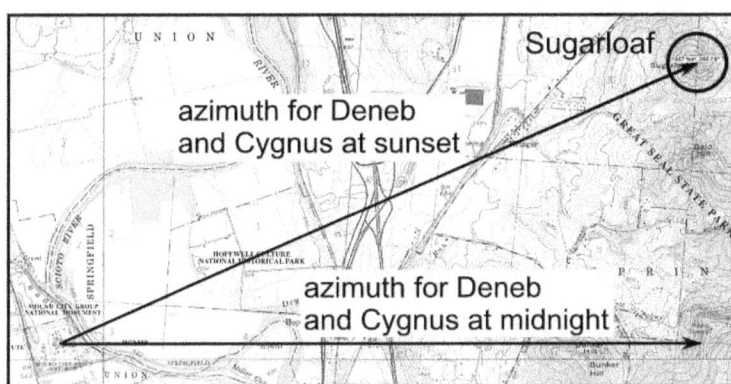
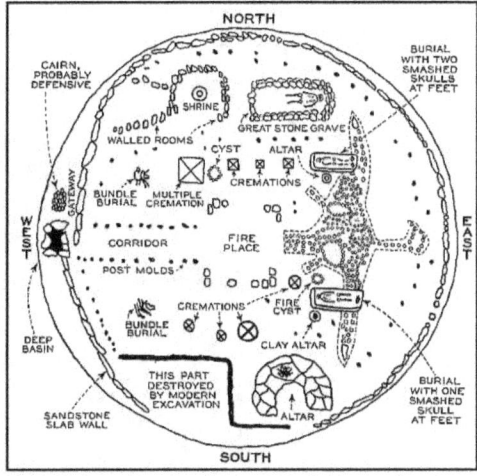

Above left: azimuths for Deneb and Cygnus relative to Mound City for A.D. 100 on night of summer solstice at sunset (nautical twilight) and midnight. Map by My Topo, annotation added.

Above right: North Benton Mound (Magrath 1940:fig.1). This Hopewell burial mound was located near North Benton, Ohio. The bird effigy is oriented due east. If the burials on its wings were to stand-up, they too would be facing east. Perhaps the bird effigy was intended to escort the souls of the dead along the Milky Way Path (note 9).

Figure 5.10. Chillicothe Earthworks: Mound City and the HMU

Mound City is not a square. Its corners are rounded and its walls are ballooned-out (note 10). The images below, however, suggest that the structure was based on the concept of a square having a diagonal equal to 1 HMU (or 1,054 feet).

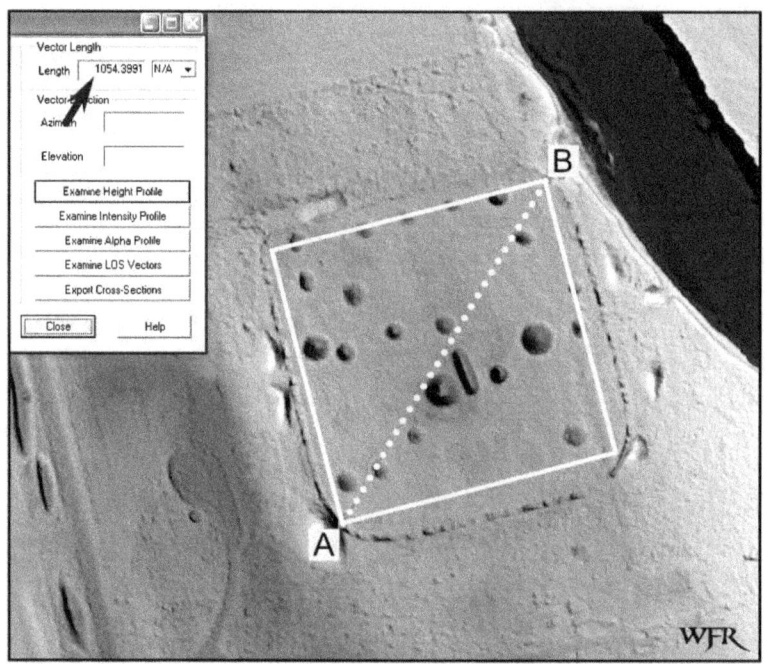

Above: the LiDAR-measured distance from point A to point B is 1 HMU (or 1,054 feet). A square is constructed using line A-B as its diagonal. The resulting square closely corresponds to the inside dimensions of the earthwork.

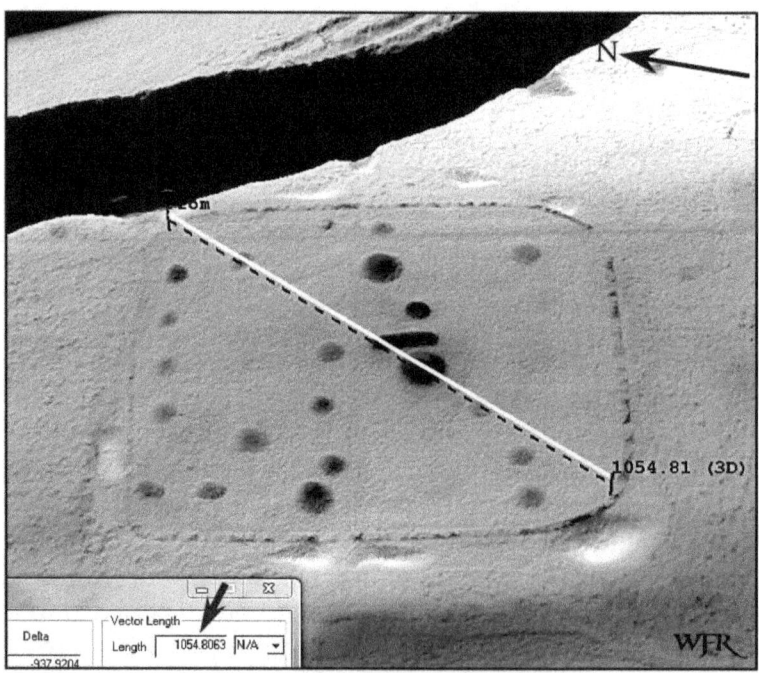

Above: oblique LiDAR view showing how the length of the southwest to northeast diagonal of the earthwork is equal to 1 HMU (or 1,054 feet).

Figure 5.11. Chillicothe Earthworks: Mound City: Visual Heirophanies

One of the things that makes Mound City special is its location relative to the sun, moon, and mountains to the east (note 11).

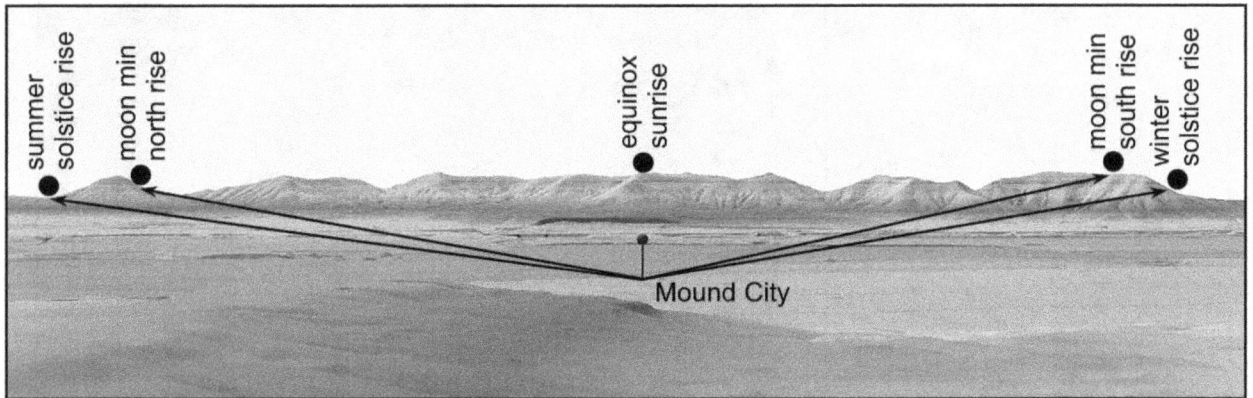

Above: LiDAR view of horizon east of Mound City, vertical exaggeration 1.3. Viewed from Mound City, the summer and winter solstice sunrises frame the mountain range while the equinox sunrise marks its center. Further, the moon's minimum north and south oscillation is marked by Sugarloaf Mountain and Mount Logan.

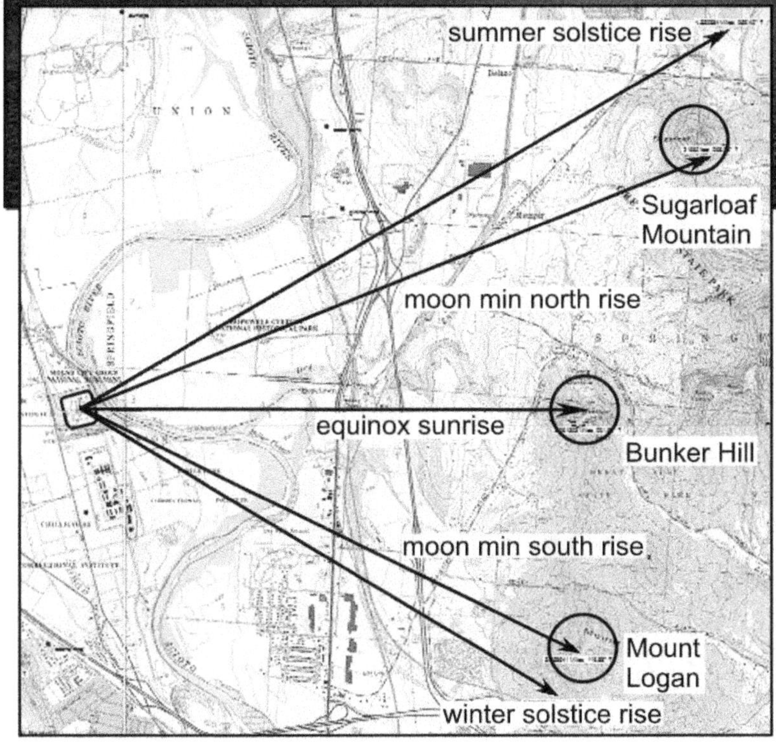

Above: photo of the equinox sunrise over Bunker Hill, March 20, 2013 (note 12). Left: celestial azimuths plotted from Mound City. Map provided by My Topo, annotation added.

123

Figure 5.12. Chillicothe Earthworks: Shriver Circle

The Shriver Circle is located about one-half mile southwest of Mound City. The earthwork has been seriously degraded but is still visible in aerial photos (note 13).

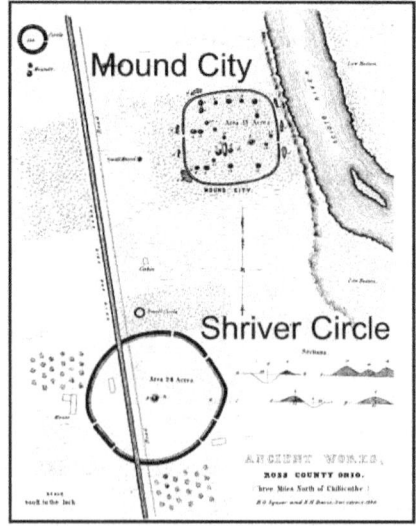

Above left: Squier and Davis (1848:Pl. XIX) map of Mound City and Shriver Circle (annotation added).
Above right: Google Earth aerial photo showing Shriver Circle.

Below: magnetic gradiometer survey results. Image courtesy of Jarrod Burks and Rob Cook.

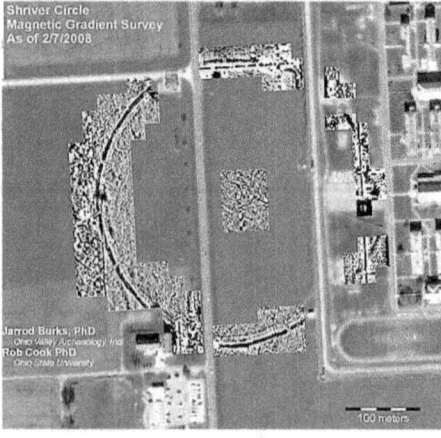
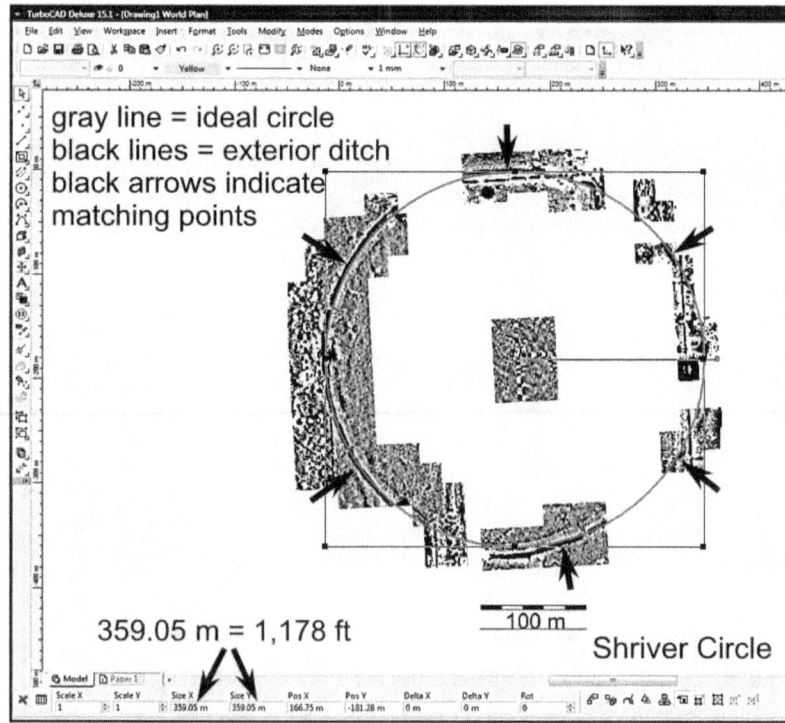

Above right: the size of the Shriver Circle is based on the 1,178-foot length - which in turn, derives from the HMU length of 1,054 feet. In the above figure, the Burks-Cook magnetic survey map is imported into TurboCAD. An ideal circle 1,178 feet in diameter is drawn and superimposed over the survey map. The ideal circle closely matches the size of the Shriver Circle as established by the outside ditch.

Figure 5.13. Chillicothe Earthworks: Shriver Circle Alignment
The Shriver Circle is situated in a unique location so that viewed from its center, the summer solstice sunrise will appear over Sugarloaf Mountain (note 14).

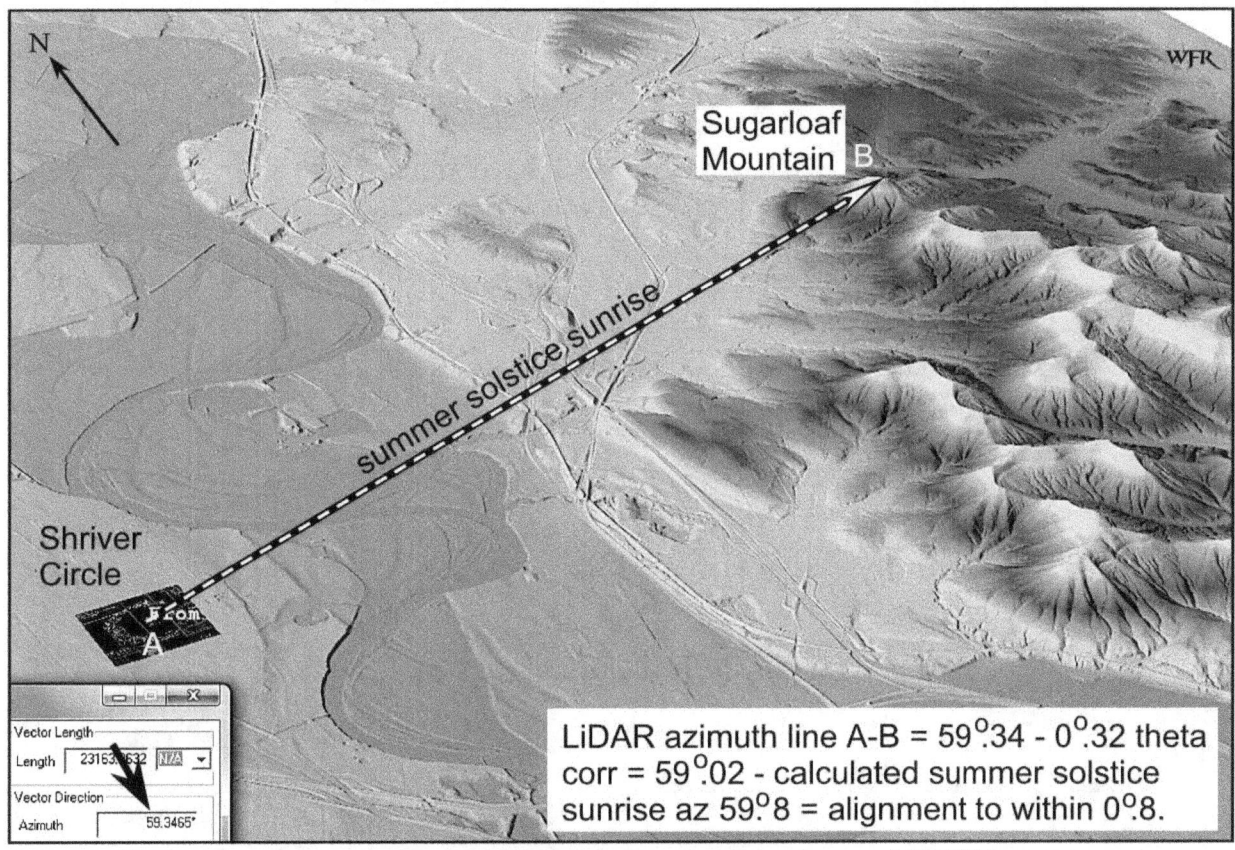

LiDAR azimuth line A-B = $59°.34 - 0°.32$ theta corr = $59°.02$ - calculated summer solstice sunrise az $59°.8$ = alignment to within $0°.8$.

Above: Burks-Cook magnetic survey map superimposed over LiDAR image.

Right: View of Sugarloaf Mountain from State Route 23, north side of Chillicothe.

Left: Enlarged detail of Squier and Davis (1848: Pl. XIX) map of Shriver Circle superimposed over Burks-Cook magnetic survey map. Solstice azimuth added. Solstice sightline coincides with one of the gateways into the enclosure (note 15).

Figure 5.14. Chillicothe Earthworks: Hopeton

The Hopeton Earthworks are situated on the east side of the Scioto River, diagonally across from Mound City. The complex originally included a large circle and square, several small circles, at least three mounds, and a set of parallel walls - nearly 2,500 feet in length.

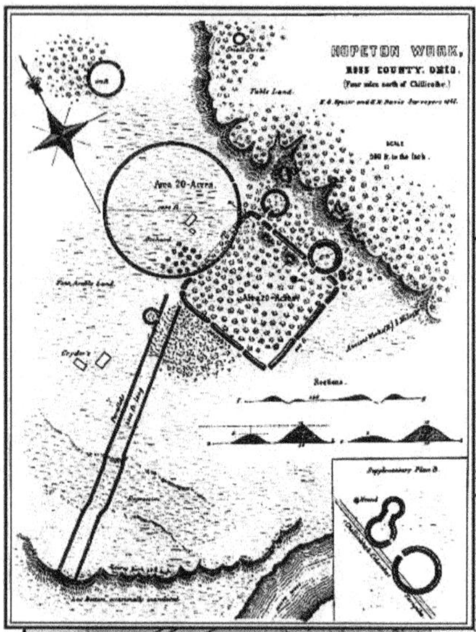

Left: Squier and Davis (1848: Pl. XVII) map of the Hopeton Earthworks.
Below: Squier and Davis map of Hopeton superimposed over LiDAR image.

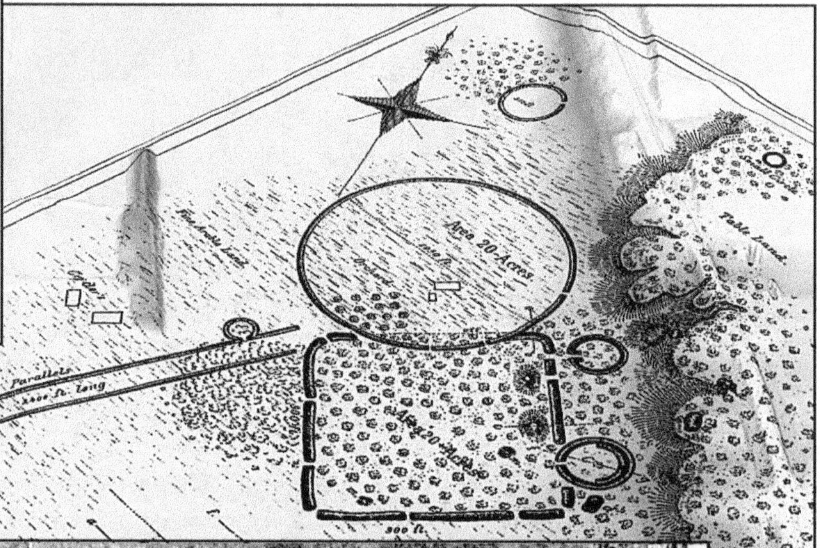

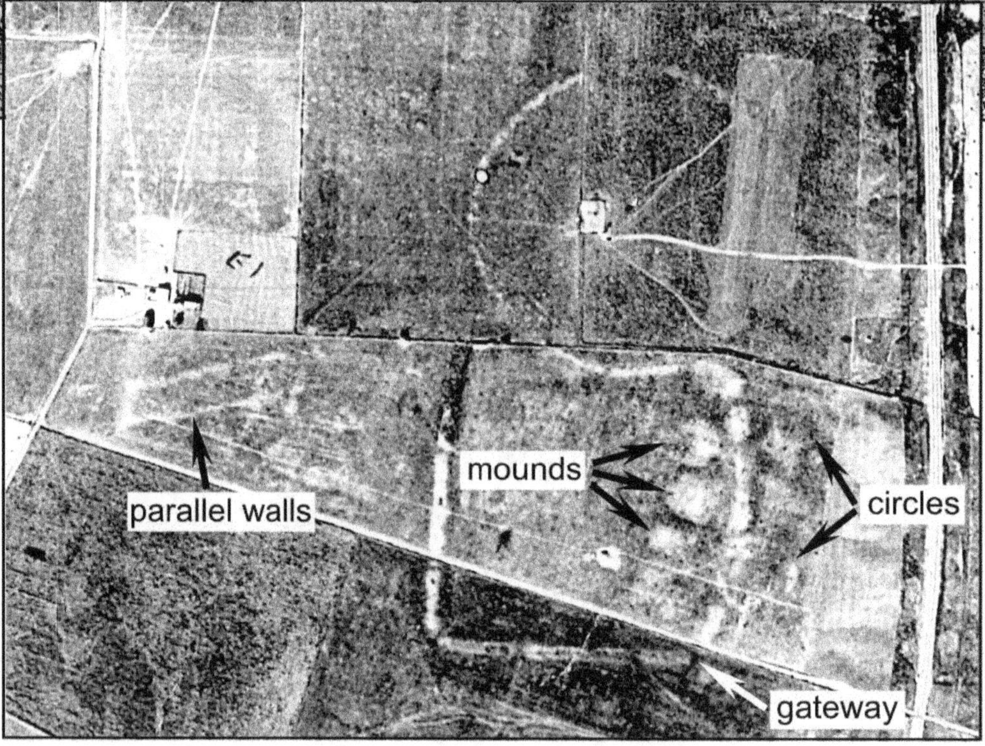

Above: US Dept of Agriculture 1938 aerial photo of Hopeton. Annotation added. The walls of the Square were 12 feet in height when measured in the 1840s.

Figure 5.15. Chillicothe Earthworks: Hopeton Alignments

The southeast gateway into the Hopeton Square at point A appears key to the location and orientation of the Square.

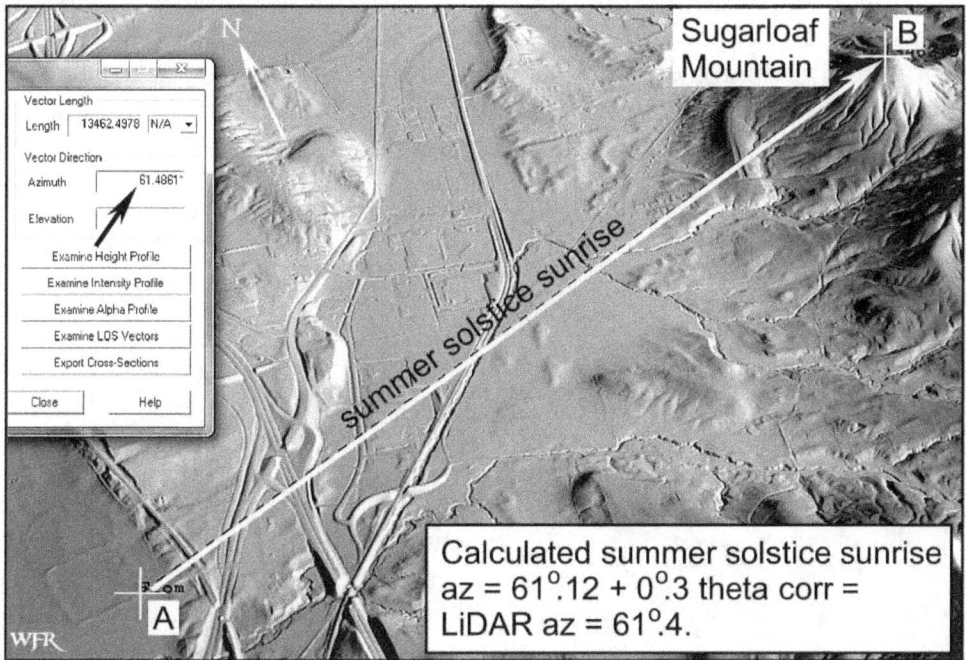

Calculated summer solstice sunrise
az = $61°.12 + 0°.3$ theta corr =
LiDAR az = $61°.4$.

Above: the Hopeton Square is situated so that viewed from the gateway at point A, the summer solstice sunrise will appear over Sugarloaf Mountain (note 16).

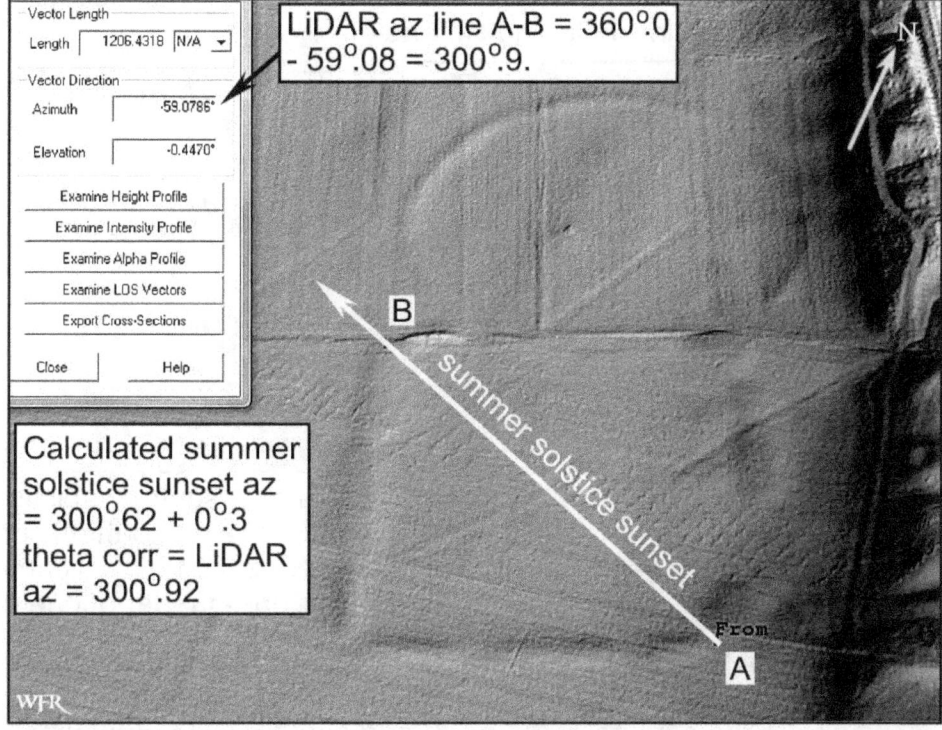

LiDAR az line A-B = $360°.0 - 59°.08 = 300°.9$.

Calculated summer solstice sunset az = $300°.62 + 0°.3$ theta corr = LiDAR az = $300°.92$

The Hopeton Square is aligned to the summer solstice sunset along a sightline that extends from the gateway at point A through the northwest corner at point B (note 17).

Figure 5.16. Chillicothe Earthworks: Hopeton Parallel Walls

The long parallel walls at Hopeton are aligned to the winter solstice sunset (note 18). The walls are no longer visible by LiDAR or geomagnetic survey. They are visible, however, in early aerial photos. In the below figure, a 1938 USDA aerial photo is georeferenced and superimposed onto a LiDAR image. The LiDAR program is then used to determine the trajectory of the parallel walls. As shown, the trajectory of the parallel walls matches the calculated azimuth of the winter solstice sunset.

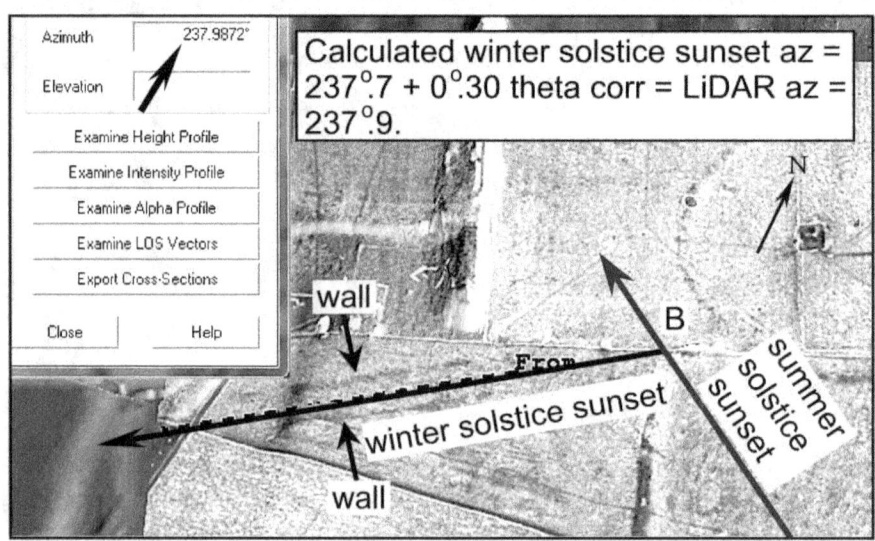

Notice that if the parallel wall - winter solstice sightline is extended, it will intersect the earlier noted summer solstice sunset sightline. The intersection occurs at point B - i.e., the northwest corner of the Square. This may account for the unique location of the parallel walls relative to the main Circle and Square.

Above left: lithograph from the 1800s showing the two circle earthworks on the east side of the Hopeton Square. From Squier 1860:figure 1.

Above right: viewed from the centers of the north and south circles, the summer solstice sun will rise over the base and peak of Sugarloaf Mountain, respectively. LiDAR image shows the summer solstice sunrise azimuth from the southernmost circle (note 19).

Figure 5.17. Chillicothe Earthworks: Hopeton and the HMU

The Hopeton Circle and Square are not geometrically perfect shapes. Both enclosures, however, are based on ideal shapes and both use the HMU to establish their size.

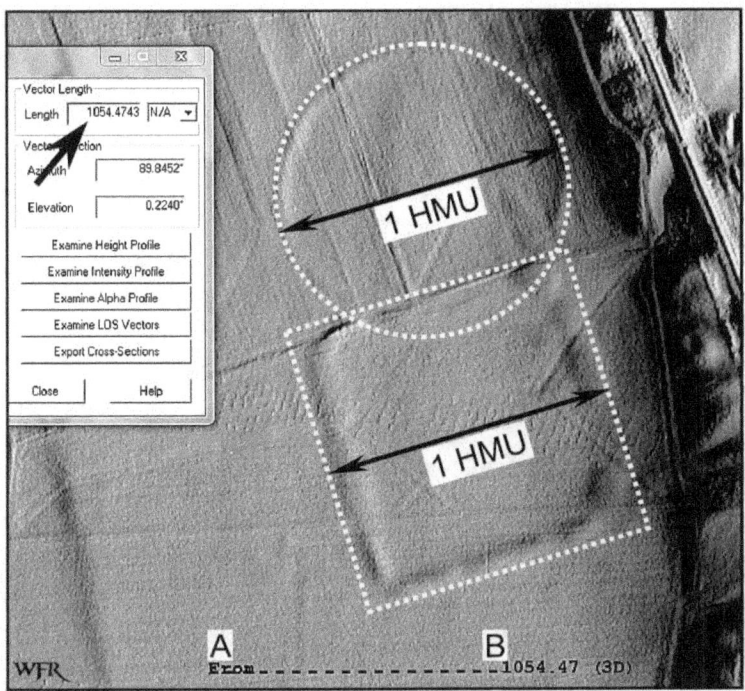

Above: the LiDAR measured length of line A-B is 1 HMU (or 1,054 feet). Line A-B is used to construct circle and square shapes. As shown, the dimensions of the constructed shapes closely match the outside dimensions of the earthwork figures (note 20).

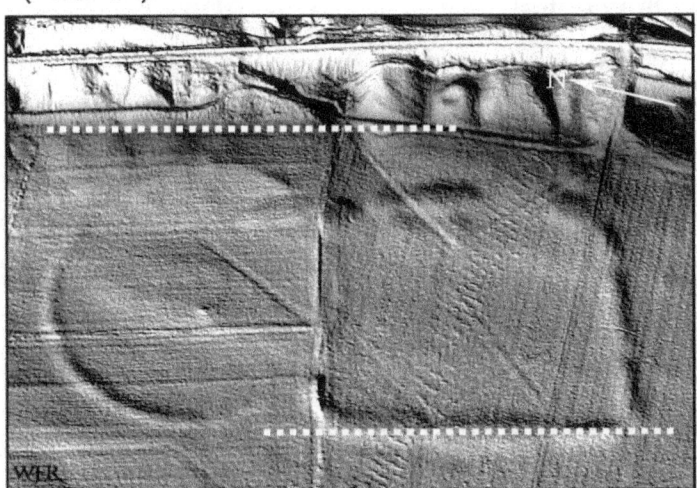 

Above left: the major axis of the Hopeton Square extends parallel to the lay of the land as established by the bottom edge of the nearby terrace. In the LiDAR image shown, the two dotted ines are parallel.

Above right: National Park Service interpreter standing on the high terrace overlooking Hopeton. What is left of the Hopeton Square is behind him.

Figure 5.18. Chillicothe Earthworks: The Adena Mound

The Adena Mound was about 26 feet high and 142 feet in diameter (Mills 1902:5-6). More than 30 burials were found in the mound, as well as the well-known Adena dwarf effigy pipe (Romain 2009:49-51). Recent radiocarbon dating of organic materials found with the central burial suggest the mound was built sometime "...between the end of second century B.C and the beginning of the first century A.D." (Leone et al. 2013:218).

Above: view of the Adena Mound ca. 1901 (Mills 1902:frontispiece).
Right: enlarged detail of USDA 1938 aerial photo showing the Adena mound. Of interest is that an earthen wall and possible ditch surround the central burial mound.

Above left: map detail showing location of Adena Mound. Map by My Topo.
Above right: Mills (1902:460) reports that subsequent to excavation, he left four feet of mound fill at the site to "show at least where it stood." Today, this mound fill is visible as a rise in the terrain, although a road now cuts across it. The area of mound fill is outlined by the dashed line.

Figure 5.19. Chillicothe Earthworks: Adena Mound Alignment

The Adena Mound is situated in a unique location. On the date of the maximum north moonrise, as viewed from the Adena Mound, the moon will appear to rise over Sugarloaf Mountain (note 21).

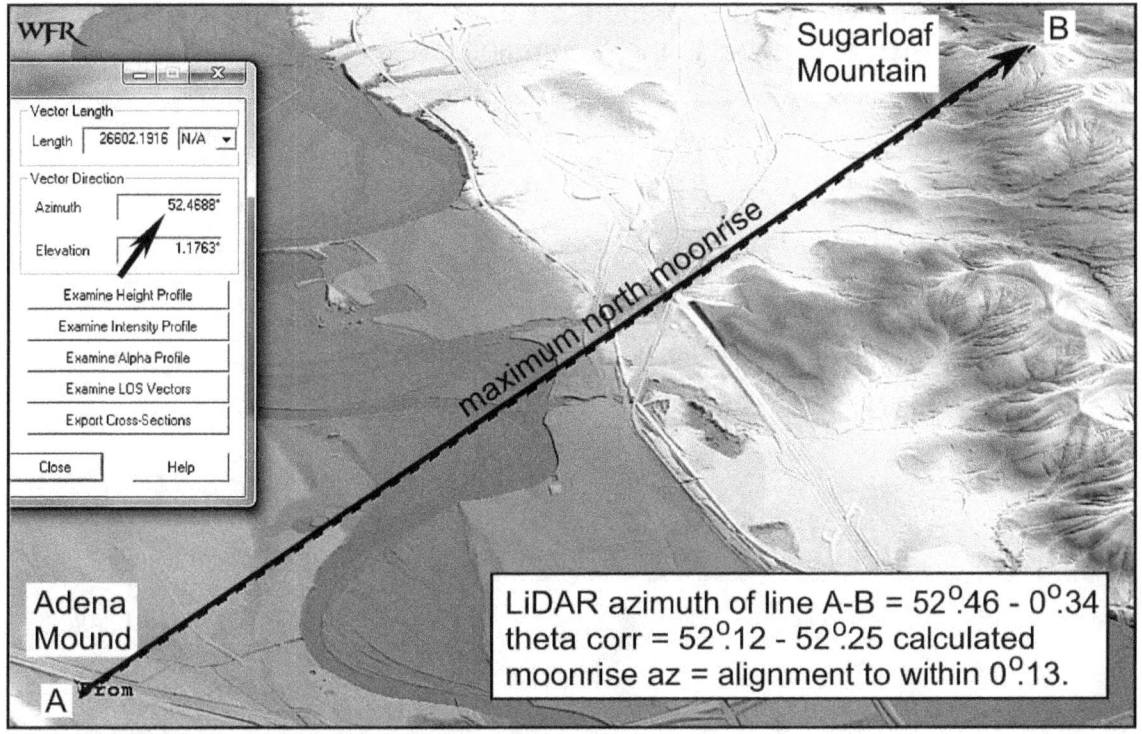

Above: LiDAR image showing relationship between the Adena Mound, Sugarloaf Mountain, and the moon's maximum north rise azimuth.
Below: USGS 7.5-minute series map details showing Adena Mound lunar azimuth plotted. Map images provided by MyTopo, annotation added.

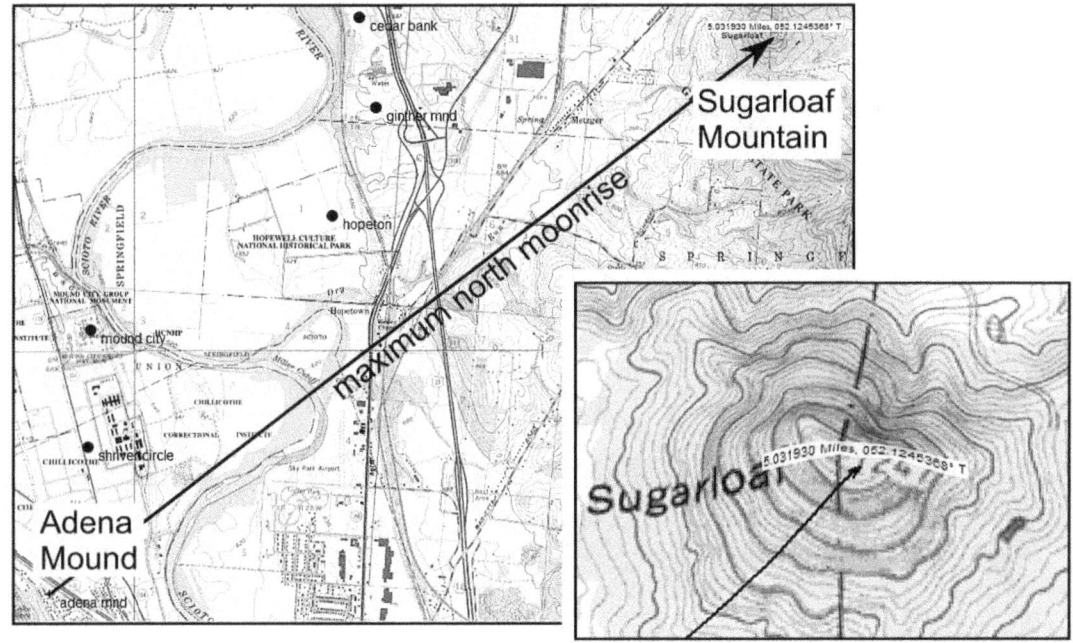

Figure 5.20. Chillicothe Earthworks: Sugarloaf Axis Mundi

As suggested earlier (e.g., Figure 3.23), Sugraloaf Mountain may have been considered the main axis mundi for the south-central Ohio Adena-Hopewell. Support for that idea comes from the observation that, in addition to the directional corridor-Milky Way Path linking the Newark Earthworks to Sugarloaf, a significant number of other Adena and Hopewell sites are also connected to the mountain along celestial azimuths.

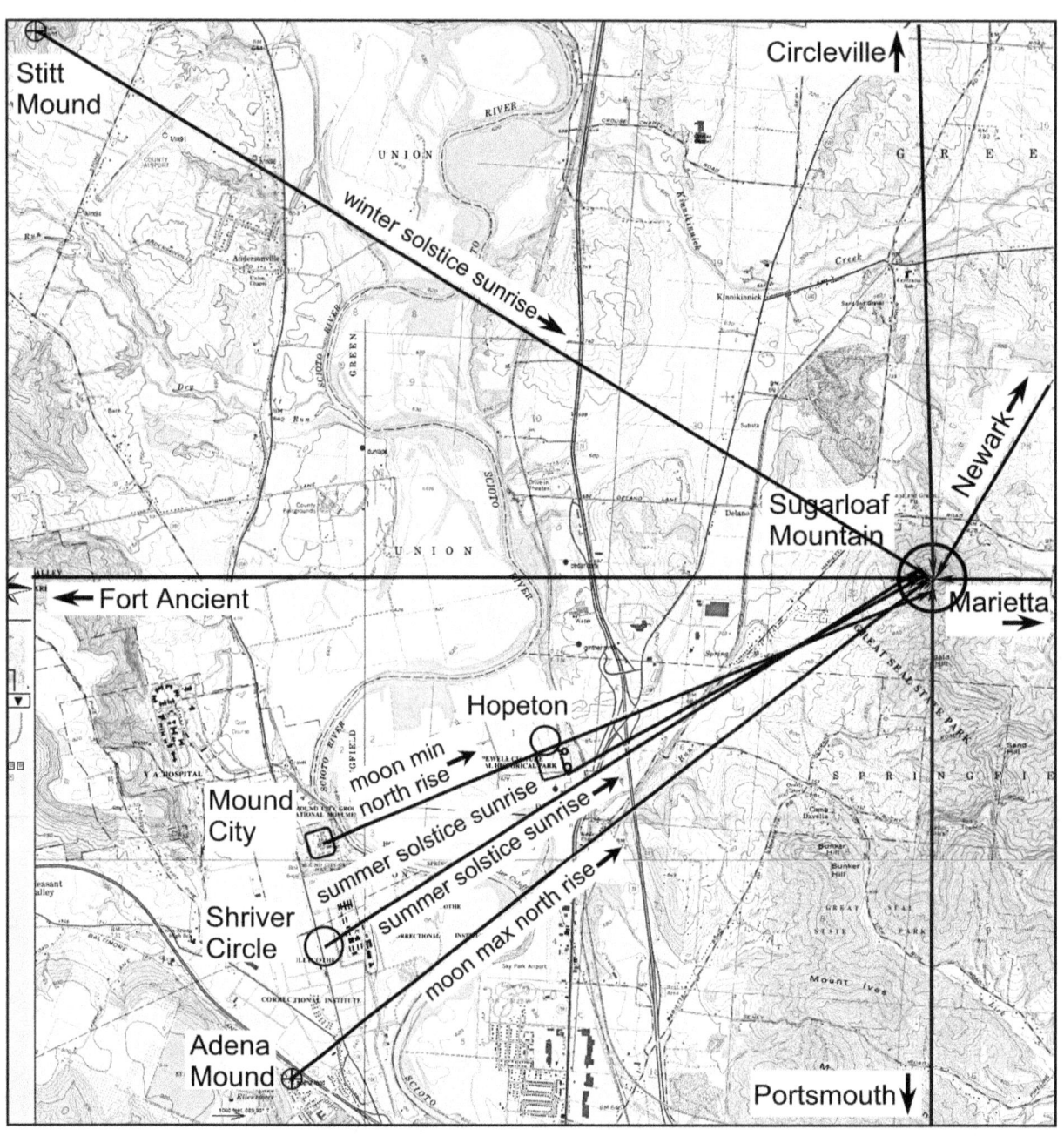

Figure 5.21. Chillicothe Earthworks: Mount Logan Mound

Of the many mounds and earthworks originally located in the Chillicothe area, several are of special interest. With reference, for example, to the Mount Logan Mound, Squier and Davis (1848:92) note: "The mound occupies the most conspicuous point in the valley; and from it is afforded the most extended view that can be obtained in that entire region." Below map provided by MyTopo, annotation added.

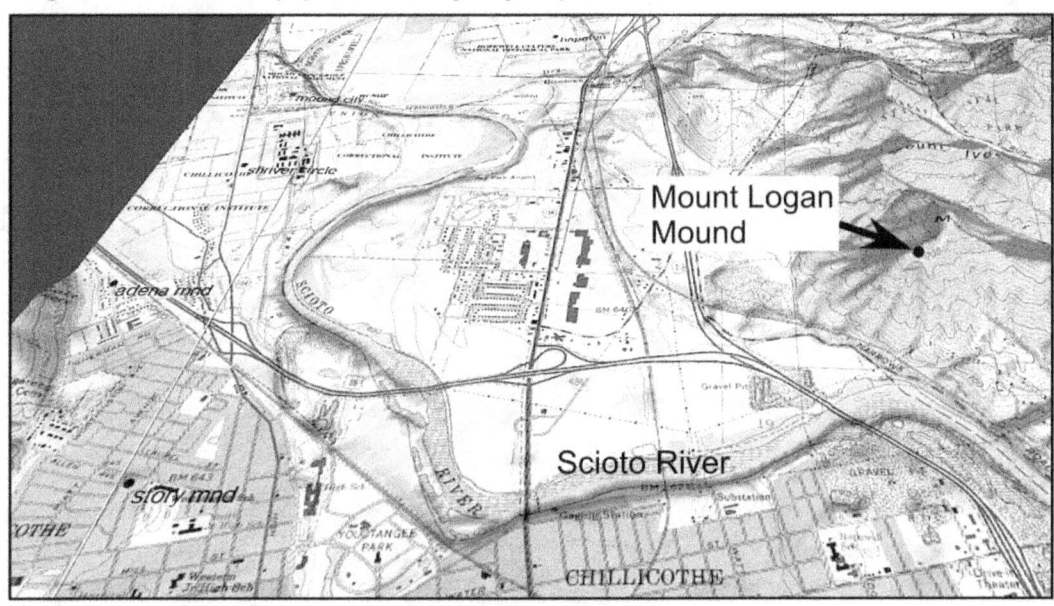

Above left: LiDAR contour map of Mount Logan Mound (10-foot contour interval). The mound is about 10 feet high and 90 feet in diameter. The mound has been dug into; however, there is no record of what was found within.

Above right: view of the Mount Logan Mound. Photo from Tonetti 2004:Pl. 2, by permission of Al Tonetti, ASC Group, Inc., and the Adjutant General's Office, Ohio National Guard.

Figure 5.22. Chillicothe Earthworks: Works East

The Works East Earthwork has been destroyed by urban growth. Its exact location and orientation have not been ground-truthed (note 22).

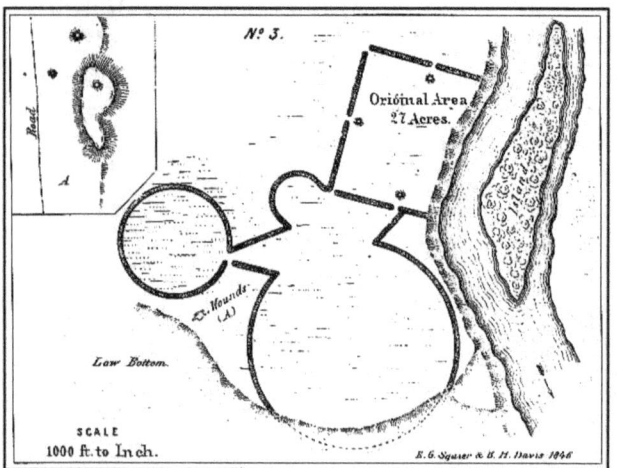
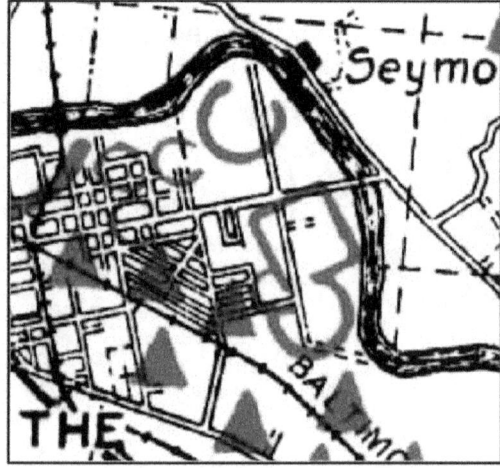

Above left: Squier and Davis (1848:Pl. XXI, no. 3) map of the site.
Above right: Mills's (1914) representation of the site. The Works East was likely destroyed by Mills's time. Also problematic is that the earthwork shown by Mills is about twice the size provided by Squier and Davis's survey. Nevertheless the Mills map is useful for helping to establish the general location for the earthwork.

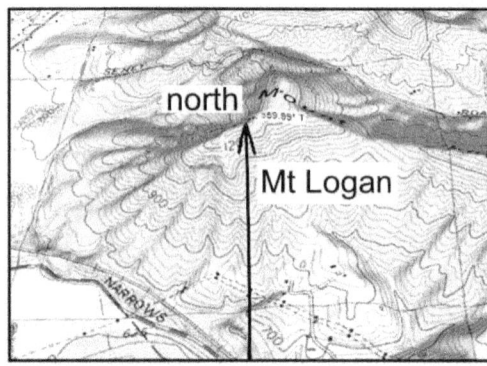

Left and below: the plotted location for Works East is estimated. If correct, then Works East was aligned to Mount Logan - due north. To the southeast, the winter solstice sunrise would have been visible over Rattlesnake Knob, about 6 miles distant (note 23).

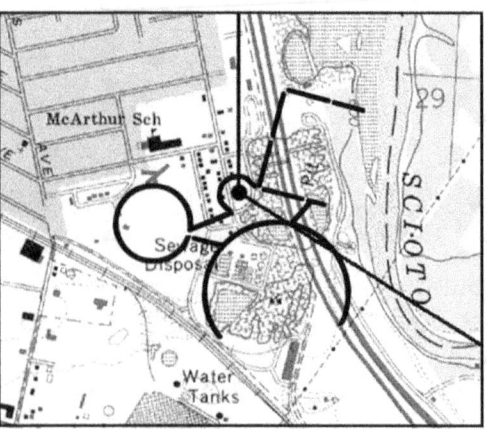
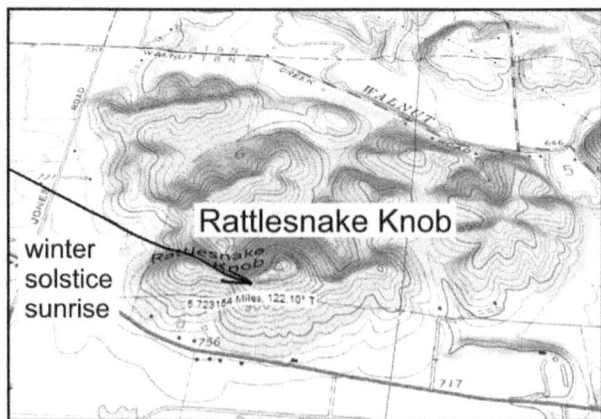

Figure 3.23. The Newark Earthworks: Great Hopewell Road Trajectory

LiDAR data do not reveal evidence for the Great Hopewell Road south of Ramp Creek in Newark-Heath. If, however, the LiDAR-determined trajectory of the GHR (az = 210.5 degrees) is extended from the woodlot location where the Road can still be seen, that trajectory will intersect the base of Sugarloaf Mountain in Ross County - 50 miles to the southwest (note 16).

Right: starting point for plotting the trajectory of the Great Hopewell Road. Map provided by MyTopo.

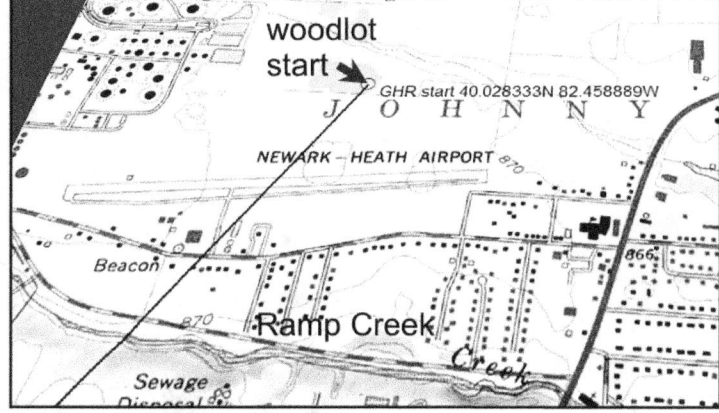

Left: end point for plotting the Great Hopewell Road. Map provided by MyTopo.

Below right: LiDAR image showing the location of Sugarloaf relative to the Scioto River Valley.

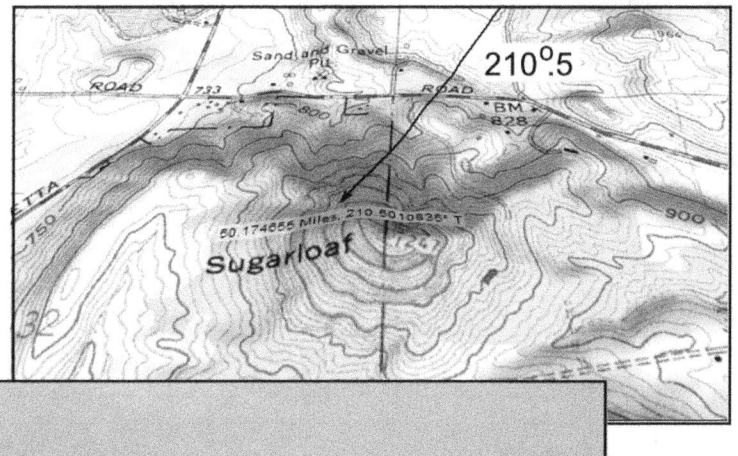

Above left and right: Sugarloaf is a prominent mountain that marks the Chillicothe earthwork area. Because Sugarloaf is an outlier, it stands out against the horizon and is visible for miles around. As discussed in Chapter 5, Sugarloaf Mountain may have been an axis mundi and portal to the Otherworld for souls of the dead.

Figure 5.24. Chillicothe Earthworks: High Bank

The High Bank earthwork complex is located about 3 miles south of Chillicothe. The complex includes one of only two octagonal-shaped earthworks built by the Hopewell. Although the 12 feet height walls have been greatly reduced by plowing, the large circle and octagon are visible at ground level.

Below: Squier and Davis (1848:Pl. XVI) map of the High Bank Works superimposed over LiDAR image (note 27).

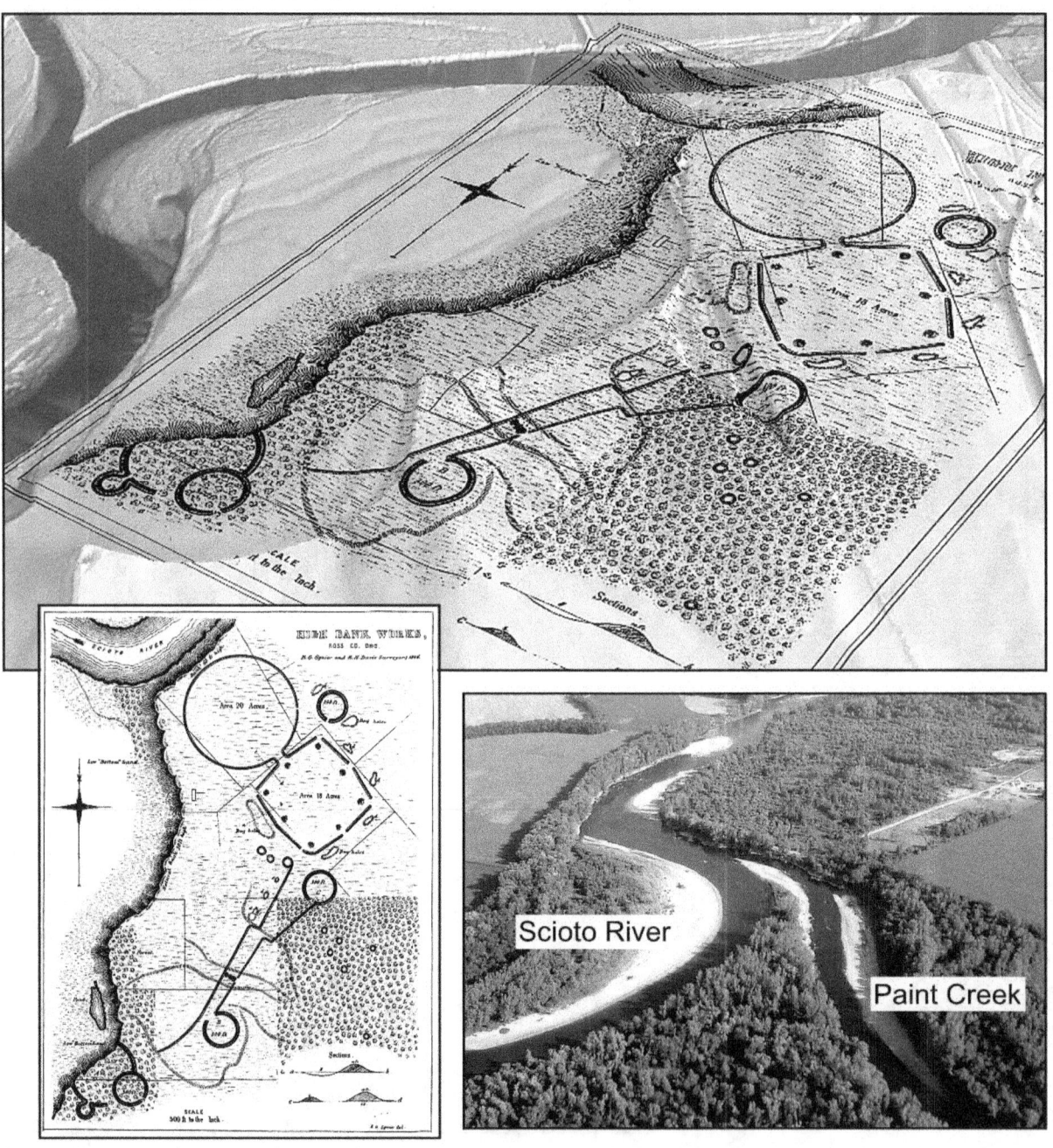

Above left: Squier and Davis (1848:Pl. XVI) map of the High Bank Works. Note the many mounds and other features located south of the main earthwork.
Above right: High Bank is located at the confluence of Paint Creek and the Scioto River, just beyond the left side of the photo.

Figure 5.25. Chillicothe Earthworks: High Bank Alignments

The High Bank earthwork incorporates the HMU in the design of both the Large Circle and Octagon (Hively and Horn 1984). As pointed out by Squier and Davis (1848:71), the High Bank, Newark Observatory, and Hopeton circles are the same size

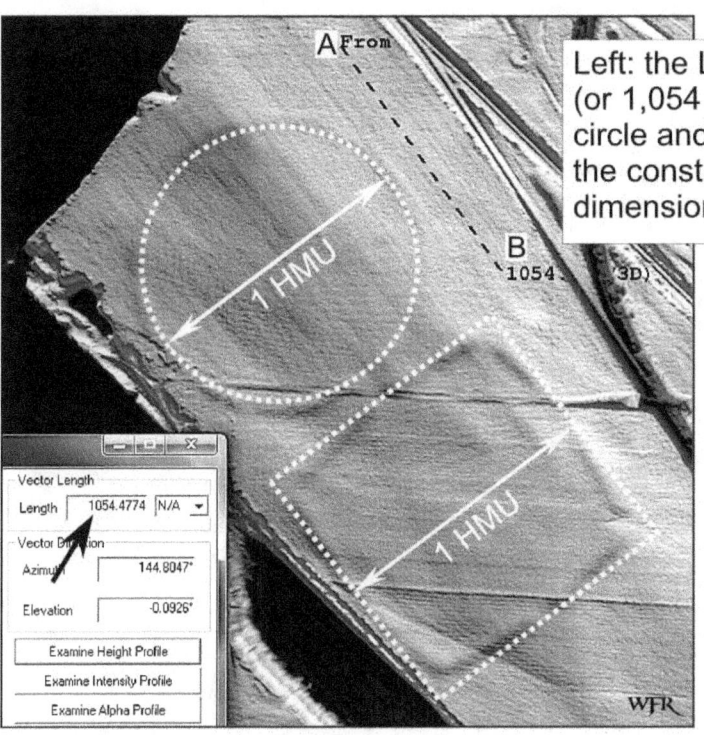

Left: the LiDAR length of line A-B is 1 HMU (or 1,054 feet). Line A-B is used to contruct circle and square shapes. The dimensions of the constructed shapes closely match the dimensions of the earthworks.

Below: High Bank also incorporates celestial alignments in its design. The alignments below are those posited by Hively and Horn (1984).

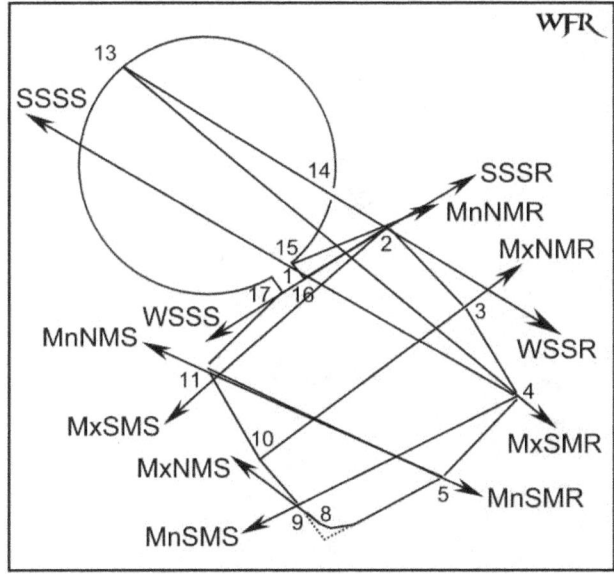

10-3 MxNMR - max north moon rise 53°.4
15-2 MnNMR - min north moon rise 67°.5
1-2 SSSR - summer solstice sun rise 59°.6
13-2 WSSR - winter solstice sun rise 121°.7
13-4 MxSMR - max south moon rise 130°.0
11-5 MnSMR - min south moon rise 115°.4
8-9 MxNMS - max north moon set 307°.3
5-11 MnNMS - min north moon set 293°.2
4-16 SSSS - summer solstice sun set 301°.0
2-17 WSSS - winter solstice sun set 238°.3
2-11 MxSMS - max south moon set 229°.4
4-9 MnSMS - min south moon set 244°.4
Left: schematic drawing of alignments as proposed by Hively and Horn (1984) (note 28).

Figure 5.26. Chillicothe Earthworks: High Bank and Scioto River
Below: of the posited High Bank alignments, the controlling one was the alignment to the moon's maximum north rise. The orthogonal (or perpendicular) of that sightline yields the azimuth for the longitudinal axis of the site (note 29).

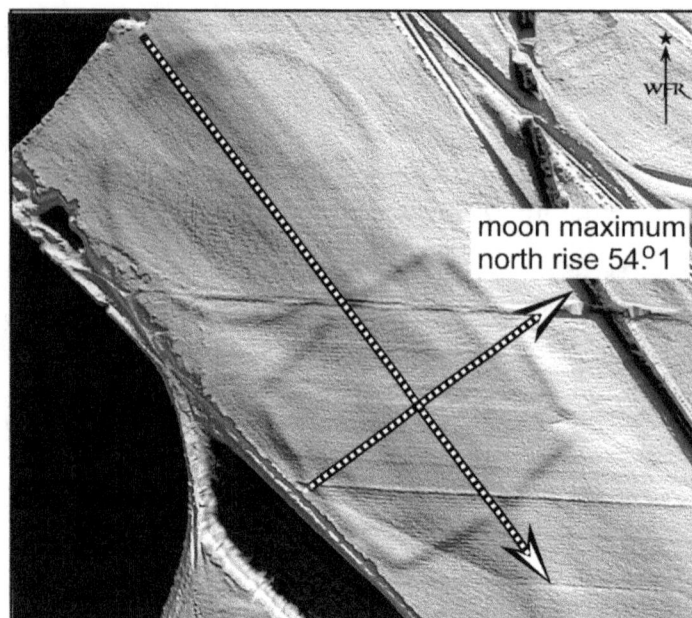

Left: LiDAR image rotated 0.26 degrees to correct for theta value. True north as indicated.

moon maximum north rise 54.°1

Below: the significance of the longitudinal axis for High Bank is that it causes the orientation of the earthwork to extend parallel to this section of the Scioto River (note 30).

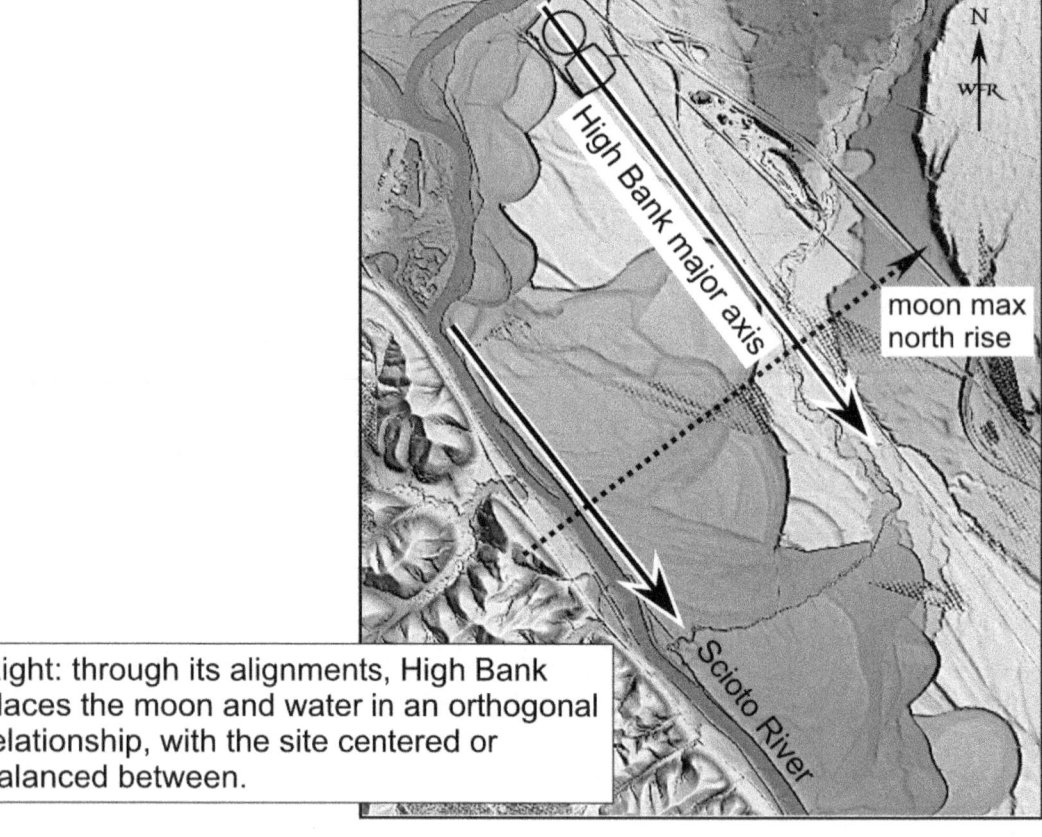

Right: through its alignments, High Bank places the moon and water in an orthogonal relationship, with the site centered or balanced between.

Figure 5.27. High Bank - Newark Relationships

There are several interesting relationships between the Newark Octagon and High Bank earthworks (see e.g., Squier and Davis 1848). The large circles are the same size, both earthworks are based on the HMU, and both octagons are aligned to the same event - i.e., the moon's maximum north rise. The way this occurs, however, differs for the two sites. At Newark the alignment is along the major axis, while at High Bank it is through the minor axis of the Octagon. The result is that the two earthworks are orthogonal to each other (note 31). As if to emphasize the concept of orthogonality, Newark and High Bank are situated on either end of a 57-mile imaginary line that extends orthogonal to the moon's minimum south rise.

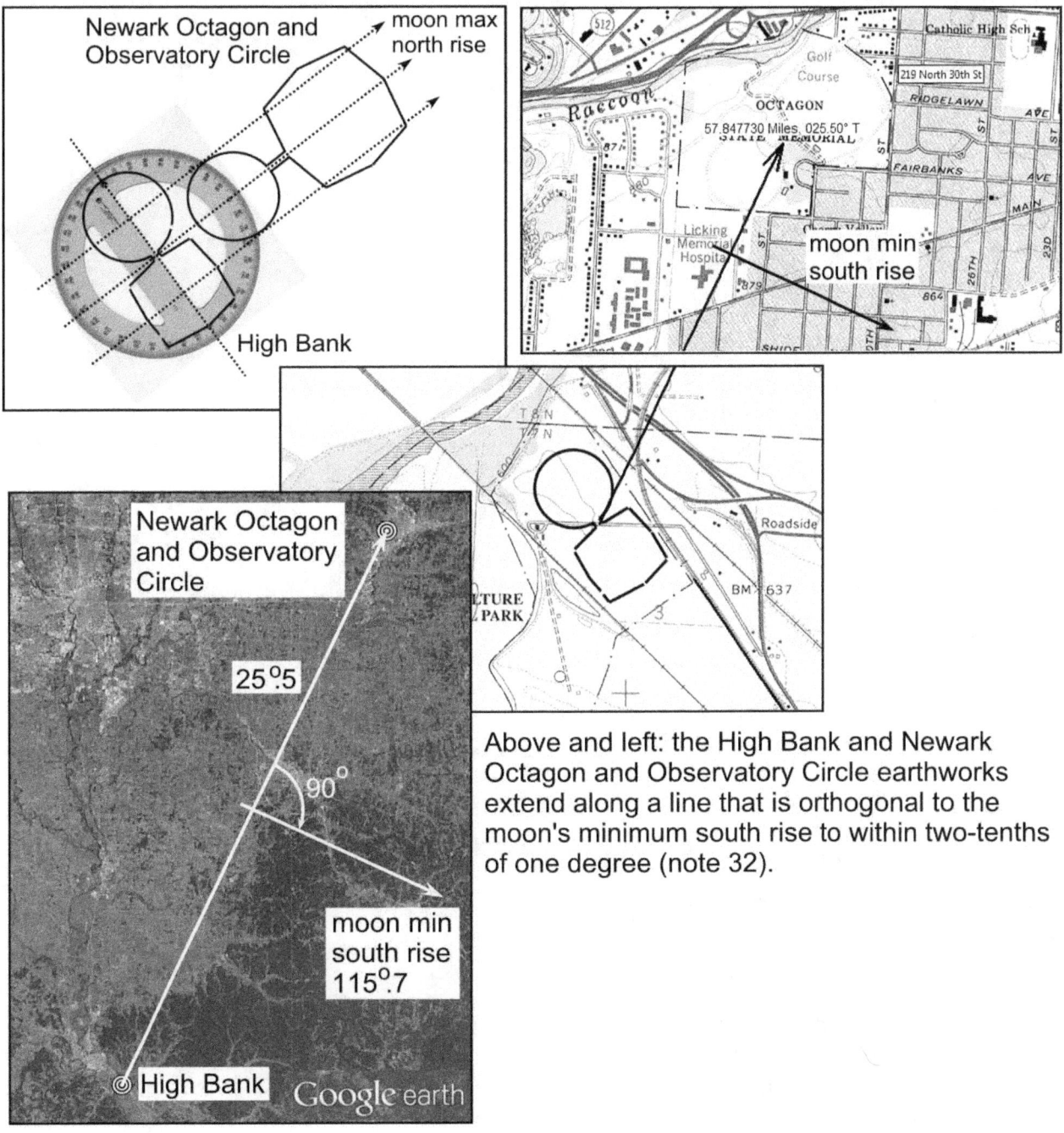

Above and left: the High Bank and Newark Octagon and Observatory Circle earthworks extend along a line that is orthogonal to the moon's minimum south rise to within two-tenths of one degree (note 32).

Figure 5.28. Chillicothe Earthworks: Liberty

The Liberty Earthworks are located about 3 1/2 miles southeast of High Bank. The site is situated on the east side of the Scioto River near the entrance to the Walnut Creek Valley. Due to farming, the earthworks are no longer visible by naked eye. The Harness Mound was located within the Large Circle, near the Square (note 33).

Above: Squier and Davis (1848:Pl. XX) map of Liberty correctly oriented and superimposed over LiDAR image.

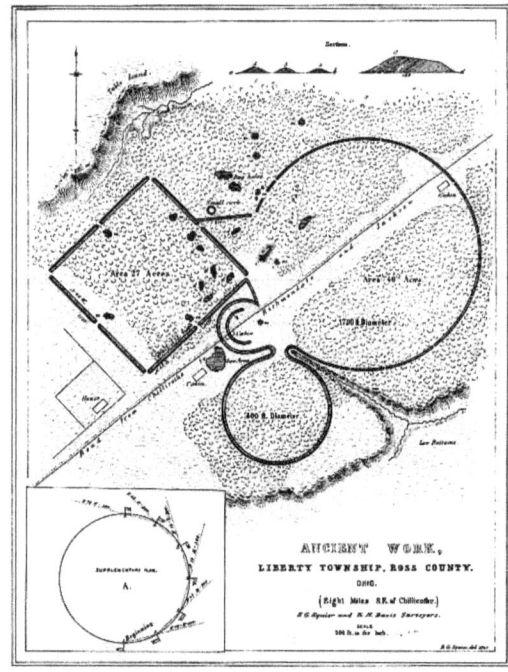
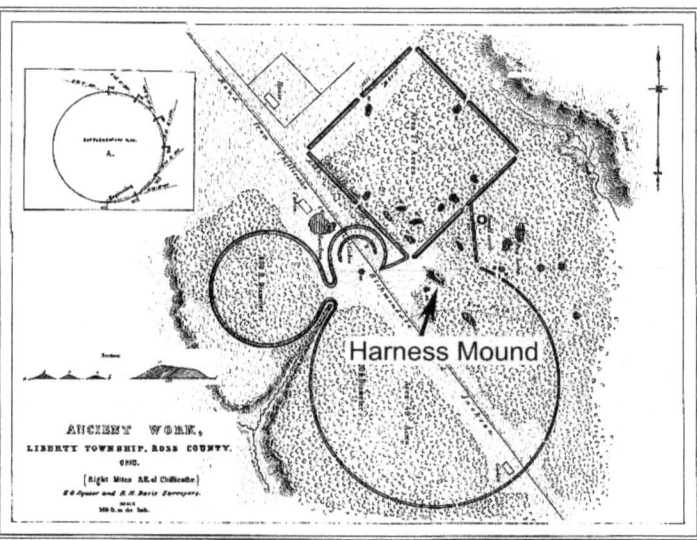

Above left: in their drawing of Liberty, Squier and Davis (1848:Pl. XX) mistakenly positioned their north arrow pointing east.
Above right: Liberty earthworks correctly oriented - as Squier and Davis might have drawn (annotation added).

Figure 5.29. Chillicothe Earthworks: Liberty and the HMU

Although not visible at ground level, faint traces of the Liberty Square and Small Circle can be seen in LiDAR imagery.

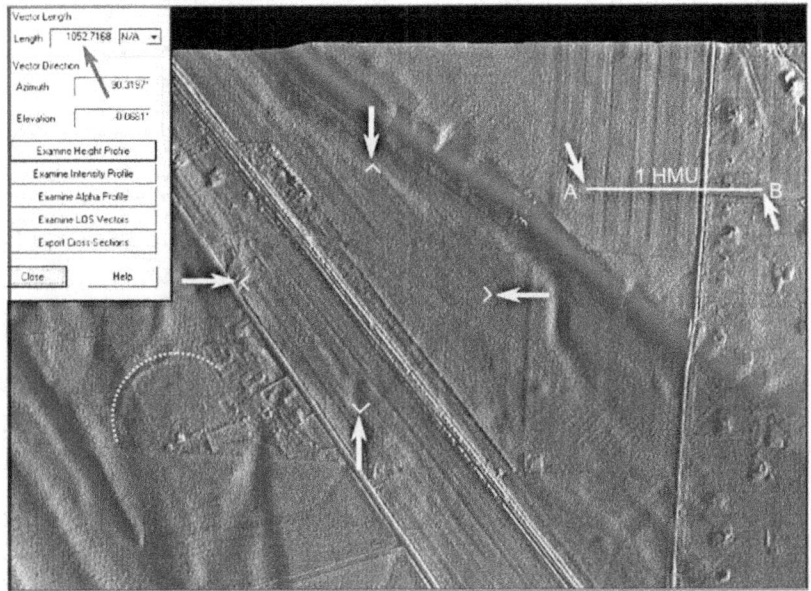

Above: the sides of the Liberty Square closely match the HMU length of 1,054 feet. Corners of Square indicated by arrows; remnant of Small Circle indicated by dotted line.

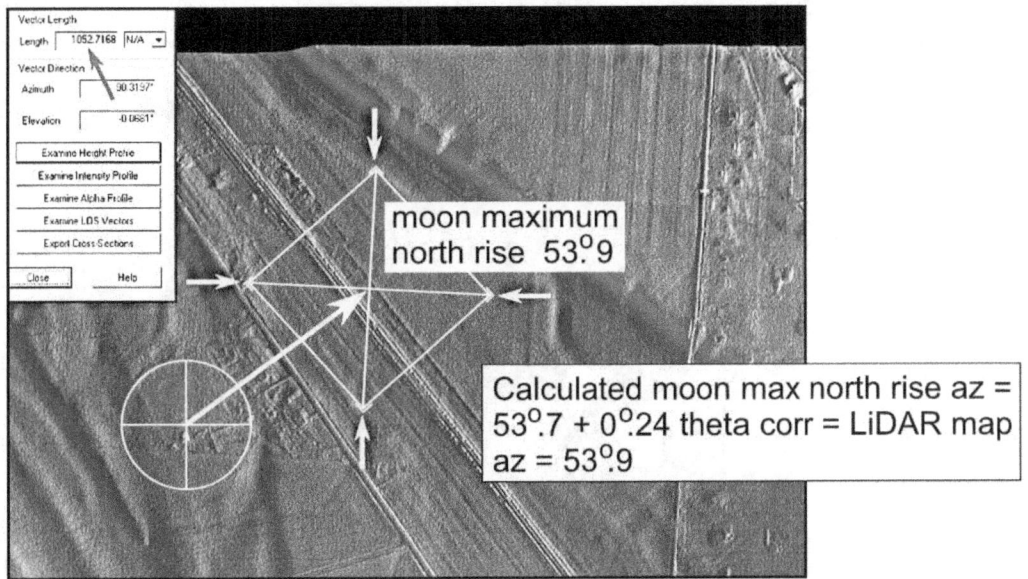

Above: the sightline between the center of the Liberty Small Circle and Liberty Square is aligned to the moon's maximum north rise (note 34).

Figure 5.30. Chillicothe Earthworks: Liberty and the Scioto River

Below right: the orientation of the Liberty Square is to the moon's maximum south set (note 35).

Below left: the orthogonal (or perpendicular) of the moon's maximum north rise generates the longitudinal axis for the site - which extends parallel to the Scioto River.

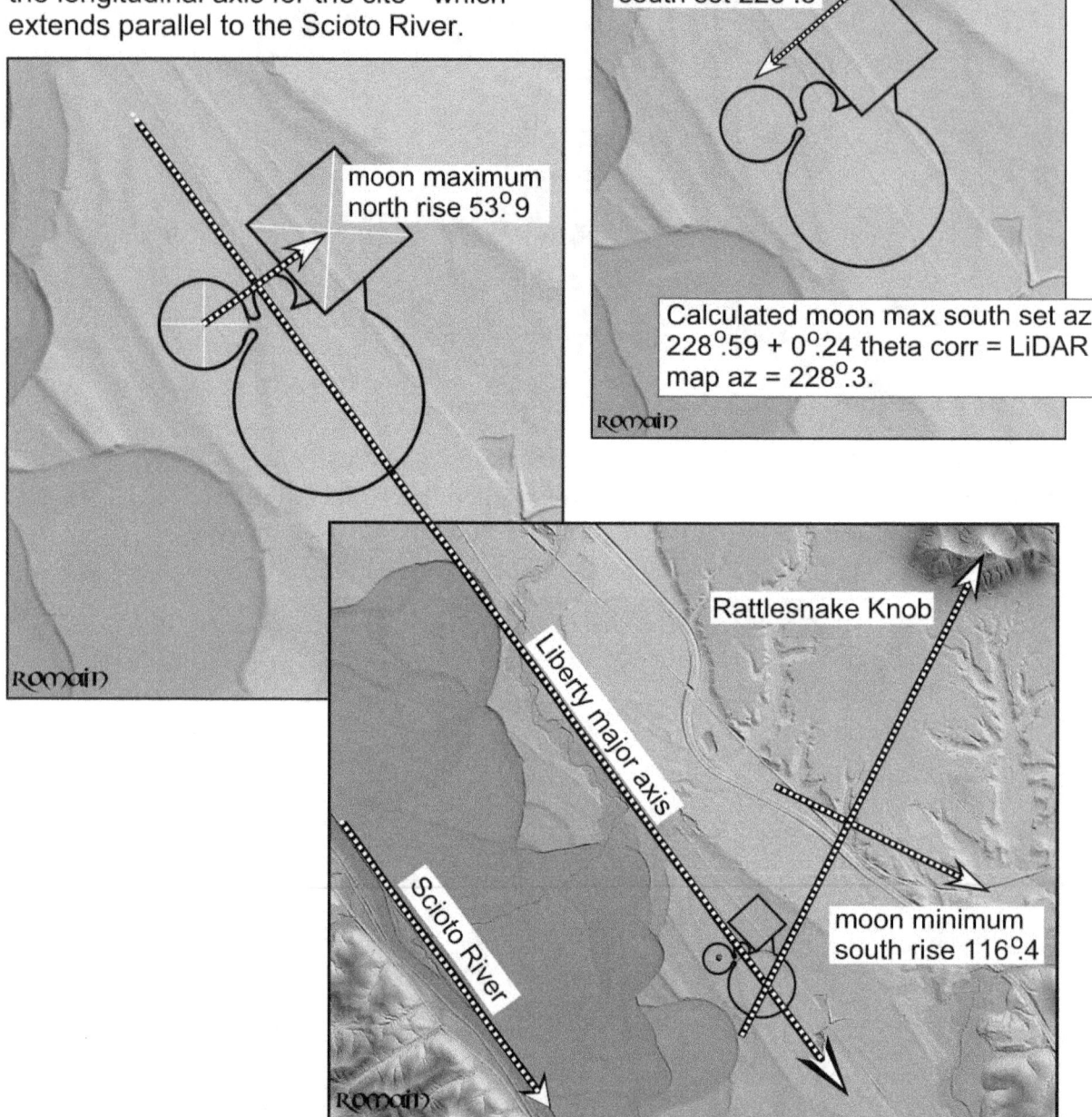

Above: the north-south location for Liberty along the Scioto River axis is established in part by he relationship of a sightline extending from the earthwork to Rattlesnake Knob. The orthogonal of that sightline is to the moon's minimum south rise (note 36).

142

Figure 5.31. Chillicothe Earthworks: Johnson Works

The Johnson Works are situated at the south end of the Central Scioto River Valley Archaeological Zone, on a bluff overlooking the confluence of the Scioto River and Salt Creek. The center mound is 3 - 4 feet high and about 40 feet in diameter. A ditch extends along the inside of the perimeter walls.

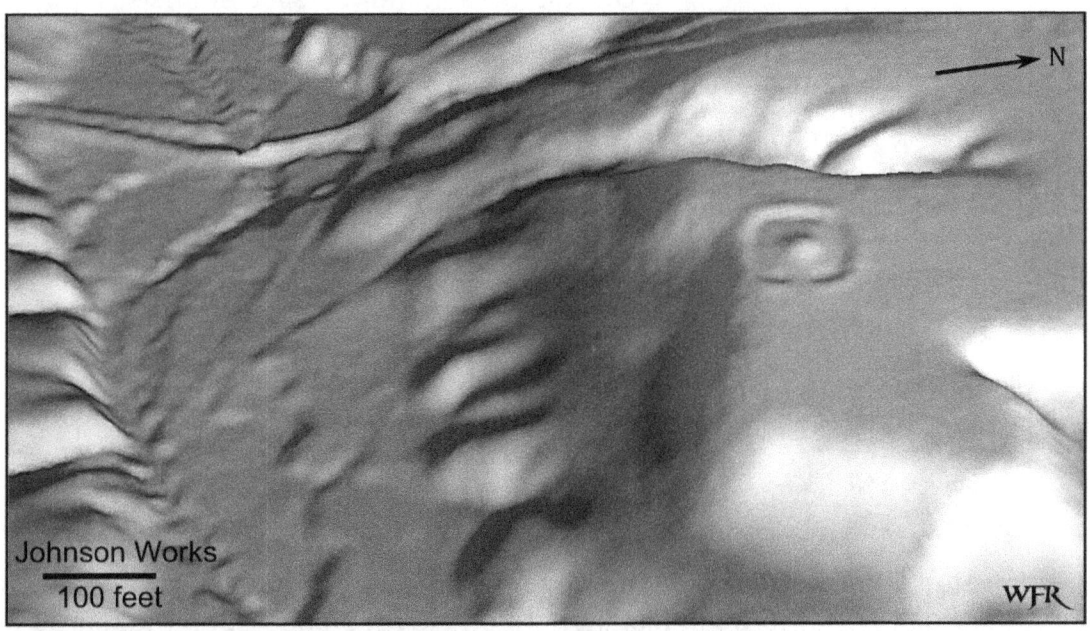

Above: LiDAR image of the Johnson Works.

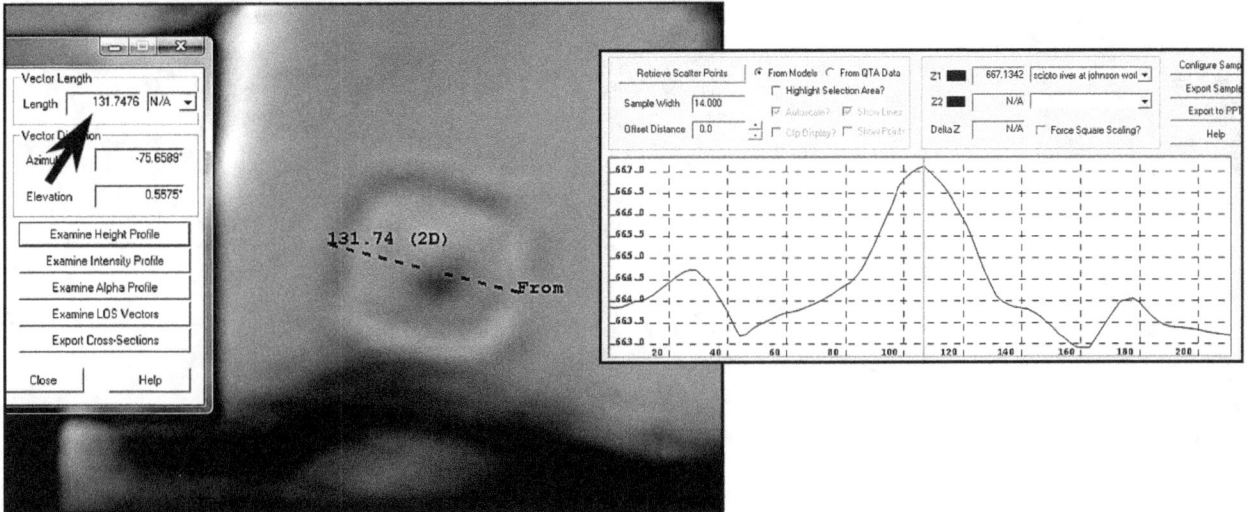

Above left: the Johnson Works uses the 1/8 HMU lesser multiple in its design. The LiDAR-measured distance from the center of the entrance to the crest of the northwest wall is about 131.8 feet - or 1/8 HMU.
Above right: LiDAR profile through the earthwork along the northeast to southwest diagonal axis.

Figure 5.32. Chillicothe Earthworks: Johnson Works Alignment

The Johnson Works is aligned to the summer solstice sunrise along its diagonal axis (note 37).

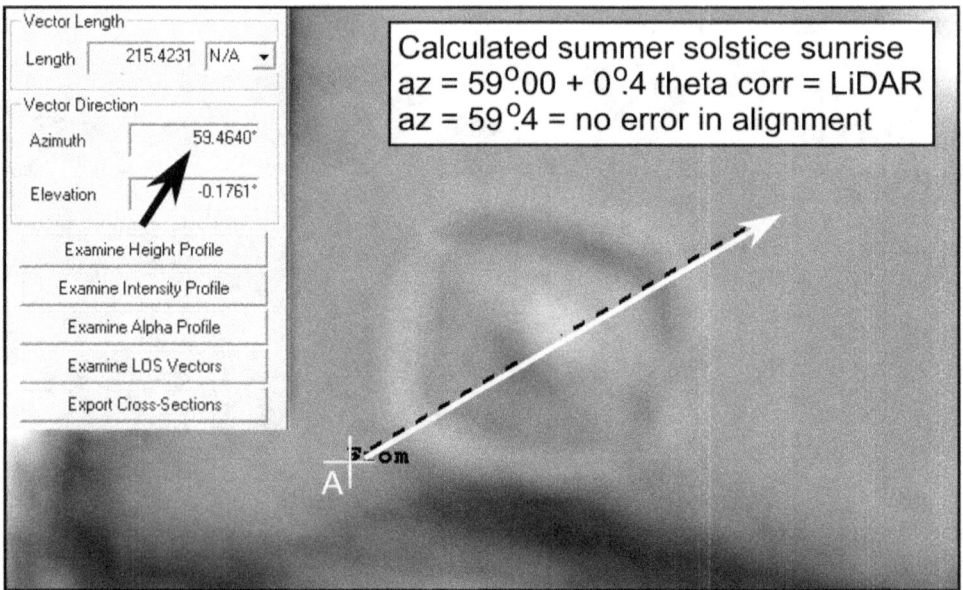

Calculated summer solstice sunrise
az = 59°.00 + 0°.4 theta corr = LiDAR
az = 59°.4 = no error in alignment

Above: calculated summer solstice azimuth for the Johnson Works superimposed over LiDAR image. Point A is where the distant horizon elevation has been measured from.

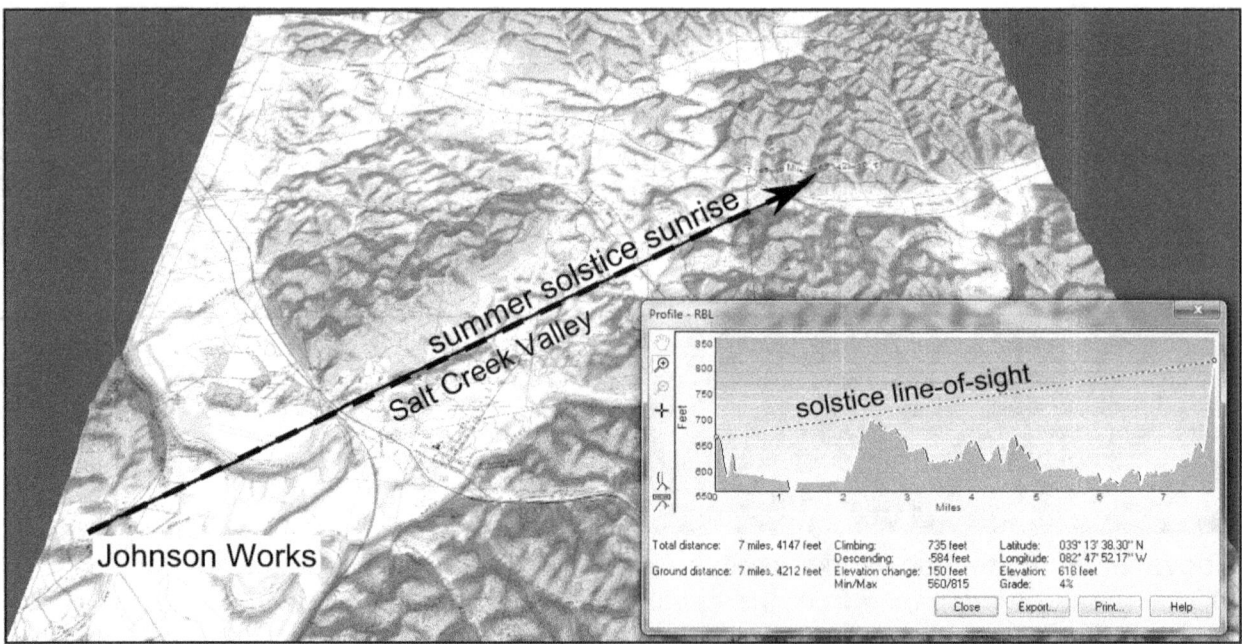

Above: viewed from the Johnson Works, the summer solstice sun will rise along the course of the Salt Creek Valley. Inset figure shows the line-of-sight profile along the solstice sightline. Map provided by MyTopo, annotation added.

Figure 5.33. Seal Township Works

The Seal Township Works (note 38) are located about 23 miles south of Chillicothe, on the east bank of the Scioto River. The site originally included a large circle and square, at least six smaller geometric earthworks, and several mounds. Most of these features have been destroyed by gravel mining and farming and are not visible in LiDAR imagery.

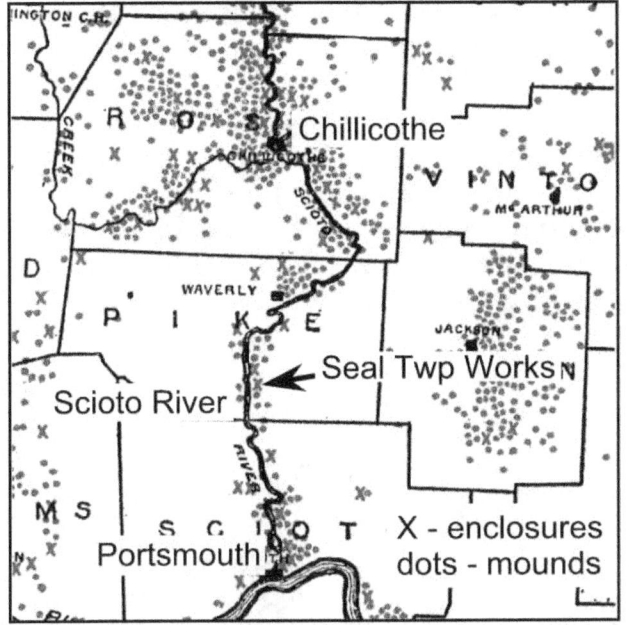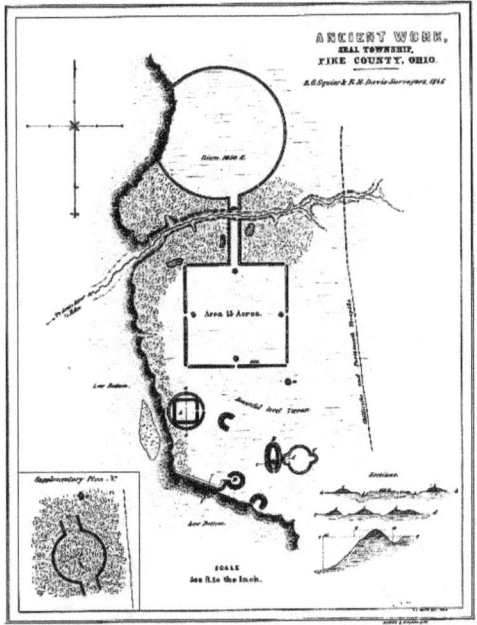

Above left: enlarged detail from Mills (1914) showing location of Seal Township Works.
Above right: Squier and Davis (1848:Pl. XXIV) map of the Seal Township Works.

Above left: USDA 1938 aerial photo showing the Seal Twp. Square. The walls of the Square appear as faint white lines. Of interest is that the parallel walls conecting the Square to the Large Circle do not extend north-south as represented by Squier and Davis - but rather, extend along an azimuth of about 348.5 degrees.
Above right: schematic drawing of the earthwork with azimuth of the parallel walls correctly shown. Squier and Davis (1848:66) give the diameter of the Circle as 1,050 feet - equal to about 1 HMU. Thomas (1880:15) gives the sides of the Square as about 853 feet in length.

Figure 5.34. Seal Township Works and the HMU

The dimensions of the Seal Township Works are based on the Hopewell Measurement unit (HMU).

As noted previously, the diameter of the Seal Twp. Circle as given by Squier and Davis is 1,050 feet - which is equal to 1 HMU to within 4 feet. The sides of the Seal Square average 853 feet in length (Thomas 1880:15). The size of the Seal Square is related to the HMU in the following way.

Below left: recall that a 30-60-90 degree angle triangle having sides that are 1,054 ft. (1 HMU) and 527 ft. (1/2 HMU) will have a hypotenuse of 1,178 feet.
Below center: if the HMU-derived 1,178-foot length is used to construct a 45-45-90 degree right triangle then, sides A-B and A-C of that triangle will be 832.9 feet. The sides of the Seal Square are equal to the 832.9-foot length to within 2 percent.
Below right: the Seal Square is about the same size as the square enclosure at the Hopewell Mound Group site.

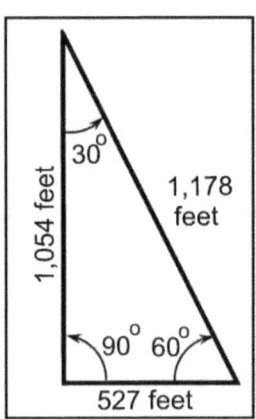
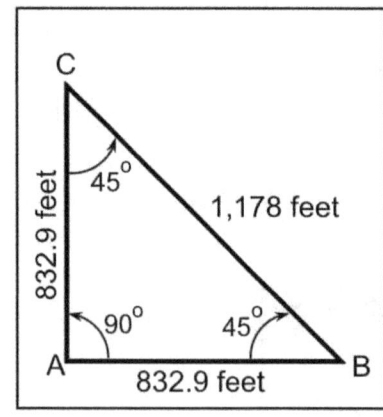

Seal Township Square
Hopewell Mnd Grp Square

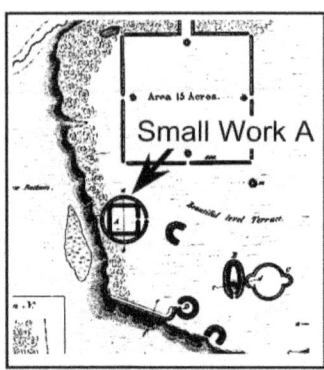

Left: enlarged detail from Squier and Davis (1848:Pl. XXIV) showing location of Small Work A.
Right: Squier and Davis (1848: Fig. 11) drawing of Small Work A.

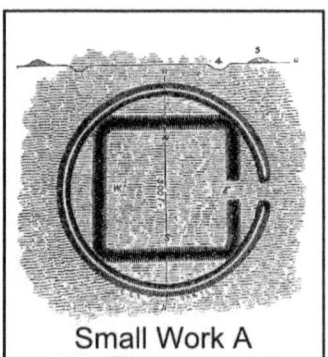

Small Work A

Above: the notion that the Hopewell were cognizant of the relationships between circles and squares is suggested by an intriguing earthwork located to the immediate southwest of Seal - referred to as Small Work A by Squier and Davis (1848:fig. 11). This earthwork is no longer visible, but if Squier and Davis's representation is even approximately accurate, then in this design we find the concept of nested geometric figures (note 39).

5.35. Seal Township Works Alignments
The Barnes Earthwork incorporates alignments to earth and sky.

Above: in this figure, a schematic drawing of the Seal Twp. Works is superimposed over a LiDAR image. The parallel walls that connect the Circle to the Square extend parallel to the trajectory of the four hill summits to the northeast.

Below left: at the same time, as shown by LiDAR contour map, the orientation of the Seal Twp. Square is parallel to the base of the hills to the northeast.

Below right: further, the Seal Twp. Square is oriented to the cardinal directions (USDA 1938 aerial photo, annotation added) (note 40).

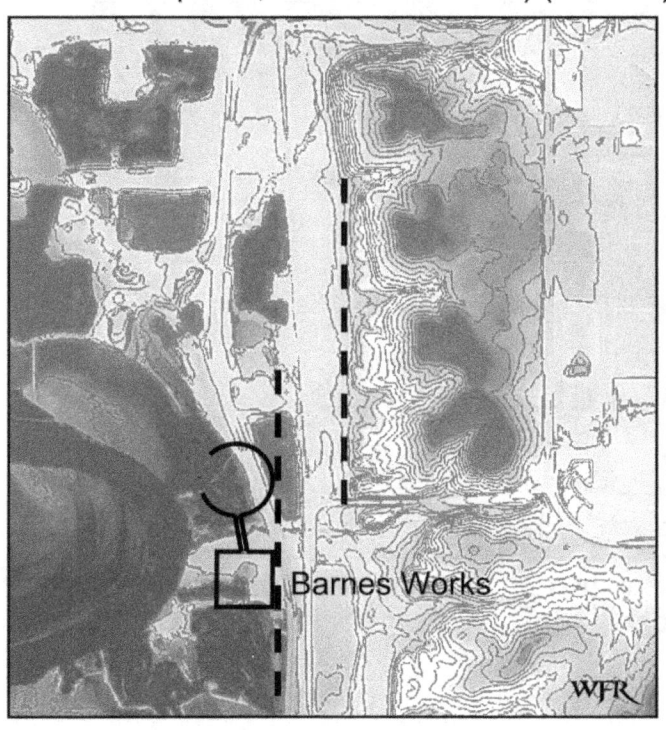
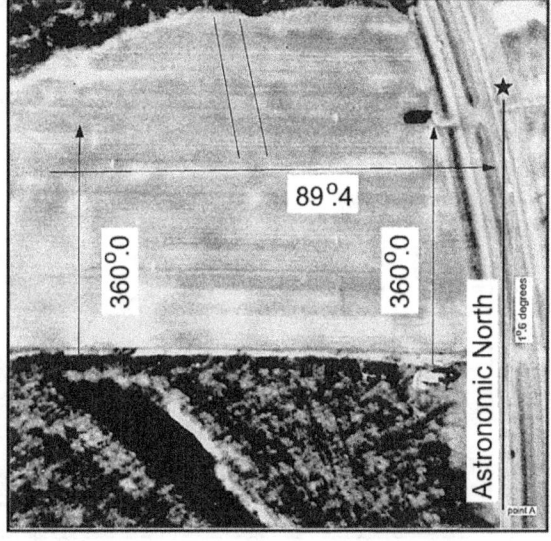

Figure 5.36. Tremper

The Tremper Earthwork is located about five miles north of Portsmouth on the west bank of the Scioto River. Excavation of the center mound (Mills 1916) found cremated human remains and hundreds of grave goods to include more than 130 stone pipes (note 41). Many of these pipes are exquisitely carved to show various animals.

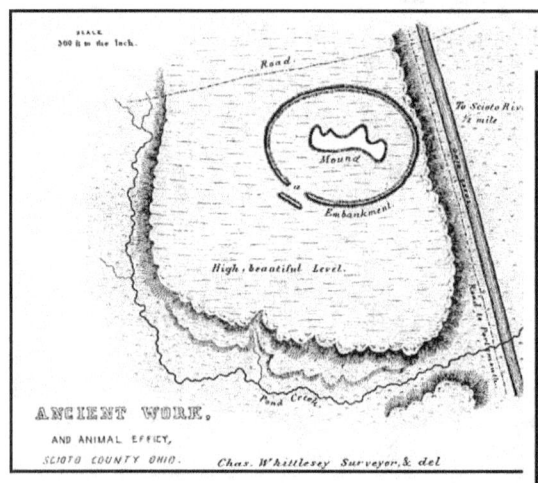

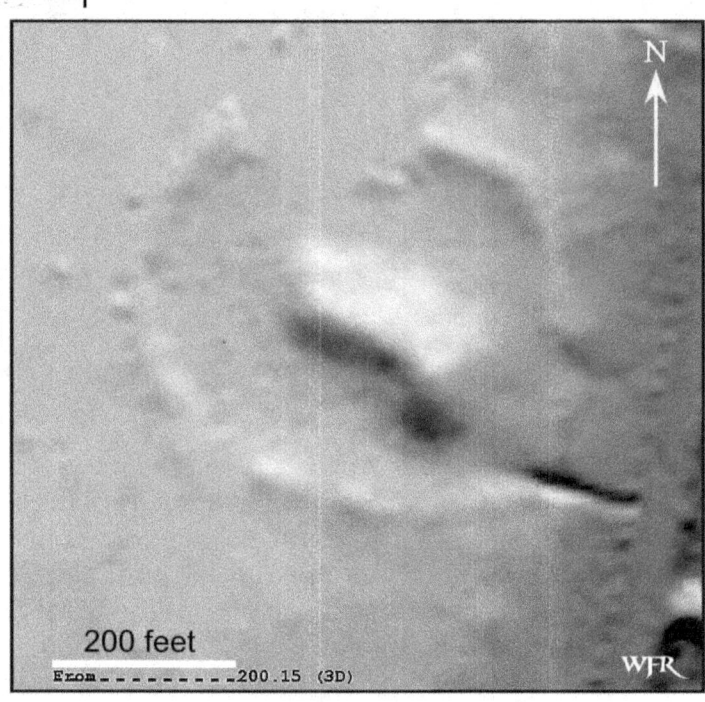

Above left: Squier and Davis (1848: Pl. XXIX) map of the Tremper Earthwork.
Right: LiDAR image of Tremper.

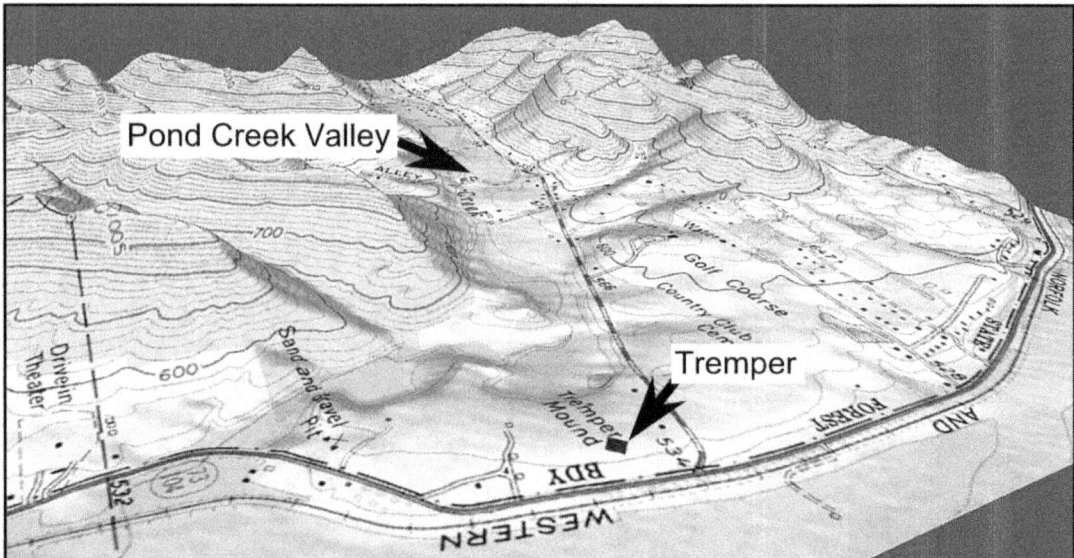

Above: the Tremper Earthwork is situated at the entrance to Pond Creek Valley. Map provided by MyTopo, annotation added.

5.37. Tremper Alignment

There are no readily apparent celestial alignments at Tremper. The Tremper Mound, however, appears oriented to Pond Creek Valley.

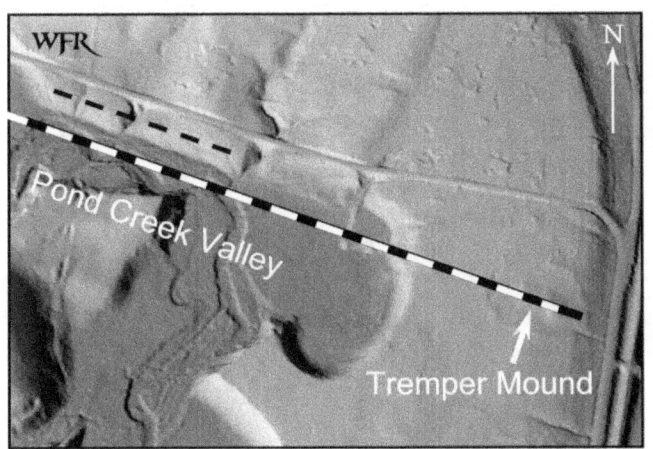

Left and below: LiDAR imagery shows that the major axis of the Tremper Mound extends along the same trajectory as the north edge of Pond Creek Valley.

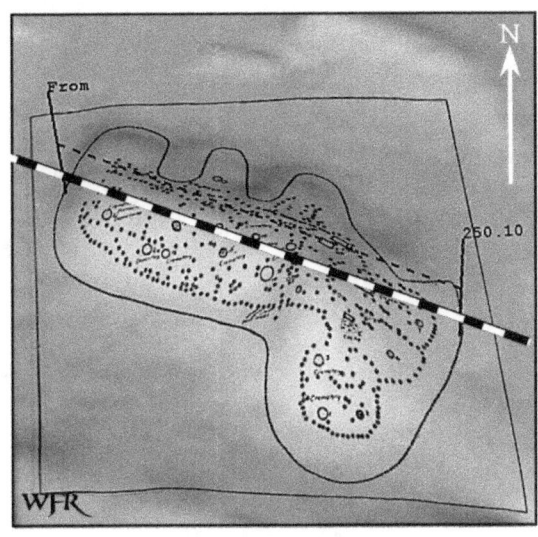

Above: the two dashed lines are parallel.
Right: Mills (1916:Fig. 3) plan of the Tremper Mound superimposed over LiDAR image. Mills (1916:270) gives the length of the mound as 250 feet (note 42).

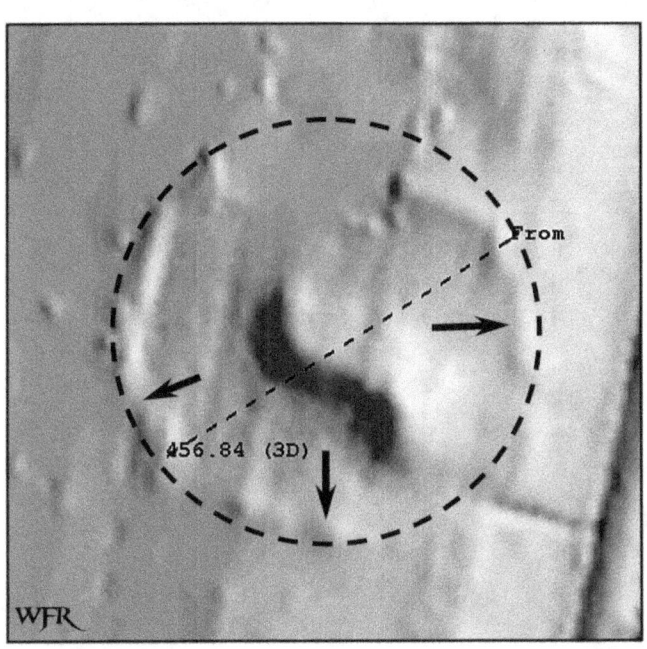

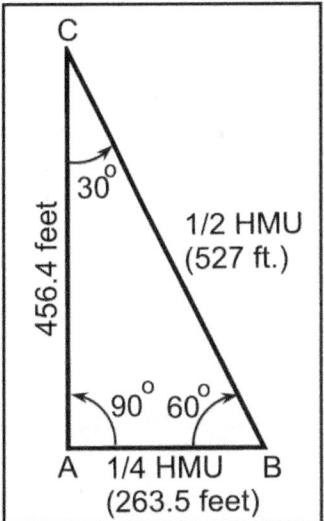

Above left: the Tremper Earthwork is somewhat elliptical in shape. Its size, however, appears based on a unit of length derived from the 1/2 HMU length. In the above LiDAR image, the computer program has drawn a line that is about 456.4 feet in length. Using this length, a circle is drawn. The 456-foot diameter circle closely matches the dimensions of the earthwork's perimeter wall.
Above right: in a 30-60-90 degree triangle, if side A-B is 1/4 HMU and side B-C is 1/2 HMU then, side A-C will be 456.4 feet in length.

Figure 5.38. The Portsmouth Earthworks
The Portsmouth Earthworks are located at the confluence of the Scioto and Ohio rivers.

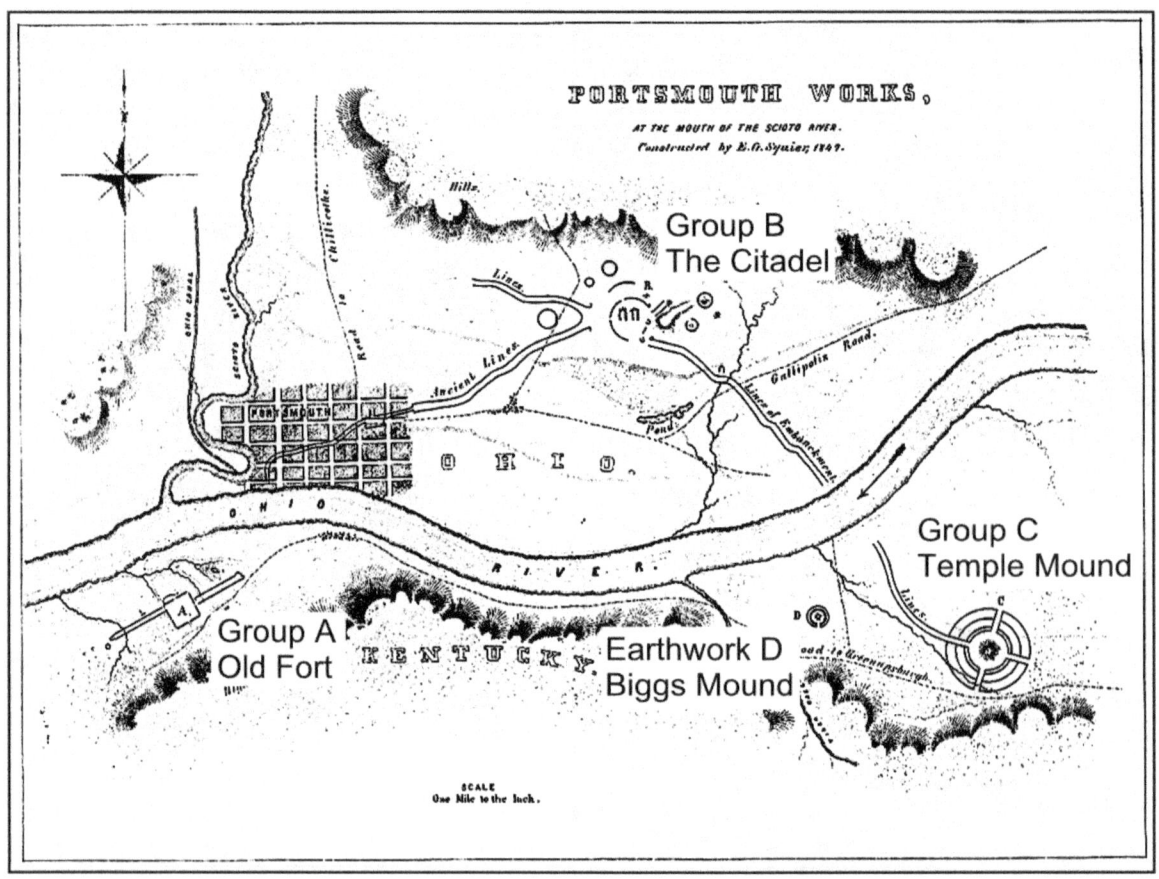

Above: Squier and Davis (1848:Pl. XXVII) map of the Portsmouth Works (annotation added). Still visible today are the Group A Old Fort, one Group B, U-shaped mound, remnants of the Temple Mound, the Biggs Mound, and several small mounds.

Left: aerial view of Portsmouth.

Figure 5.39. The Portsmouth Earthworks: Group A Square

The Portsmouth Group A earthworks includes several mounds, geometric earthworks, and a large square enclosure known as the Group A Square or Old Fort (note 43).

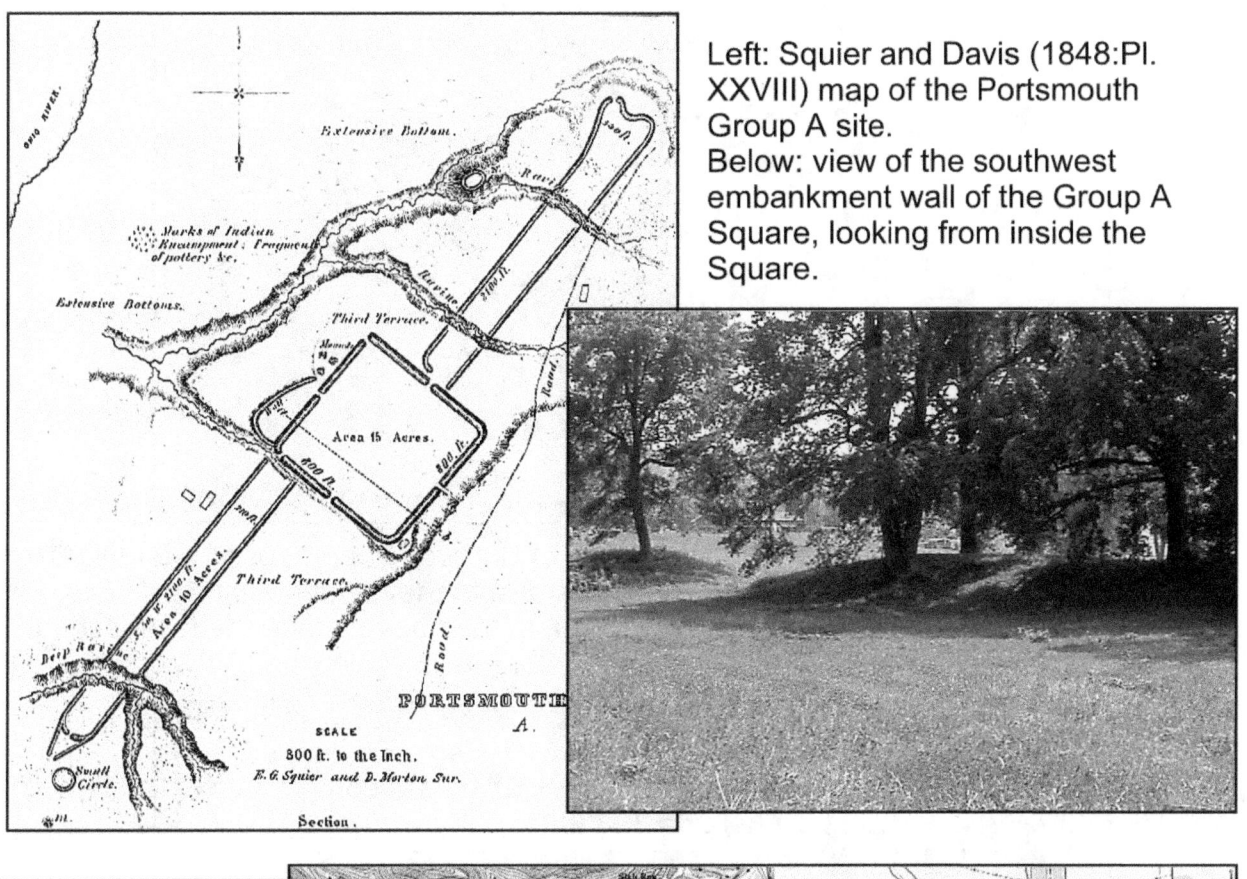

Left: Squier and Davis (1848:Pl. XXVIII) map of the Portsmouth Group A site.
Below: view of the southwest embankment wall of the Group A Square, looking from inside the Square.

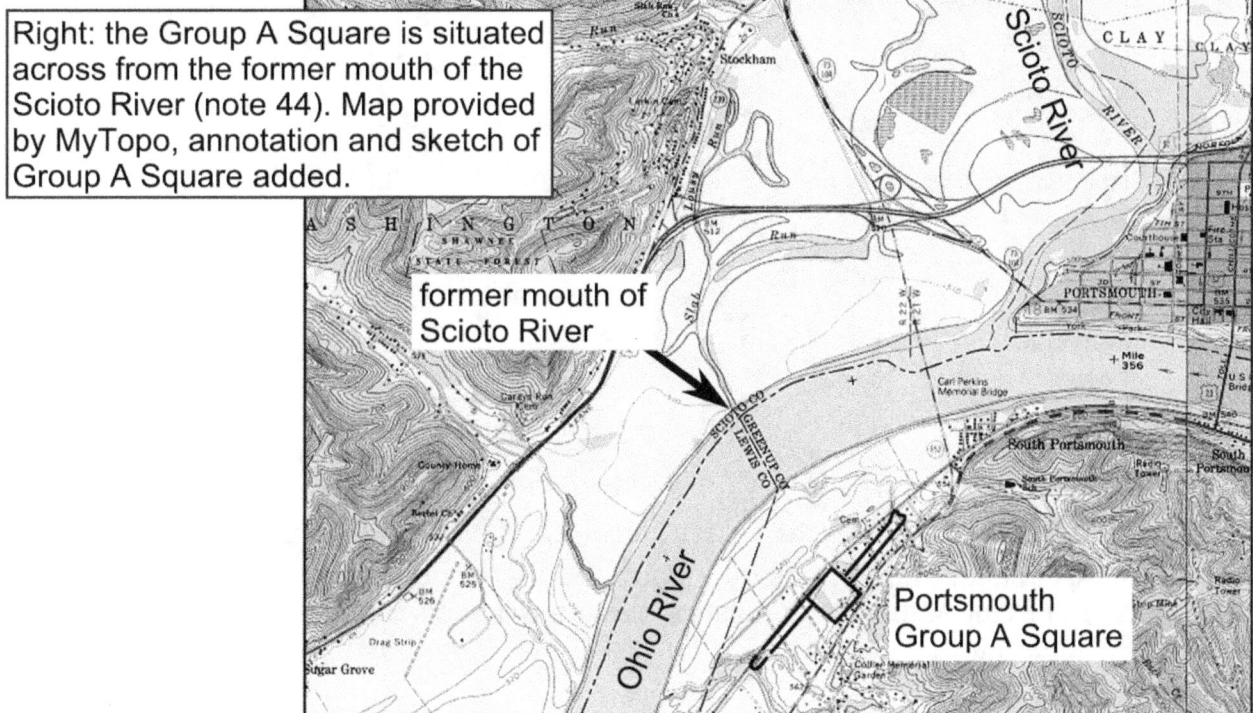

Right: the Group A Square is situated across from the former mouth of the Scioto River (note 44). Map provided by MyTopo, annotation and sketch of Group A Square added.

Figure 5.40. The Portsmouth Earthworks: Group A Square Alignments
The Portsmouth Group A Square is aligned to both earth and sky.

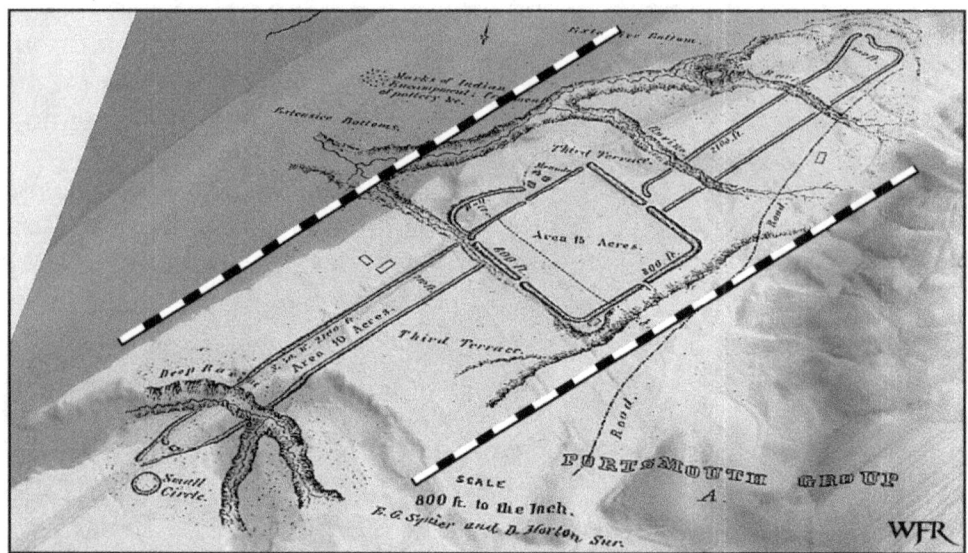

Above: Squier and Davis (1848:Pl. XXVIII) map of Square superimposed over LiDAR image. The major axis of the Square extends parallel to the lay of the land as established by the terrace edge to the northwest. In the above image the two dashed lines are parallel.

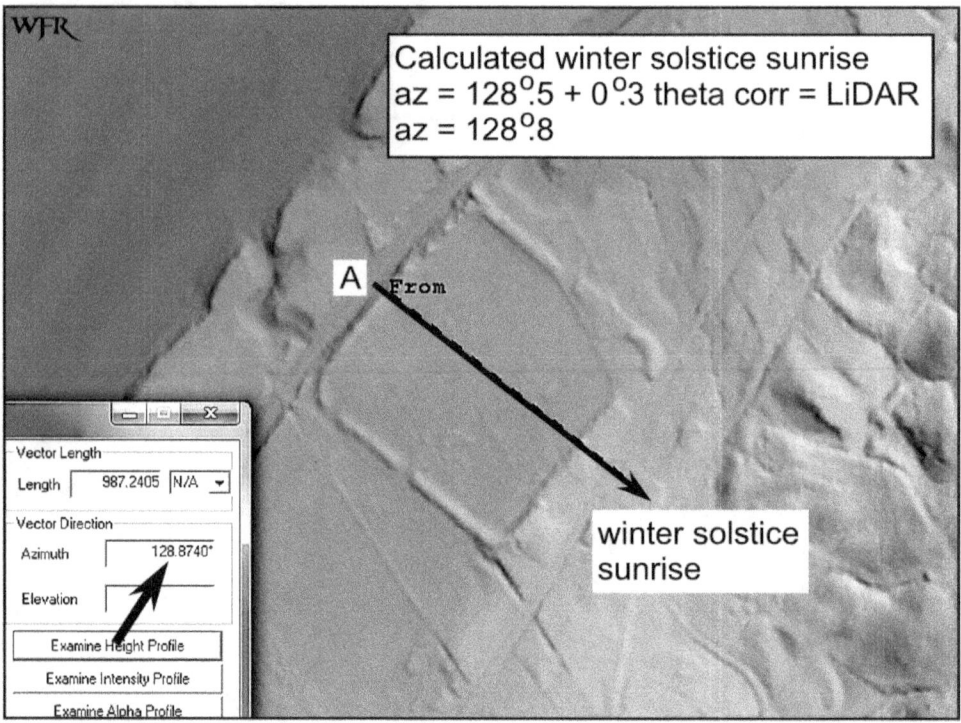

Above: LiDAR image. The Portsmouth Group A Square is oriented so that viewed from the gateway at Point A, the winter solstice sunrise will occur in alignment with the minor axis of the Square (note 45).

Figure 5.41. The Portsmouth Earthworks: Group A Square and the HMU

The Portsmouth Group A Square has rounded corners. As shown below, however, the structure was based on the concept of a square having a diagonal equal to 1 HMU (or 1,054 feet). In its size and shape, the Portsmouth Group A Square closely resembles Mound City.

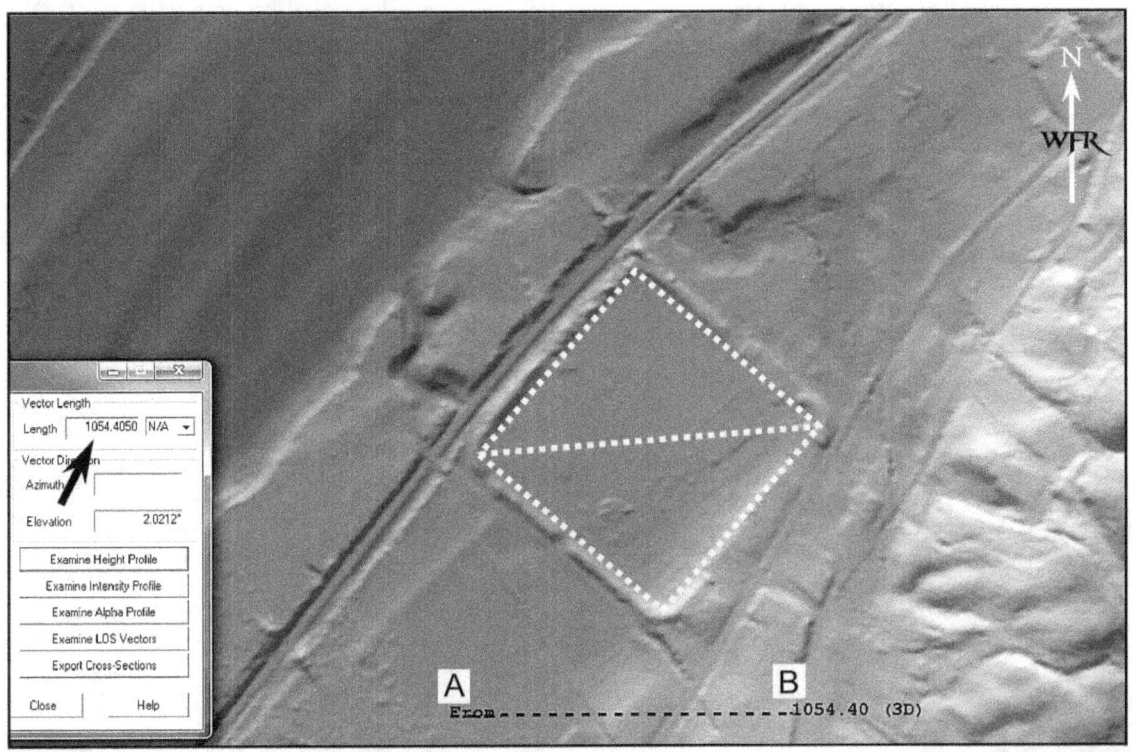

Above: the LiDAR-measured distance from point A to point B is 1 HMU (or 1,054 feet). A square is constructed using line A-B as its diagonal. The resulting square closely marks the inside dimensions of the earthwork.

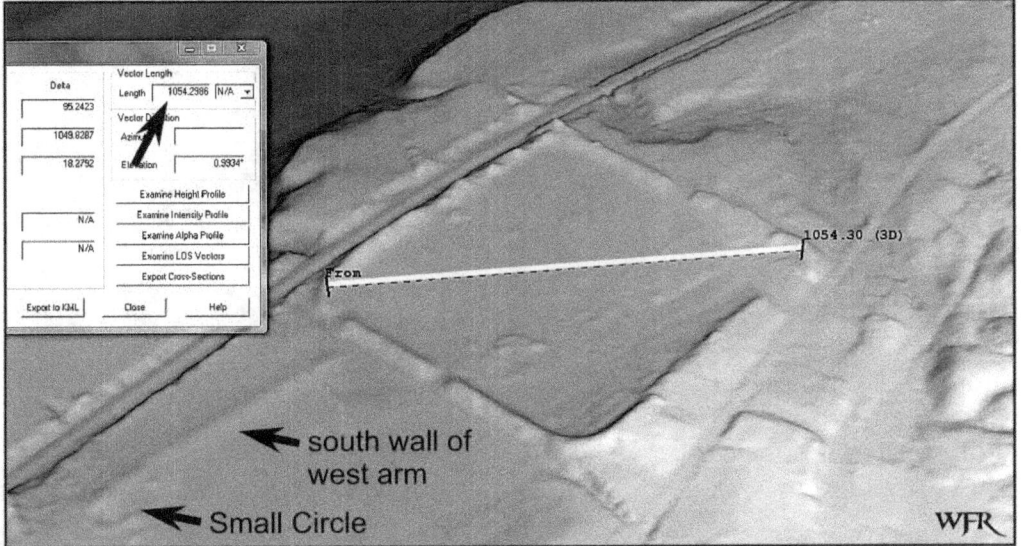

Above: oblique LiDAR view showing the length of the east-west diagonal of the earthwork - equal to 1 HMU (or 1,054 feet). Also visible in this view is a small circle near the far end of the west arm and faint traces of both the east and west arms.

Figure 5.42. The Portsmouth Earthworks: Group B: Citadel

The Portsmouth Group B earthworks are almost entirely obliterated. Remnants of what may be a couple of mounds and one U-shaped mound are preserved in a city park.

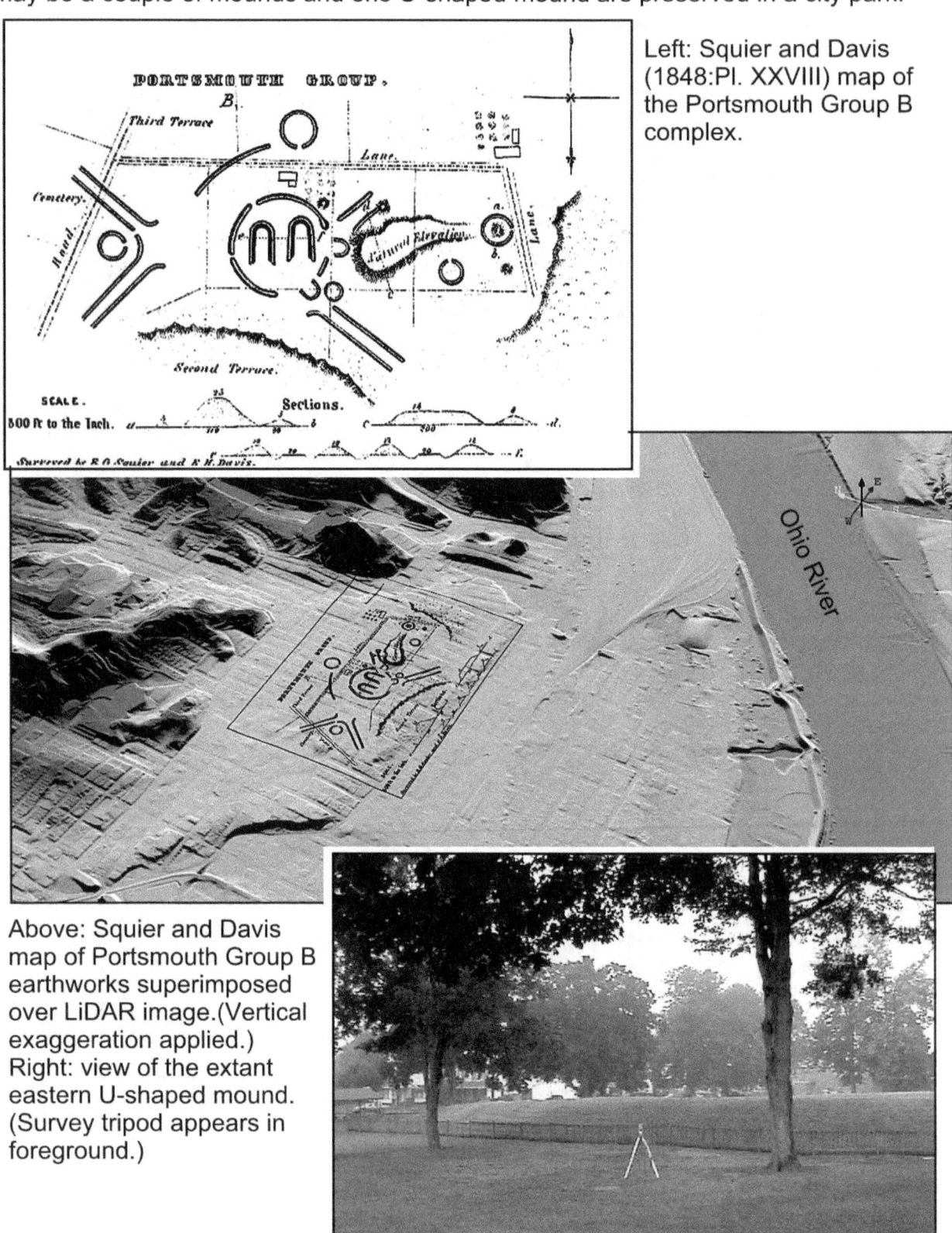

Left: Squier and Davis (1848:Pl. XXVIII) map of the Portsmouth Group B complex.

Above: Squier and Davis map of Portsmouth Group B earthworks superimposed over LiDAR image. (Vertical exaggeration applied.)
Right: view of the extant eastern U-shaped mound. (Survey tripod appears in foreground.)

154

Figure 5.43. The Portsmouth Earthworks: Group B: Alignment and Size

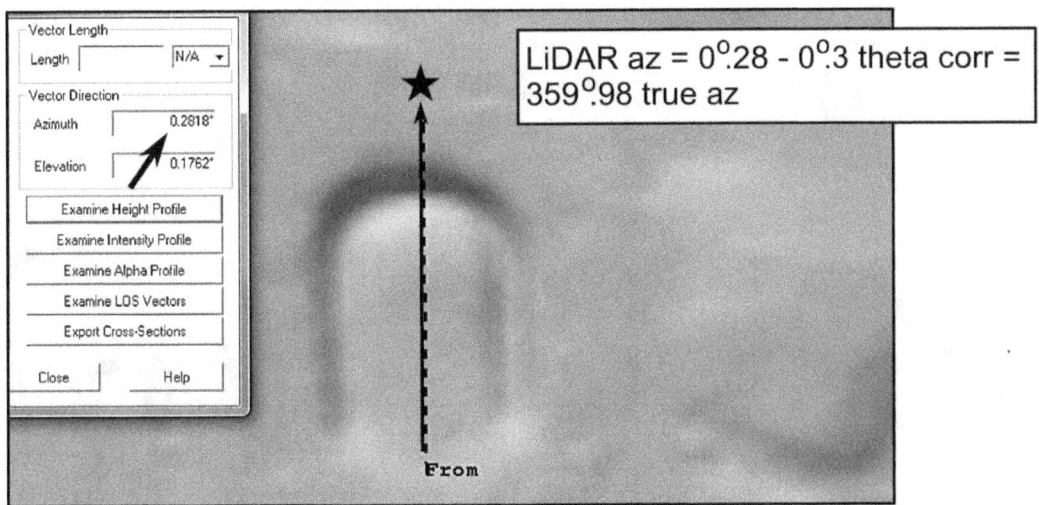

The survivng U-shaped mound is not perfectly symmetrical - its west side is a bit skewed. All things considered, however, it appears the earthwork is oriented along a north-south axis.

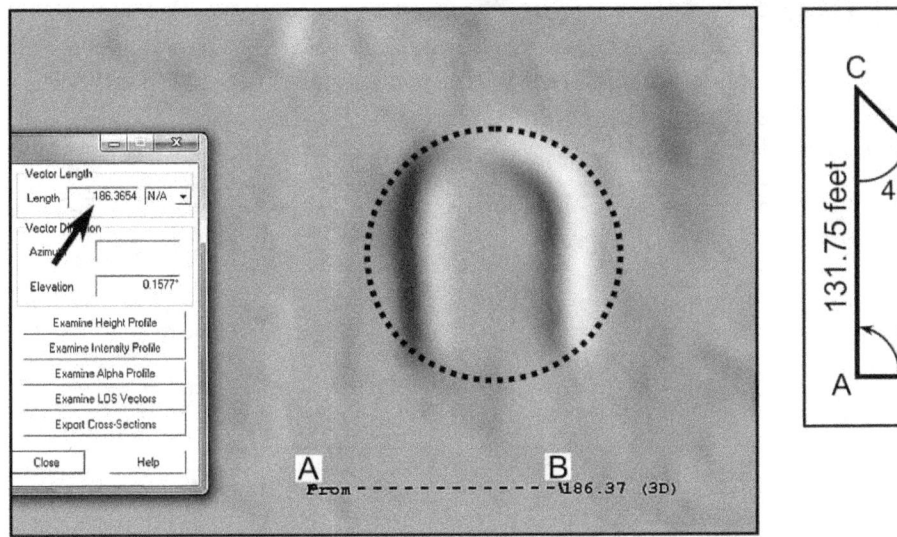
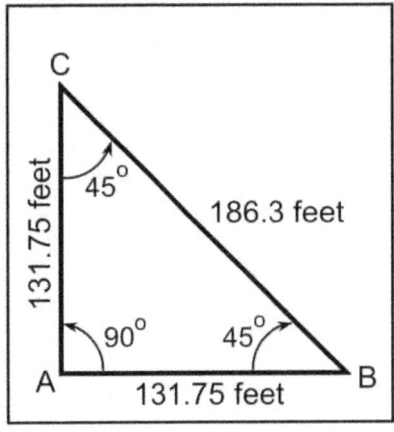

Above left: the size of the U-shaped mound appears based on the diameter of a circle having a diamter of 186.3 feet. In the above figure, the diameter of the dotted line circle is 186.3 feet.

Above right: the 186.3-foot unit of length is related to the HMU in the following way. In a 45-45-90 degree triangle, if sides A-B and A-C are each 131.75 feet (or 1/8 HMU) then, side B-C will be 186.3 feet in length.

Figure 5.44. Portsmouth U-Shaped Valley

The Portsmouth U-shaped mounds may have been intended to mimic the shape of the valley to the immediate north.

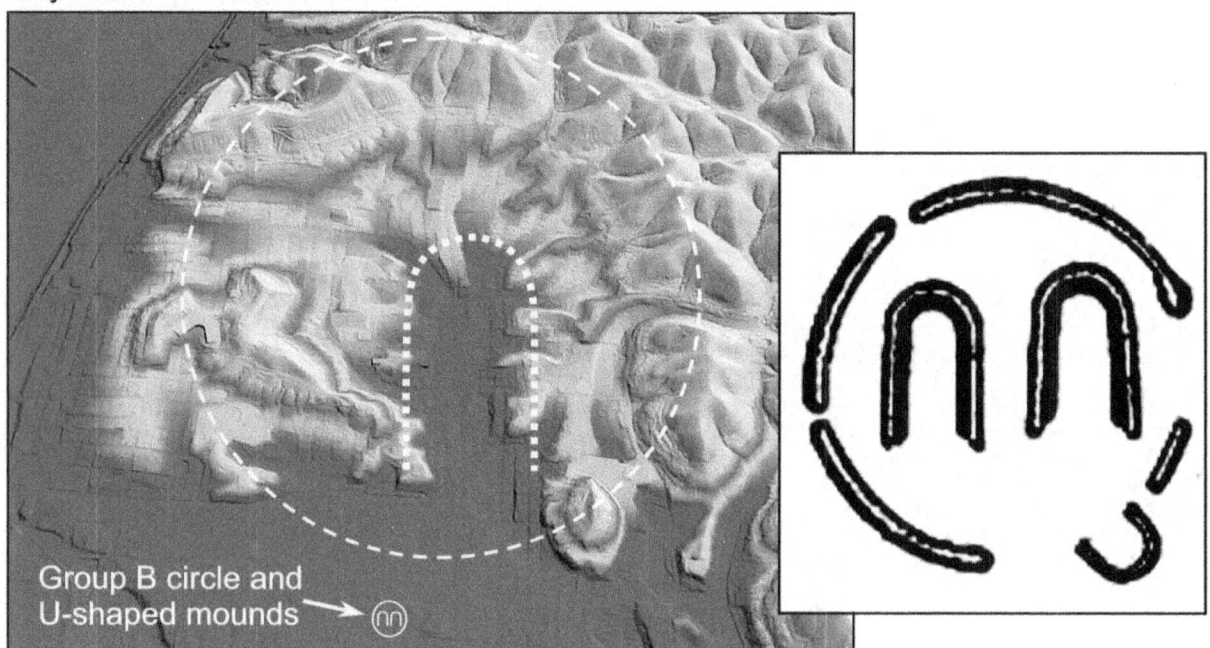

Above left: LiDAR image of north Portsmouth area with location and size of U-shaped earthworks plotted to scale. U-shaped valley outlined by dotted line.
Above right: Group B earthwork detail from Squier and Davis (1848:Pl. XXVIII).

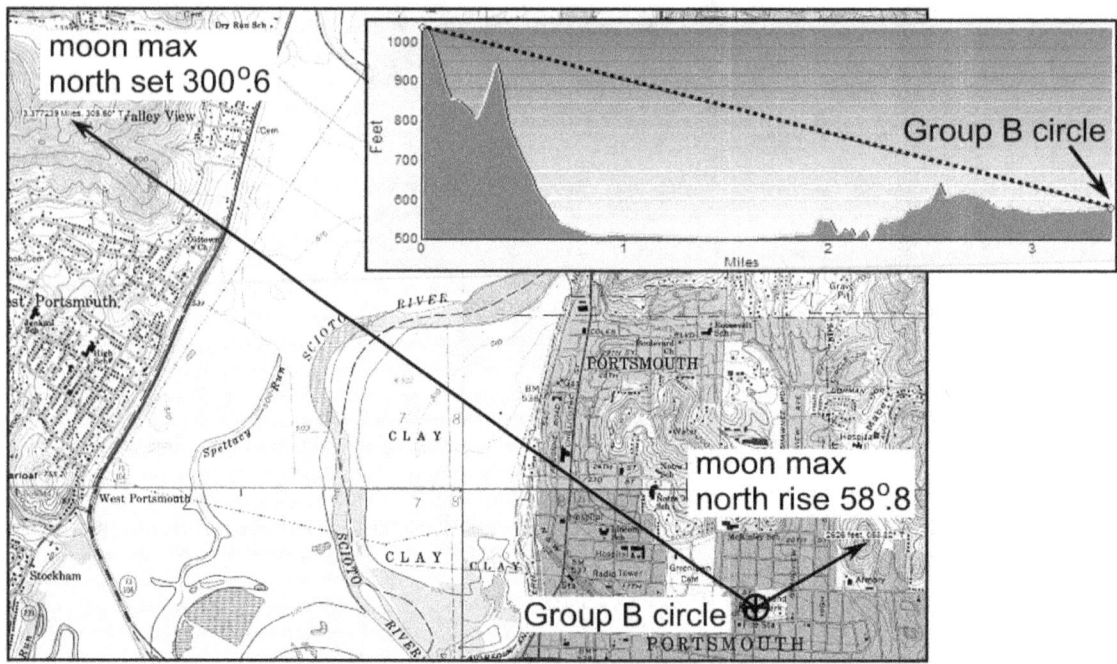

Above: the Group B earthwork is situated so that as viewed from its center, the moon's maximum north rise and maximum north set will appear over nearby hills (note 46). Inset figure shows horizon profile and line of sight for moon max north set. Maps by MyTopo, annotation added.

Figure 5.45. The Portsmouth Earthworks: Group C: Temple Mound

The Portsmouth Group C earthwork is located on the Kentucky side of the Ohio River, about four miles southeast of Portsmouth. The earthwork consisted of concentric embankments surrounding a central mound. The rings were broken by four parallel-walled passageways leading to the center mound. The earthwork has been seriously degraded by plowing and is very difficult to see at ground level. To date, LiDAR data are not available for this part of Kentucky.

Above: view from near the Group C center mound looking toward Portsmouth.

Above left: Squier and Davis (1848:Pl.XXVIII) map of Portsmouth Group C.

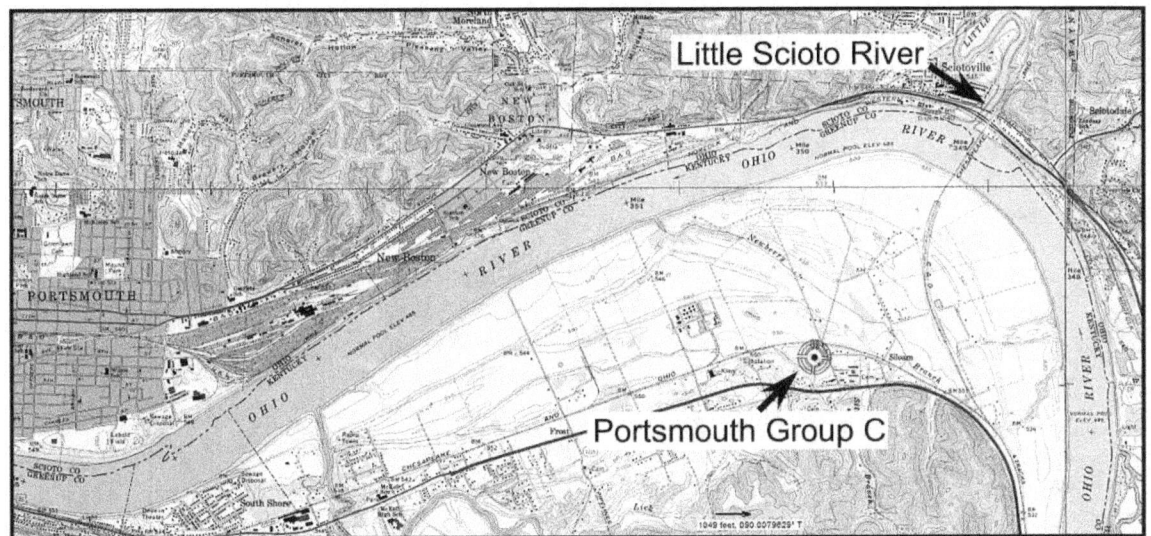

Above: the Portsmouth Group C earthwork is situated at a major bend in the Ohio River, across from the confluence of the Little Scioto and Ohio rivers.

Figure 5.46. The Portsmouth Earthworks: Parallel Wall Embankments
Squier and Davis (1848:78) described the Portsmouth parallel-walled embankments as totalling 8 miles in length, averaging 4 feet high and 20 feet wide at their base. Today, only a few sections of the walls are visible in LiDAR imagery or old aerial photographs.

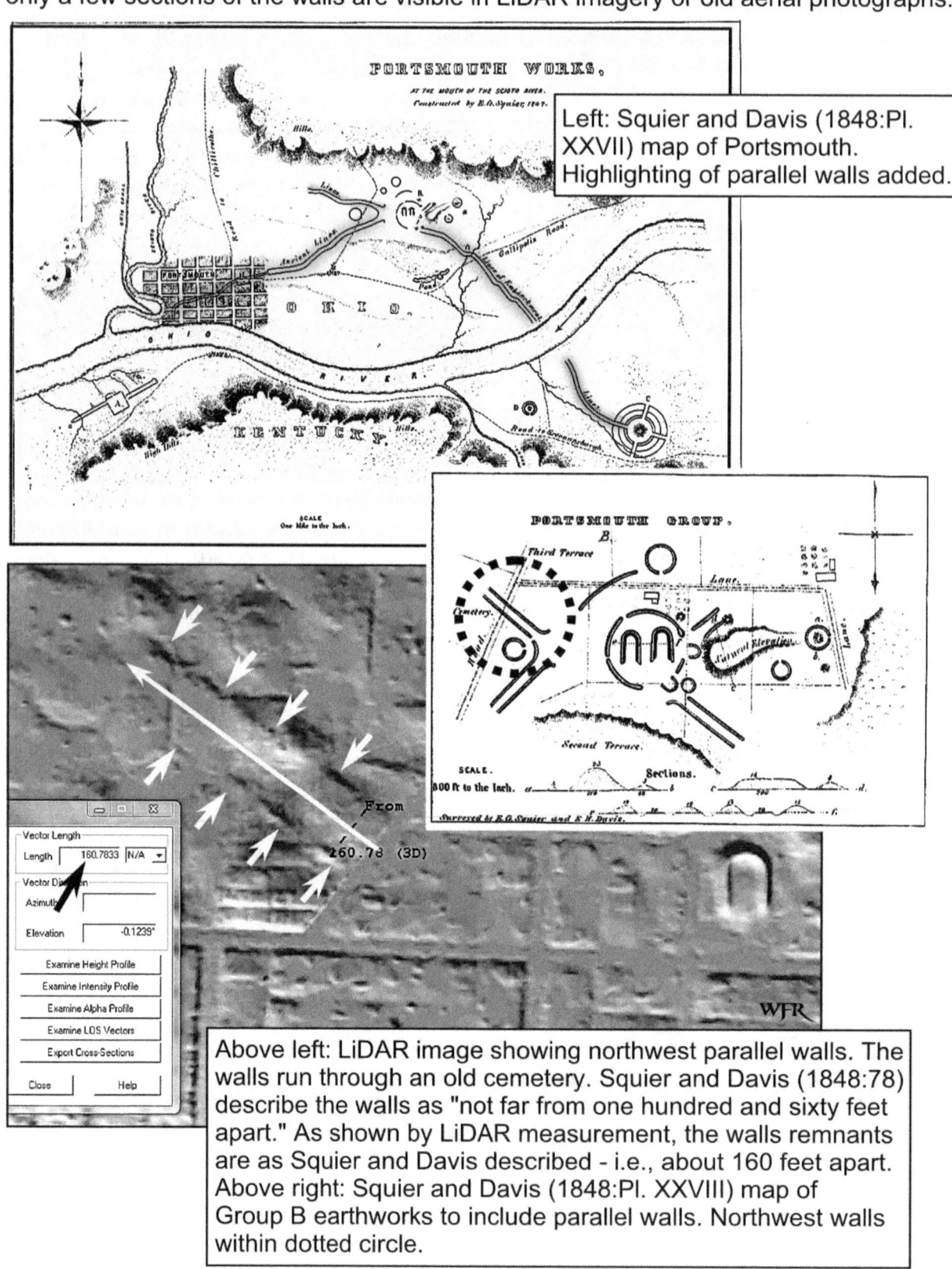

Left: Squier and Davis (1848:Pl. XXVII) map of Portsmouth. Highlighting of parallel walls added.

Above left: LiDAR image showing northwest parallel walls. The walls run through an old cemetery. Squier and Davis (1848:78) describe the walls as "not far from one hundred and sixty feet apart." As shown by LiDAR measurement, the walls remnants are as Squier and Davis described - i.e., about 160 feet apart. Above right: Squier and Davis (1848:Pl. XXVIII) map of Group B earthworks to include parallel walls. Northwest walls within dotted circle.

Figure 5.47. The Portsmouth Earthworks: Parallel Wall Embankments II

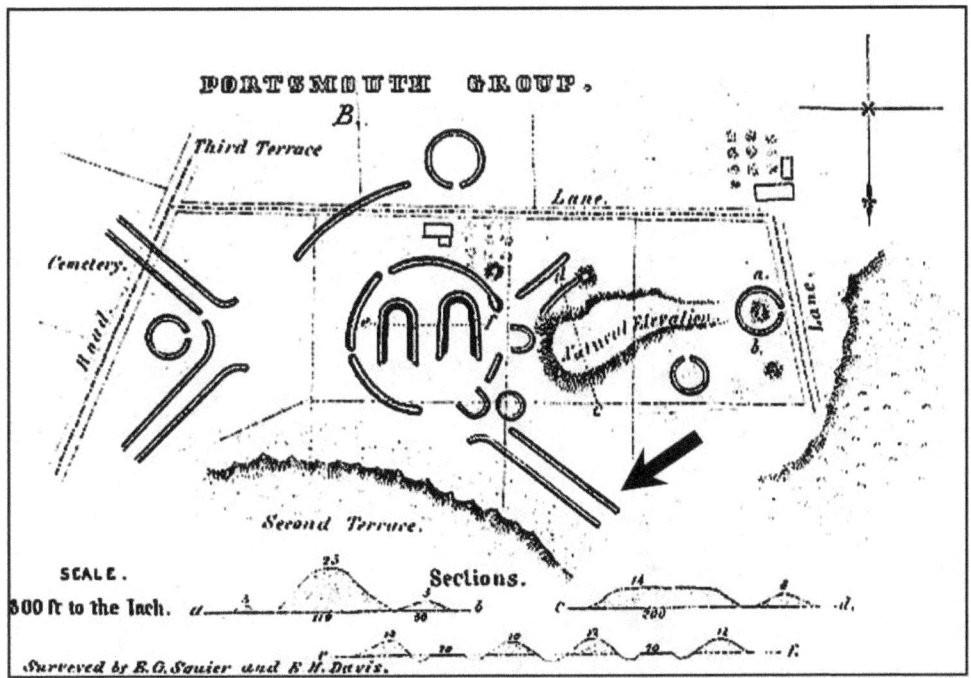

Above: Squier and Davis (1848:Pl. XXVII) map of Portsmouth Group B earthworks. Southeast parallel walls shown by added arrow.
Below: remnants of the southeast parallel walls are just barely visible in LiDAR imagery. In this case, the walls are about 140 feet apart.

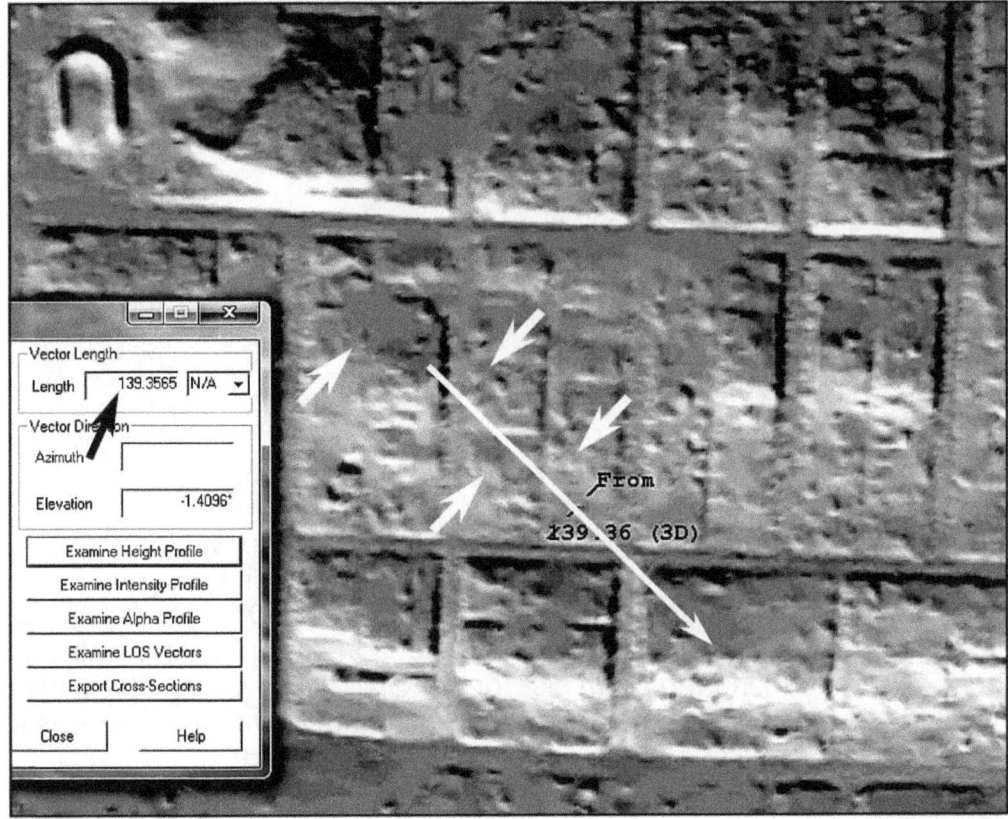

Figure 5.48. The Portsmouth Earthworks: Parallel Wall Embankments III
The Kentucky side of the Group C parallel walls are not visible in LiDAR imagery. However, they are visible in the 1930s USDA aerial photo shown below.

Left: the aerial photo shown here has been cropped, enlarged, and correctly oriented to true north. The prehistoric parallel walls are indicated by arrows (note 47).

Below: in this image the aerial photo is superimposed over the USGS 7.5 minute series map for the area (map provided by MyTopo). Next, the trajectories of the Group B southeast parallel walls and the Group C parallel walls are superimposed (white lines).

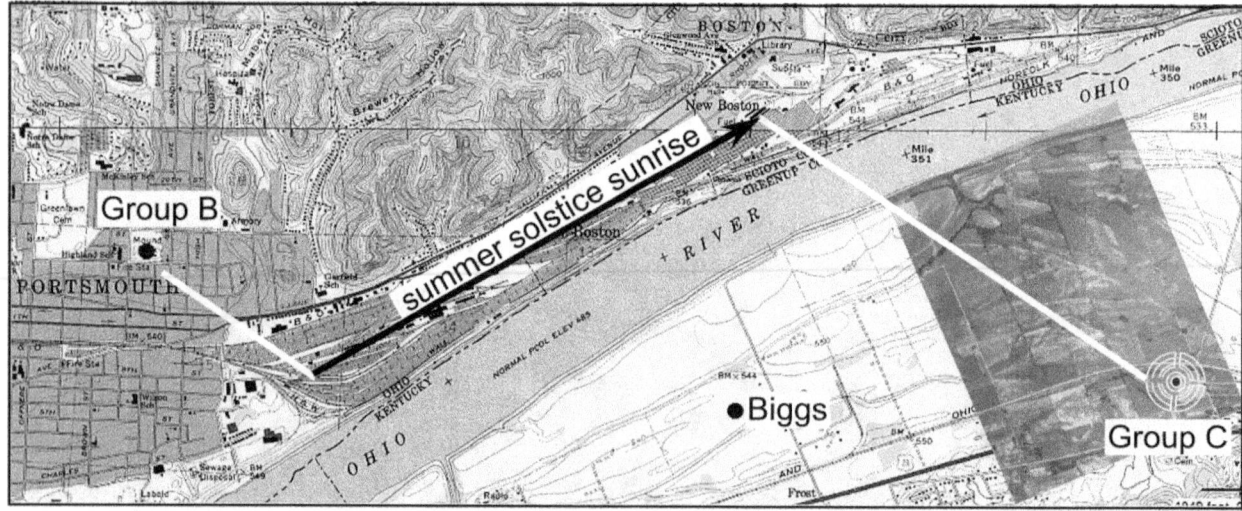

Above: the results of this exercise show that: 1) contrary to the Squier and Davis (1848: Pl.XXVII) map, for the Group B and Group C parallel walls to link-up, an intervening stretch of parallels would have to have proceeded along the Ohio River for about 2 miles; and 2) if the trajectory of this posited 'lost' section of connecting walls extended parallel to the hills that define the north side of the Ohio River Valley, then, that trajectory would simultaneously extend along the azimuth of the summer solstice sunrise (note 48). Map provided by MyTopo, annotation added.

Figure 5.49. The Portsmouth Earthworks: Biggs Earthworks

The Biggs site is located on the Kentucky side of the Ohio River, about 2 miles southeast of Portsmouth. The earthwork consists of a central mound surrounded by a ditch and embankment. Excavation of the central mound revealed cremated remains on a clay platform, with associated fire basin. Grave goods included a tubular pipe, mica, four celts, and hematite and quartz-tempered ceramics (Railey 1996).

Above left: view of the Biggs earthwork. Courtesy of the William S. Webb Museum of Anthropology, University of Kentucky.
Above right: Squier and Davis (1848:fig. 19) sketch of the Biggs Earthwork.

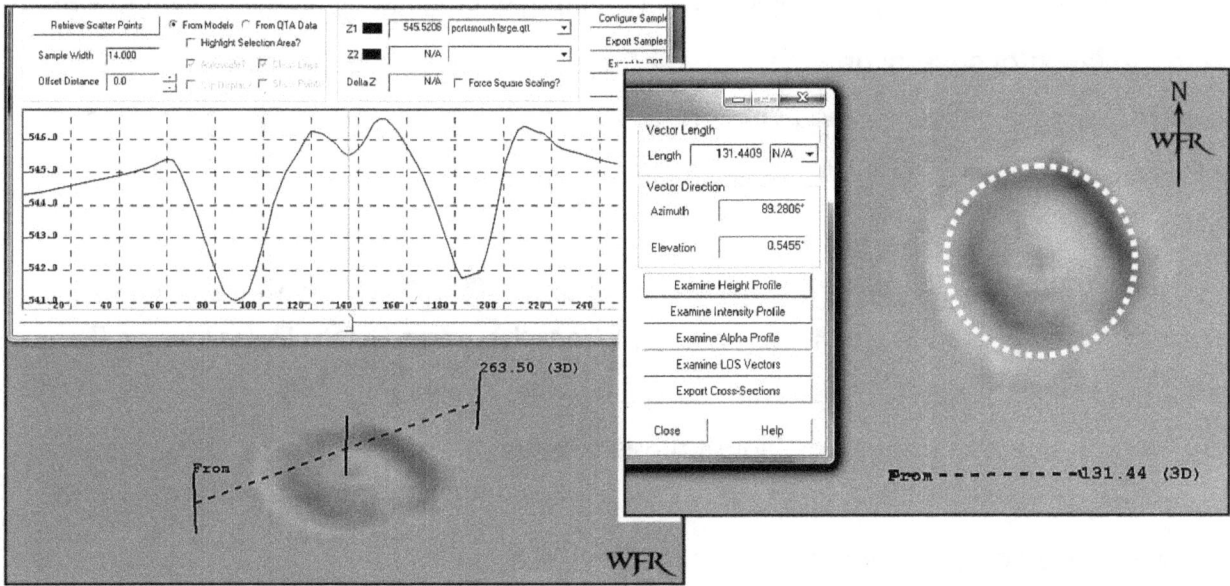

Above left: LiDAR image and profile of Biggs Earthwork. The depression in the center of the mound is from modern digging.
Above right: the size and shape of the Biggs Earthwork closely resembles that of the Johnson Works. Its size appears based on the 1/8 HMU (or 131.75 feet). The above figure shows the relationship between the earthwork and a 1/8 HMU-diameter circle.

CHAPTER 6
PAINT CREEK WATERSHED

Paint Creek is a tributary of the Scioto River. It is about 95 miles in length. From its confluence with the Scioto River just south of Chillicothe, Paint Creek extends southwest for a few miles before it turns northwest, toward the town of Washington Courthouse. The watercourse drains about 1,100 square miles.

North Fork Paint Creek is a tributary of Paint Creek. Its confluence with Paint Creek is on the west side of Chillicothe. The North Fork Paint Creek is about 47 miles long and drains an area of about 236 square miles. Several of the best-known Hopewell earthworks are situated in the two valleys formed by these watercourses (Figure 6.1).

Connector Mounds

In Figures 6.1 - 6.6, several interesting resemblances and relationships occur between mounds and earthwork groups. Specifically, the Steel, Junction, and 33RO257 groups bear similarities to each other in how they are situated on flat terraces overlooking Paint Creek. More striking is that each group is comprised of small circles, with the addition of several unusually-shaped enclosures. Moreover, Steel, Junction, and 33RO257 are at special locations - at the entrances to the three separate forks of the Paint Creek and North Fork Paint Creek valleys. Each earthwork group commands its own valley fork.

Two mounds - i.e., the Story and Country Club mounds are situated in locations that mark the way into the Paint Creek valleys from the north Scioto River Valley. The Scioto and Paint Creek valleys are separated from each other by fairly rugged hills. The Story Mound, however, is within sight of a passageway through the hills. Once into this passageway, the Country Club Mound marks a divide in the route, with one direction leading to the Steel Group and the other to the Junction Group.

In total then, five earthwork groups and mounds are linked through terrestrial pathways. As a practical matter, this network of earthworks and mounds established important lines of communication and access between the Scioto and Paint Creek river valleys. My suspicion, based solely on earthwork morphology is that the earthwork groups are most likely Adena. If that is the case, then an argument can be made that the concept of inter-related earthwork constructions had its origins in Adena, with more and further elaborations exhibited by Hopewell.

Exploring this notion a bit further, it will be recalled from Chapter 6 that the Works East earthwork is situated at the intersection of two sightlines - one of which is parallel to the Scioto River Valley and the other parallel to the Paint Creek River Valley. Further it will be recalled that the Paint Creek sightline connects the Works East earthwork to site 33RO257; while the Scioto River sightline connects the Works East to High Bank and Liberty. In this configuration, another connection is made linking the Paint Creek Valley to the Scioto River Valley. If intentional – as seems likely, then this may indicate a deliberate effort of Hopewell people to connect their earthworks (i.e., Works East, High Bank, and Liberty) to earlier Adena earthworks, namely Steel, Junction, and 33RO257. Support for this scenario awaits radiocarbon dating of the Paint Creek Valley earthworks. But whatever the temporal relationships, there seems little doubt that the earthworks in the Scioto and Paint Creek River valleys were thought-of in terms of their interconnected relationships not only with the landscape - to include earth and sky, but also with each other.

The Hopewell Mound Group

The Hopewell Mound Group is one of the largest of the geometric enclosures built by the Hopewell. Within the Great Enclosure was Mound 25, the largest Hopewell burial mound ever discovered - at 500 feet in length, 188 feet in width, and 33 feet high (Shetrone 1926:203).

Several things came together at the Hopewell Mound Group that made that location special. To begin with, the winter solstice sunrise sightline from Frankfort extends through the heart of the Hopewell Mound Group. In turn, this solstice sightline is intersected by the major axis of the Hopewell Mound Group, which extends parallel to the third river terrace edge. Thus intersecting celestial and terrestrial azimuths establish the center place for the Hopewell Mound Group. Add to that the fact that Hopewell Mound Group is situated at the topographic head of the main Paint Creek Valley and along a major route of travel to Frankfort from all points east, and the significance of this location becomes apparent.

Spruce Hill

Entering into the Paint Creek Valley from the northeast, Spruce Hill commands one of the most physically imposing locations in entire valley. Very little physical evidence of an Adena or Hopewell presence is found within the enclosure itself. However, the expansive view of Paint Creek from the north edge the Spruce Hill summit leaves little doubt that in this case, the viewscape contributed to what made the site special. Indeed, the special nature of the site appears designated by the perimeter wall that encloses most of the promontory. I suspect that Spruce Hill was a vision quest site.

In any case, a second thing that may have contributed to the special status for Spruce Hill is the occurrence of large concretions at that locale that were identified by Squier and Davis (1848:Pl. IV). In particular Squier and Davis show several large concretions in Paint Creek that presumably fell from the Spruce Hill cliff. In searching for these concretions, I was not successful. Perhaps they were removed years ago by collectors. I have, however, found large concretions in the cliff overlooking Paint Creek at the nearby Copperas Mountain.

The interesting thing about the Paint Creek concretions is that when broken open, some are found to contain fossilized arthrodire fish bone. Arthrodire fishes are extinct armor-jawed fish that lived during the Devonian period. Perhaps it was the case that the discovery of fish bone inside the Paint Creek concretions led Moundbuilder peoples to think of the concretions as having once been alive and thus possessing some remnant life-force or special power. If this was the case, it might explain a large collection of concretions found in a mound at the Snake Den Group, as well as the special nature attributed to Spruce Hill - presumably a source for such concretions.

Flood Mitigation
Several sites in the Paint Creek Valley are susceptible to flooding from Paint Creek.

At Seip, Baum, and Frankfort, modern flood levees have been built to protect crops now grown in these areas. The flood levees look remarkably like earthwork walls. In any case, given the visible of old stream channels visible in the LiDAR imagery for Seip and Baum, there is no reason to think that flooding was not problematic in these areas during ancient times. This raises the possibility that, at least at some sites, some earthwork walls may have been built to help mitigate potential flood damage to things inside the enclosures. By 'mitigate' I do not mean that walls were intended to prevent water from entering a site. Anyone who has witness a major flood knows that such efforts are futile. Rather, some enclosure walls may have been built with the idea of lessening the destructive effects of erosion caused by fast moving flood waters, as in the purpose of flood levees in these very areas today.

Of course the question then becomes: What might people have wanted to protect from flood damage and erosion?' I would suggest: burial mounds that hold the ancestors, wooden house or temple structures such as found at Seip (Baby and Langlois 1979), and maybe even food sources - such as stands of chenopodium, knotweed, maygrass, and marshelder, which tend to favor disturbed habitats as found within the Hopewell enclosures (Romain 2004a). Does this mean that all enclosures were built for flood protection? No. Do I mean to suggest that plant foods were grown in all of the enclosures? No. But both possibilities need to be considered for at least some enclosures.

Inter-Site Lunar Alignments

The internal geometry of several earthworks in the Paint Creek and North Fork Paint Creek river valleys suggest direct relationships with each other and nearby Scioto River Valley earthworks. The relevant earthworks are Seip and Baum in the Paint Creek Valley, the Frankfort Works in the North Fork Paint Creek Valley and the Works East and Liberty earthworks in the Scioto River Valley. As Figure 6.22 shows, each earthwork is comprised of three basic geometric shapes situated in different ways relative to each other. I have commented on these relationships elsewhere (Romain 1996, 2000, 2004a). The reasons for, or meanings behind the resemblances are not known. The intent could be to reflect social or political unity or identity. Alternatively, it might simply be a geometric exercise on a grand scale showing the possible combinations of a two circles and a square.

Whatever the reason, it is interesting to note that the five earthworks just mentioned not only share geometric similarities, they are also connected to each other along lunar azimuths - see Figure 6.23. To these five sites, two more lunar connected sites can be added, namely Anderson and High Bank, thus bringing to seven, the number of lunar-connected sites. If we add the Newark Octagon and Observatory Circle, then the total is eight lunar inter-connected earthworks. With reference to Figure 6.23, the sightline that extends from Works East, through Anderson, to the Frankfort earthwork coincides with the moon's north minimum set. Correspondingly, the sightline that extends from Seip, through Baum, to the Works East earthwork coincides with the moon's minimum north rise (Romain 1992). As discussed later in this chapter and as pointed-out by Hively and Horn (2010:139), the Liberty, High Bank, and Works East earthworks extend in a line having an azimuth of 143.4 degrees. Not noted by Hively and Horn is that this 143.4-degree azimuth is orthogonal to the moon's maximum north rise to within one degree.

Also of interest is that, although the High Bank Earthwork is part of the web of relationships just noted, it is obviously different in its geometry from the others. With its octagon and attached circle, High Bank more closely resembles the Newark Octagon and Observatory Circle than it does any of the earthworks in the Paint Creek or Scioto River valleys. Notably, the High Bank Circle and Newark Octagon Circle are both 1 HMU in diameter; and the octagons at both sites are based on the HMU length. It is often commented that the major axis of the High Bank earthwork is oriented at a near 90-degree angle to the major axis of the Newark Octagon and Observatory Circle. In other words, the two earthworks are orthogonal to each other. The connection goes deeper than these similarities alone, however. As Figure 6.23 shows, a line drawn across the 58-mile distance between the centers of the High Bank and Newark Octagon and Observatory Circle earthworks extends along an azimuth of 25.5 degrees. The orthogonal of 25.5 degrees is 115.5 degrees. The azimuth of 115.5 degrees is to within one-third of a degree of the moon's minimum south rise at 115.2 degrees. Thus Newark and High Bank are connected along a lunar-derived azimuth and all eight sites are entangled with each other in relational web that links the Newark and Chillicothe areas.

Figure 6.1. Paint Creek Convergence Zone

Paint Creek and the North Fork of Paint Creek converge near the Steel, Junction Group, and 33RO257 earthworks before flowing into the Scioto River.

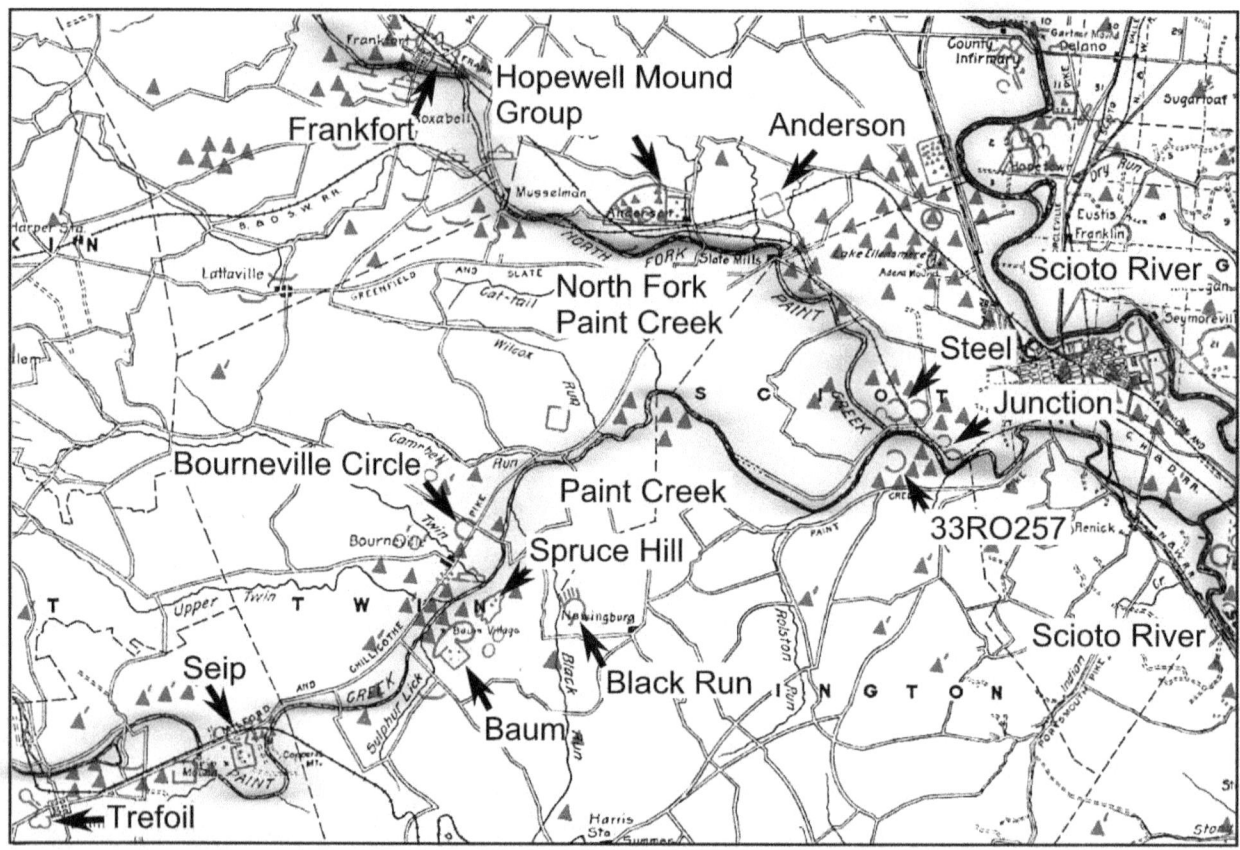

Above: enlarged detail from Mills (1914) showing Paint Creek sites. Annotation, highlighting and Anderson Earthwork added.

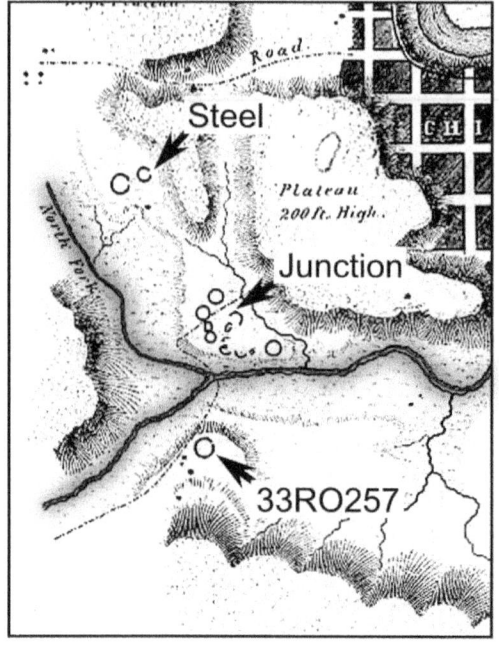

Left: three earthwork complexes stand at the threshold entrances to the sites of Paint Creek and the North Fork of Paint Creek - i.e., Junction Group, the Steel Group, and 33RO257. Very little is known about these complexes (note 1). All three sites have been largely destroyed by farming. The image to the left is an enlarged detail from Squier and Davis (1848:Pl. II), annotation added.

Figure 6.2. Paint Creek Convergence Zone: Junction Group

The Junction Group consists of several circle and other geometric enclosures and mounds. Note the unique four-lobed earthwork in the below left image.

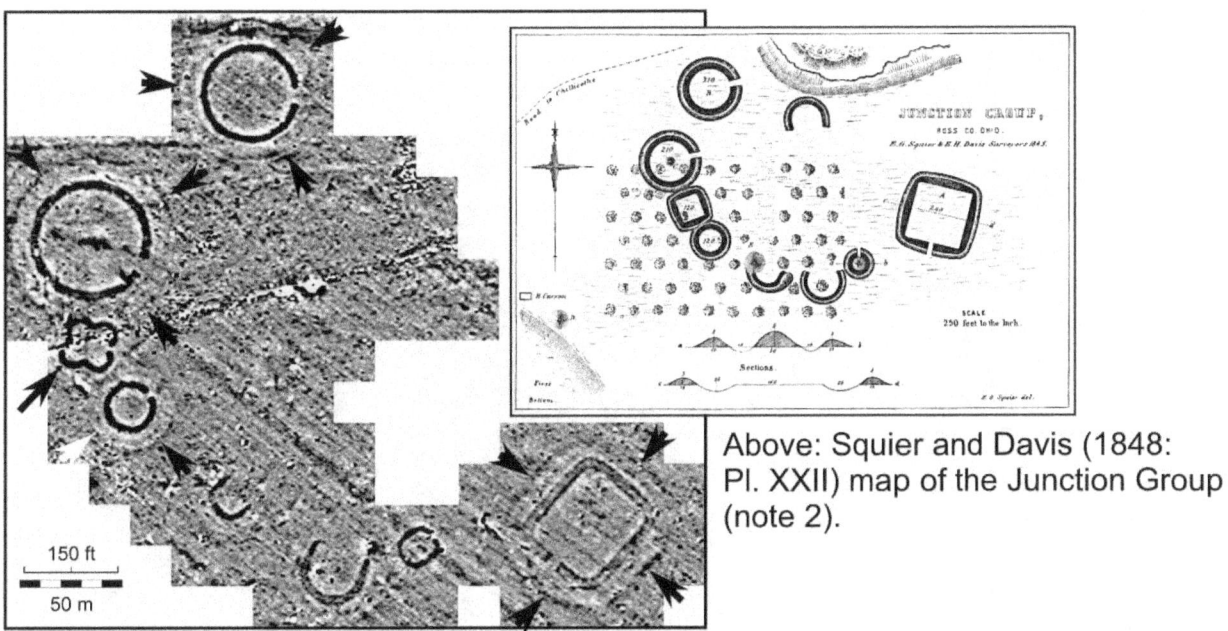

Above: Squier and Davis (1848: Pl. XXII) map of the Junction Group (note 2).

Above left: magnetic gradiometer survey of Junction Group by Jarrod Burks and associates (also see Burks 2006, 2010, 2011). Interior, filled-in ditches indicated by dark lines. Deflated perimeter walls are lighter with arrows showing same.
Below left: mag survey map over LiDAR image. LiDAR analysis suggests that at least four of the Junction Group earthworks were designed using either the 1/4 HMU or 1/8 HMU lengths. Dashed circles = 1/4 HMU diameter; dotted circles = 1/8 HMU. Mag survey images courtesy of Jarrod Burks, annotation by the present author.
Below right: Aerial view showing trefoil earthwork. Photo by Jeffrey Wilson.

Figure 6.3. Paint Creek Convergence Zone: Steel Group

Squier and Davis (1848:Pl. II) show the Steel Group as comprised of two earthworks and a mound. Magnetic gradiometer survey by Jarrod Burks and associates, however, shows the complex included at least eleven earthworks.

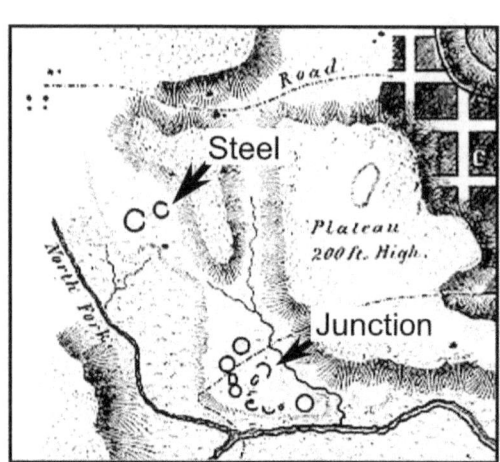

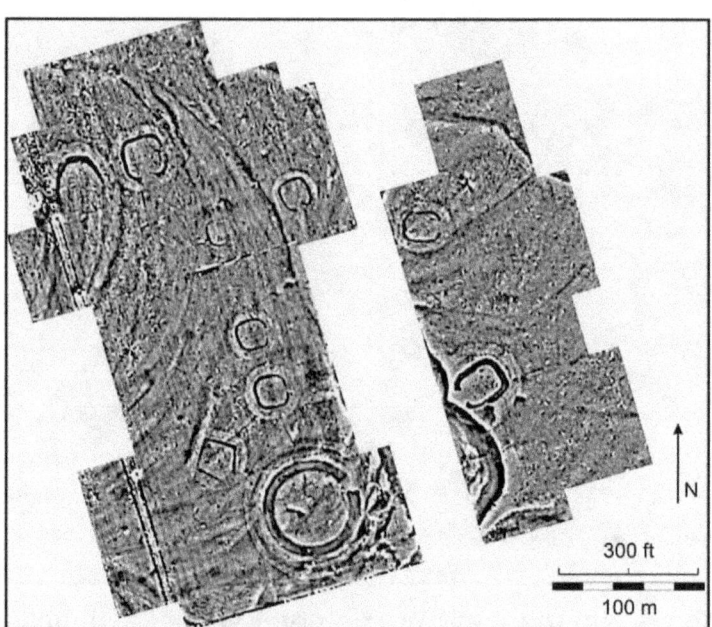

Above: the Steel Group is located about one mile northwest of the Junction Group.

Above right: magnetic gradiometer survey image of Steel Group, courtesy of Jarrod Burks.

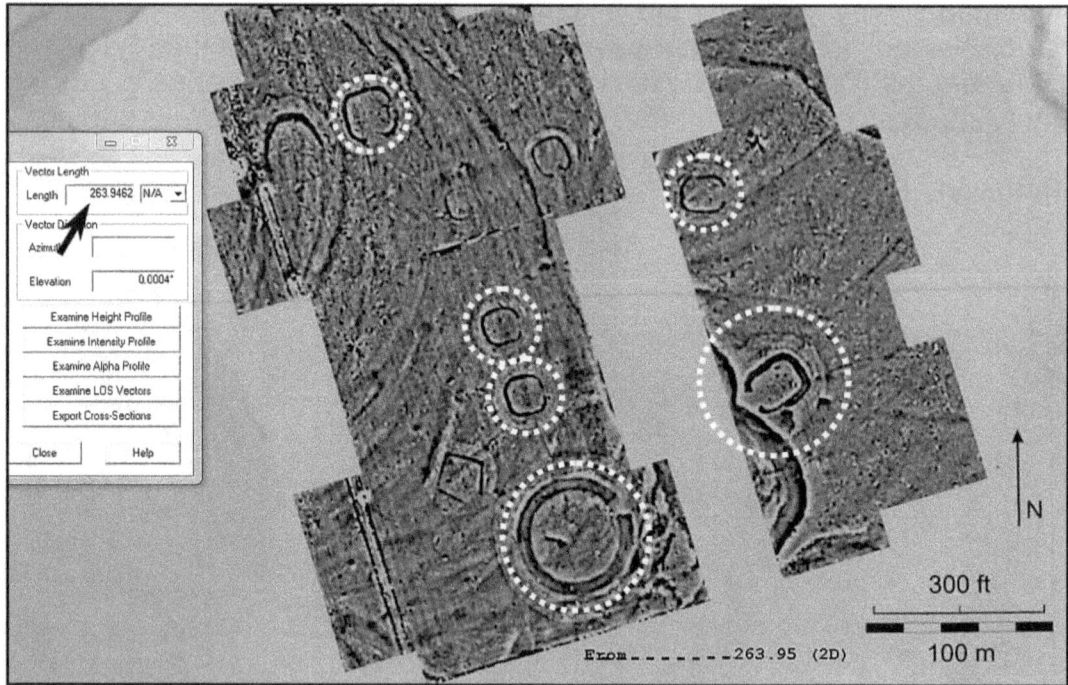

Above: mag survey map superimposed over LiDAR image. As shown, LiDAR analysis suggests that at least six of the Steel Group earthworks are designed using either the 1/4 HMU or 1/8 HMU lengths. Large white dot circles = 1/4 HMU diameter; small white dot circles = 1/8 HMU diameter.

Figure 6.4. Paint Creek Covergence Zone: Lines of Communication

Junction Group, Steel, and 33RO257 are related not only in terms of design similarities, but also in how each commands an important line of communication, direction of travel, and access to major earthworks. To these three sites the Story and Chillicothe Country Club mounds can be added as waypoints for traffic between Paint Creek and Scioto River Valley sites. White dotted lines show waterways, black dotted lines show valley passes.

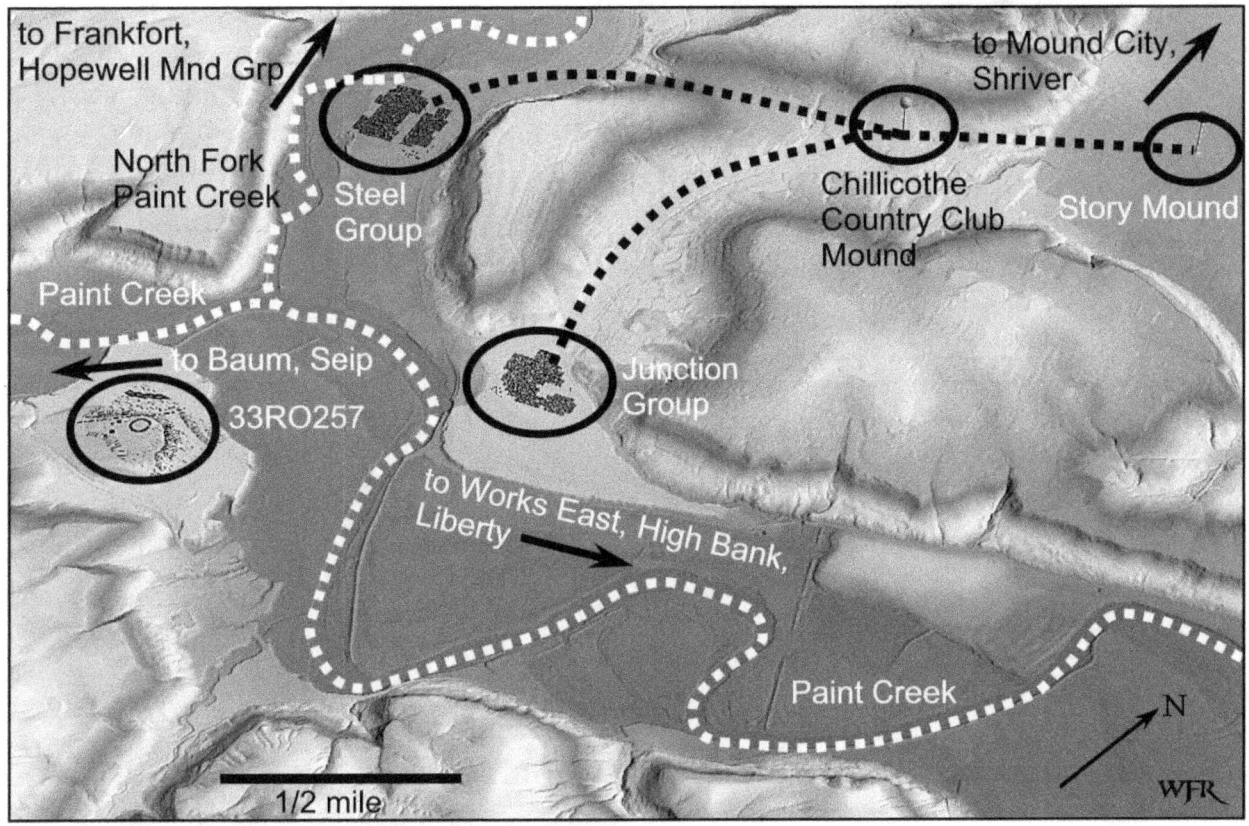

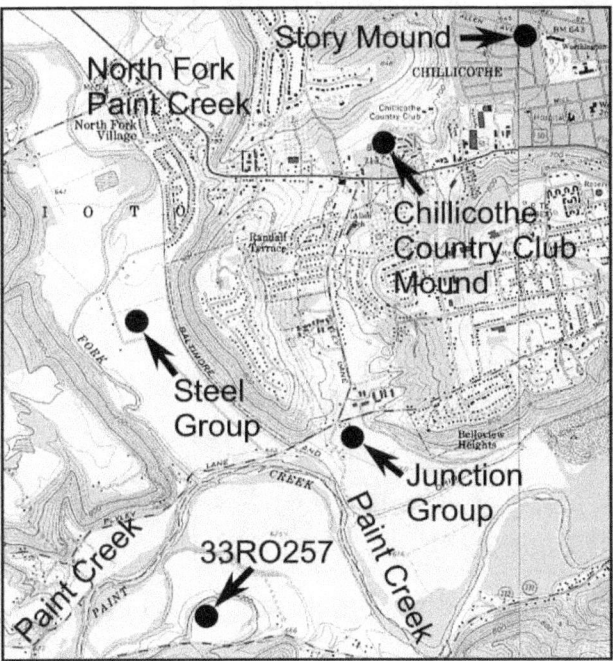

Left: USGS 7.5-minute series topographic map of area of interest. Map provided by MyTopo, annotation added.

Figure 6.5. Paint Creek Connector Mounds: Story Mound

The Story Mound is an Adena mound located in Chillicothe, about one mile southwest of the Adena Mound. The mound was reportedly about 25 feet high when excavated in 1897, which would make it about the same size as the Adena Mound. Found within the mound was an adult male burial, grave goods, and post holes forming a circular structure (Moorehead 1899:132-133).

Left: view of the Story Mound. Below: although the mound is a bit irregular in shape, LiDAR analysis suggests that its size was based on the 1/8 HMU (or 131.75 feet).

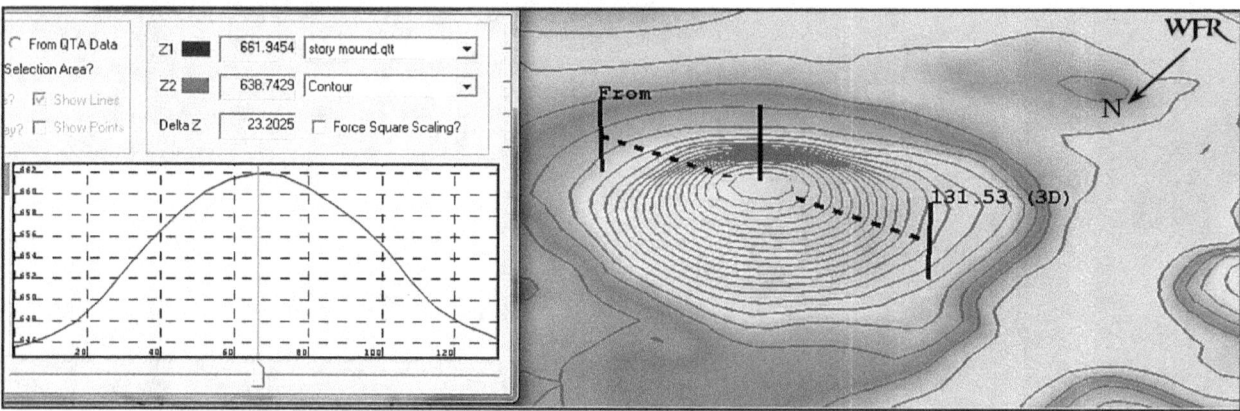

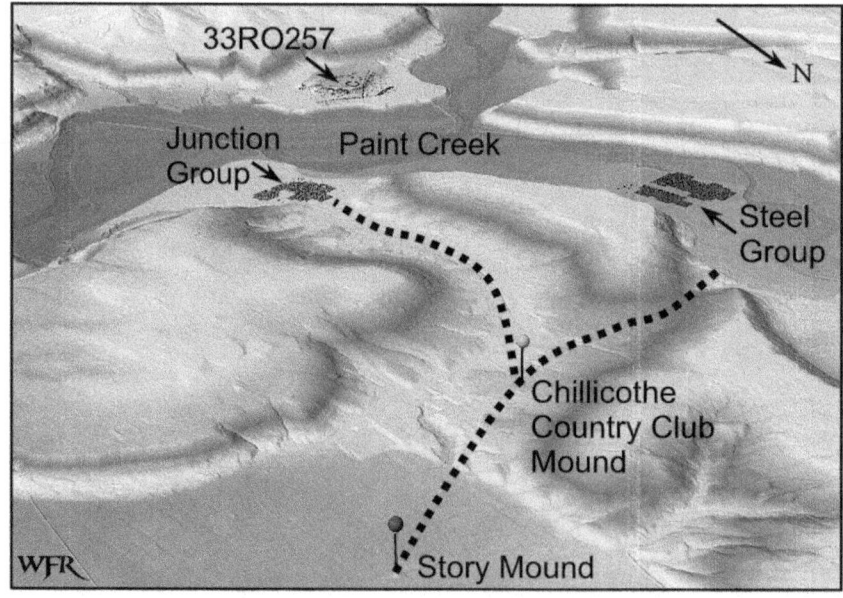

Above: the Story Mound marks the entrance to a valley that cuts through from the north Chillicothe area to Paint Creek.

Figure 6.6. Paint Creek Connector Mounds: Chillicothe Country Club Mound
Very little is known about the Country Club Mound. It may have been dug into in the 1920s - but if it was, no record of that work is known (note 3).

Above left: Detail of Mills (1914) map. Annotation added. Arrow shows what may be the Chillicothe Country Club Mound.
Above right: view of the Chillicothe Country Club Mound, 17-September-2011.

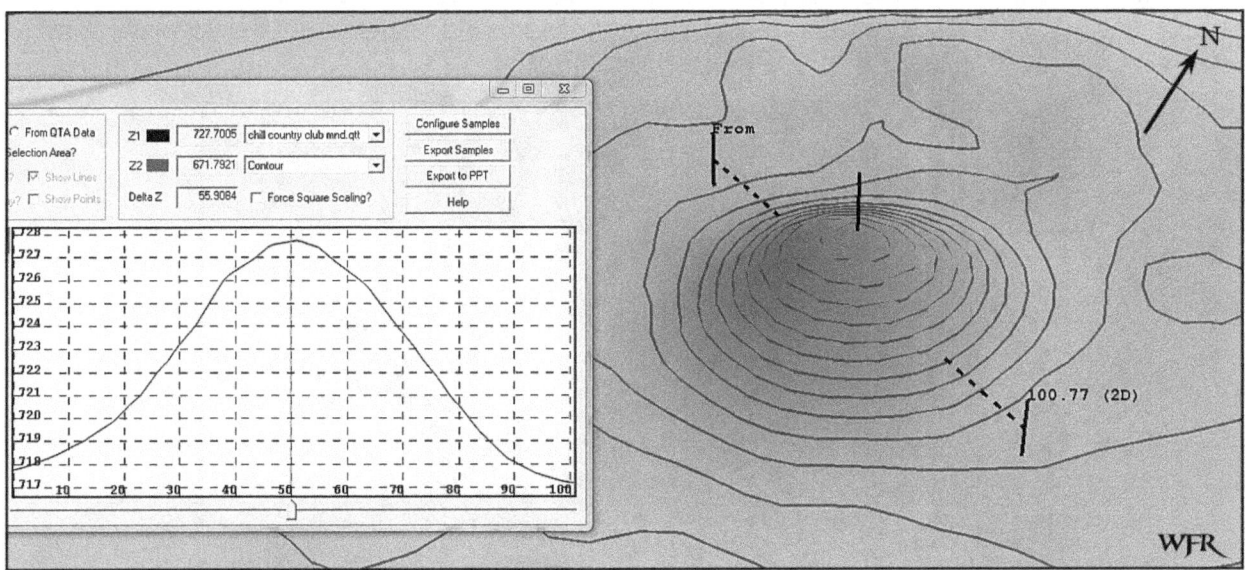

Above: LiDAR contour map and image and profile showing dimensions of Chillicothe Country Club Mound. The mound is about 10 feet in height and 90 feet in diameter.

Figure 6.7. Paint Creek North Fork: Anderson Earthwork

The Anderson Earthwork is located on a terrace overlooking the North Fork of Paint Creek. It has been severely impacted by a housing development.

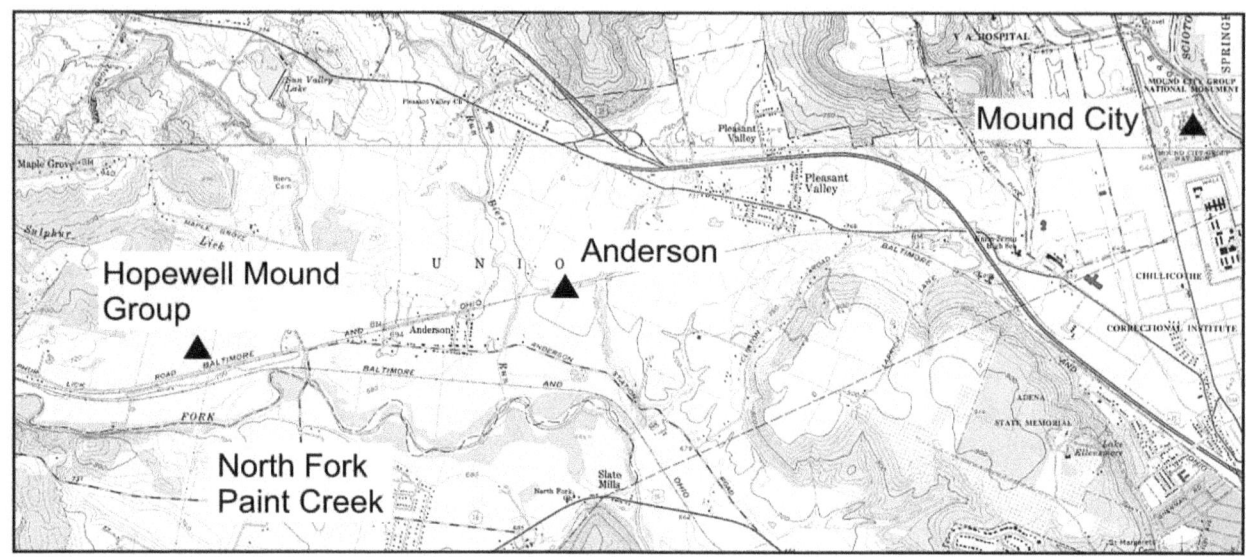

Below: the Anderson Earthwork was first reported in the literature in 1980 (Anderson 1980) - when it was identified in a 1938 USDA aerial photo. Analysis of the aerial photo shows the earthwork was an irregular square with one or possibly, two, attached small circle features. No mounds, burials, or habitation sites have been found in association with the earthwork. Limited salvage excavations through two wall sections were made in 1993 (Pickard and Weinberger 2009).

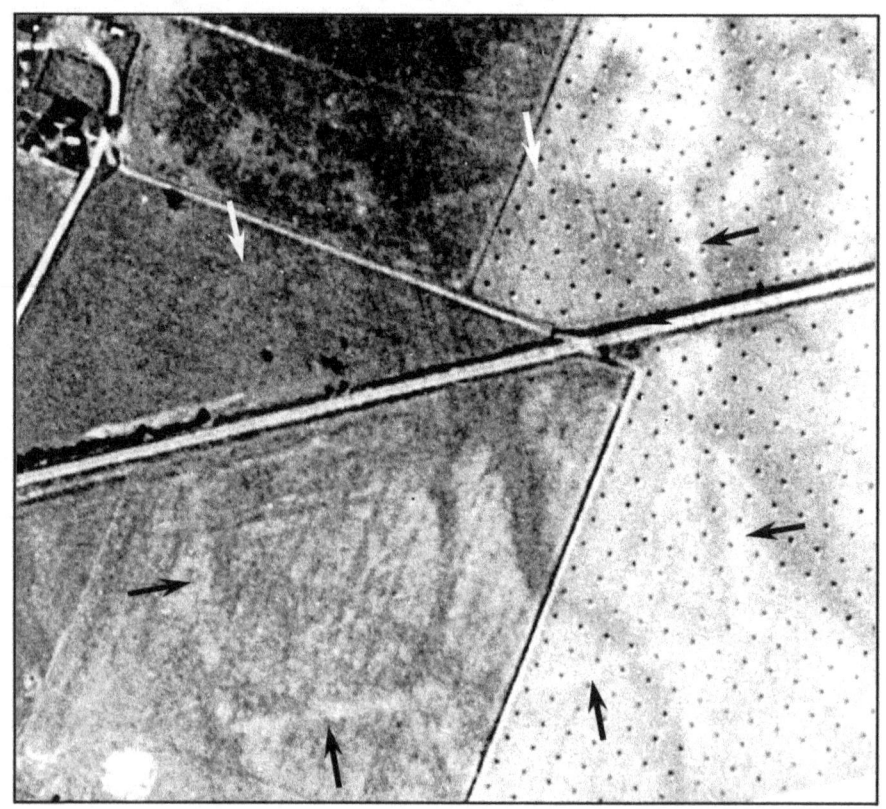

Figure 6.8. Paint Creek North Fork: Anderson: Alignment and HMU

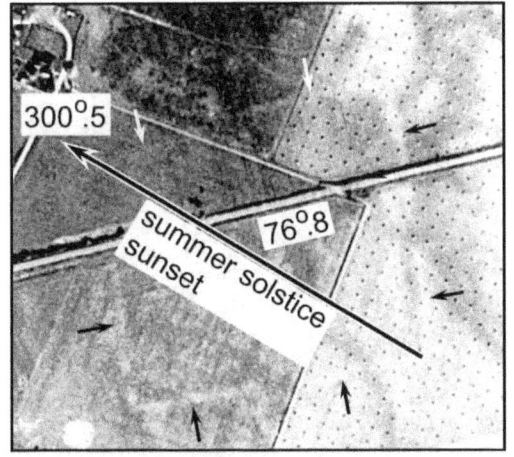
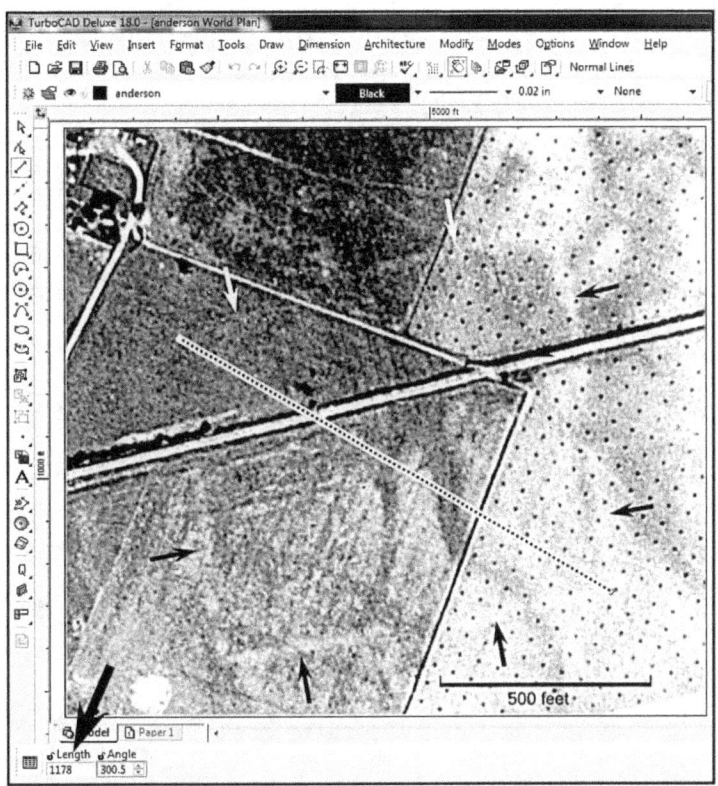

Above: the northwest corner of the Anderson Earthwork is not visible. However, it may have been oriented to the summer solstice sunset along its diagonal axis (note 4).
Right: analysis suggests that the solstice diagonal (black dotted line) was about 1,178 feet in length - making the earthwork about the same size as the Circleville Square.

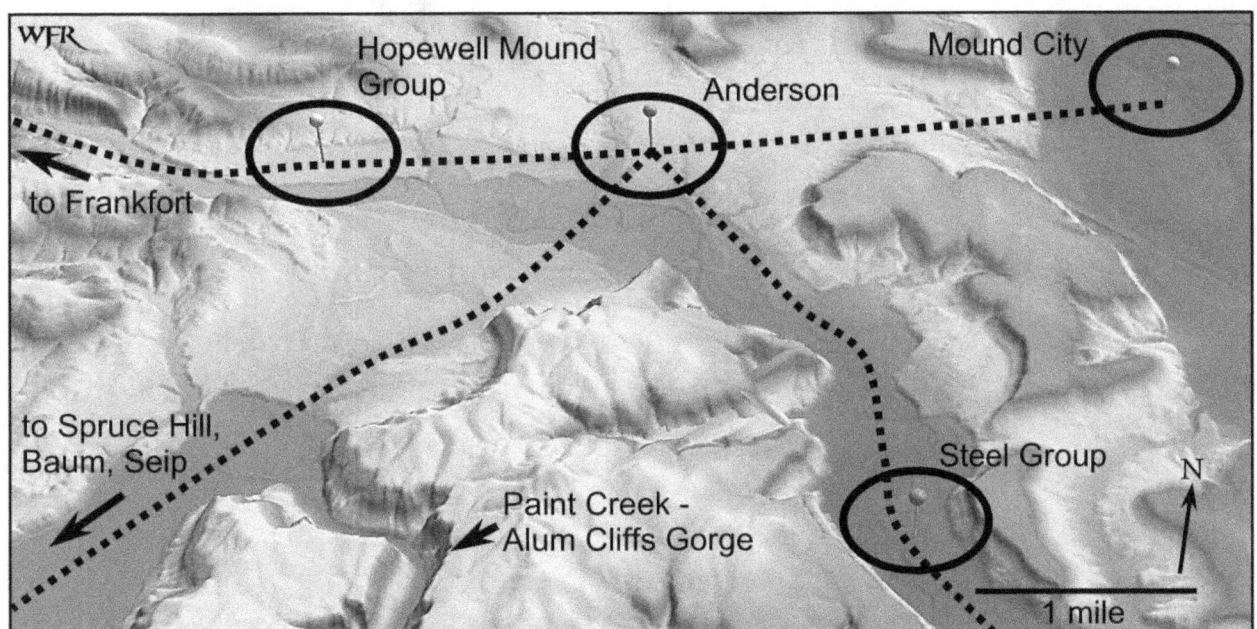

Above: LiDAR image showing location of the Anderson Earthwork. Anderson is located at a strategic location overlooking three river valley trajectories and four significant routes of travel (note 5).

Figure 6.9. Paint Creek North Fork: Hopewell Mound Group

The Hopewell Culture derives its name from this site (note 6). The Hopewell Mound Group is comprised of the Great Enclosure and attached Square Enclosure. Originally, the site included about three miles of embankment walls and more than 40 mounds (Shetrone 1926:203).

Below: Shetrone's (1926:196) map of Hopewell site superimposed over LiDAR image.

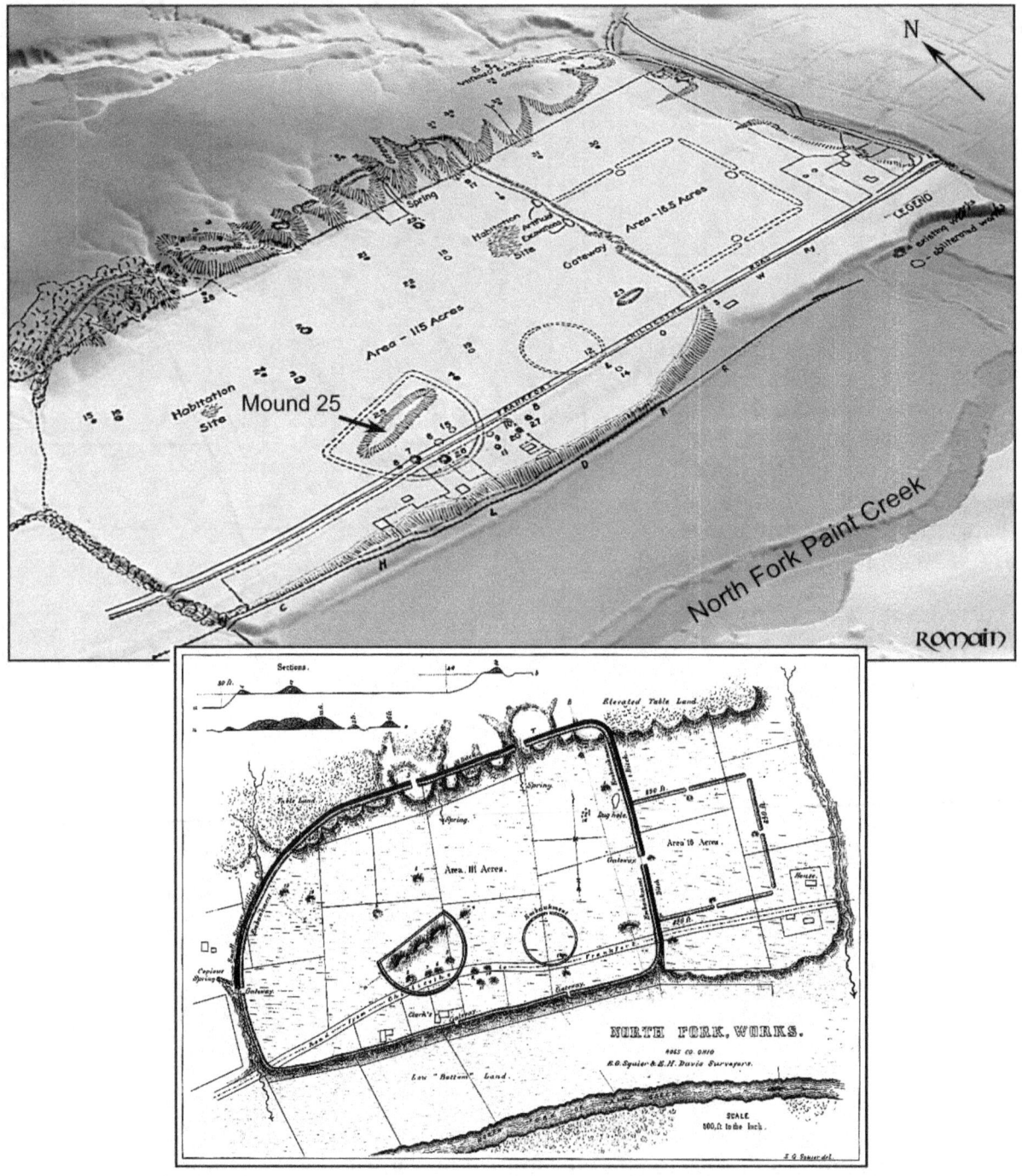

Above: Squier and Davis (1848:Pl. X) map of the Hopewell Mound Group. Although not as accurate as the Shetrone map, it provides a useful early view of the site. Note the D-shaped and circle enclosures within the Great Enclosure.

Figure 6.10. Paint Creek North Fork: Hopewell Mound Group: Lay of Land

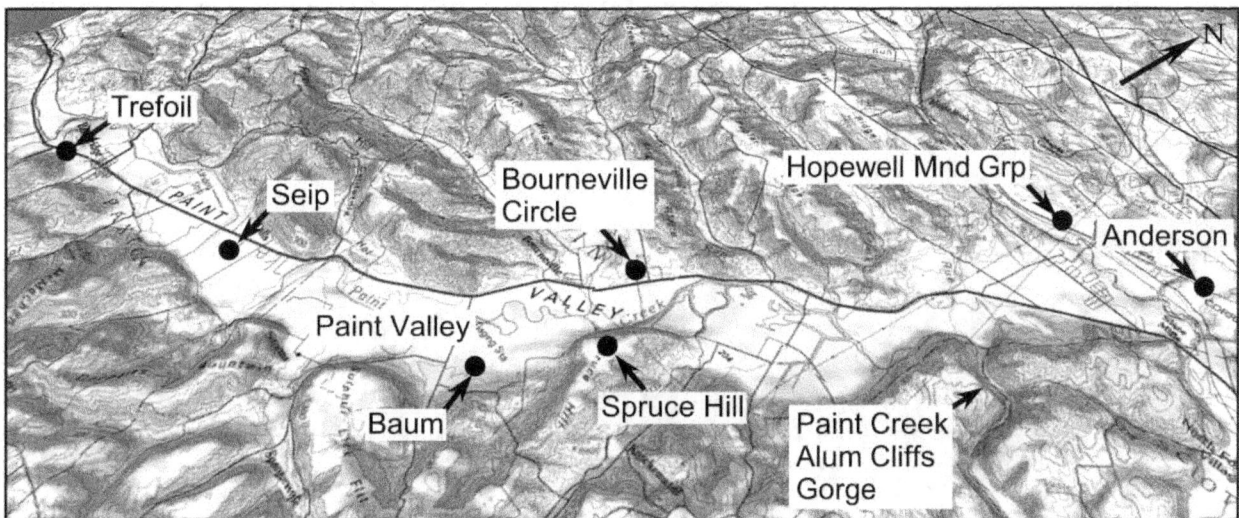

Above: the Hopewell Mound Group is strategically located at the head of Paint Valley. Note how the Hopewell Mound Group is situated at the northwest corner of the valley, while the Anderson site is located in a complementary fashion - on the northeast edge of the valley. Significant sites located in Paint Valley include: Spruce Hill, Baum, Seip, Bourneville Circle, and Trefoil. Map provided by MyTopo, annotation added.

Above: the Hopewell Mound Group extends parallel to the lay of the land established by the trajectory of the third terrace edge. The major axis (also the longest length of the enclosure) is 2,978 feet in length - which is equal to 2 X 1,489 feet. The 1,489-foot length is equal to the hypotenuse of a 45-45-90 degree triangle having two sides that are each 1 HMU (or 1,054 feet) in length.

Above right: view of the Hopewell Mound Group from the third terrace level looking across the site toward the North Fork of Paint Creek.

Figure 6.11. Paint Creek North Fork: Hopewell Mound Group: Alignment

The Hopewell Square has been severely degraded by plowing. Only sections of it are still visible on the ground or in LiDAR imagery. The Square does, however, show-up in old aerial photos. For the below illustrations, a 1938 USDA aerial photo is superimposed over a LiDAR image. Azimuths and dimensions are then determined using the LiDAR program.

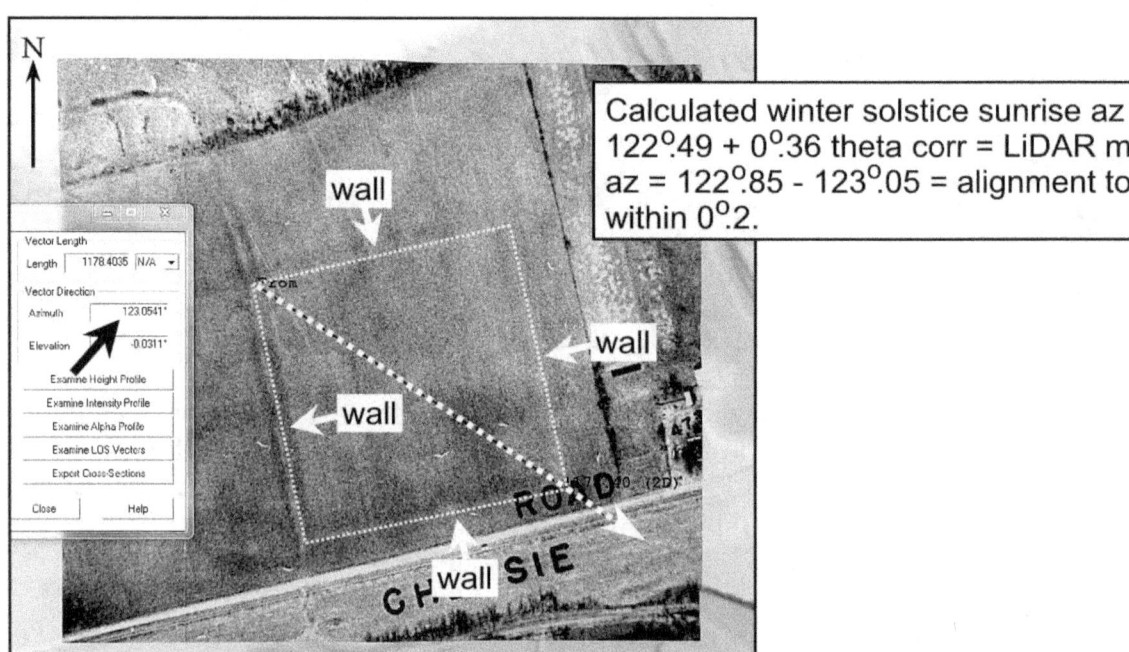

Calculated winter solstice sunrise az = $122°.49 + 0°.36$ theta corr = LiDAR map az = $122°.85 - 123°.05$ = alignment to within $0°.2$.

Above: the Hopewell Group Square Enclosure is aligned to the winter solstice sunrise along its diagonal axis (note 7).

Below: the length of the solstice diagonal across the Square is approximately 1,178 feet. As shown elsewhere, the 1,178-foot length is related to the HMU (or 1,054-foot length).

Figure 6.12. Paint Creek North Fork: Frankfort Works

The Frankfort Earthwork was located a few miles upstream from the Hopewell Mound Group, on a terrace overlooking the North Fork of Paint Creek. The earthwork has been mostly destroyed. Except for the Henneberger Mound, the earthworks are not visible on the ground level or in LiDAR imagery.

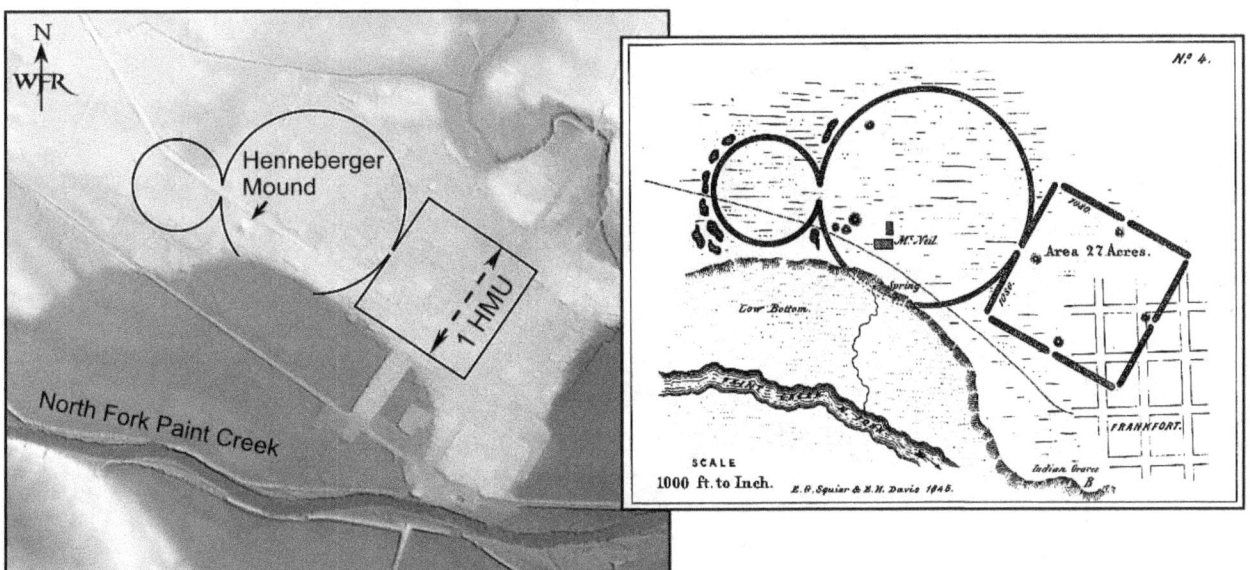

Above left: sketch of Frankfort Works superimposed over LiDAR image. When mapped in the 1800s, erosion had already cut into the Large Circle. Orientation is estimated; size is to correct scale (note 8). The dimensions of the Large and Small Circles, as well as the Square are based on the HMU.
Above left: Squier and Davis (1848:Pl. XXI, no. 4) map of Frankfort.

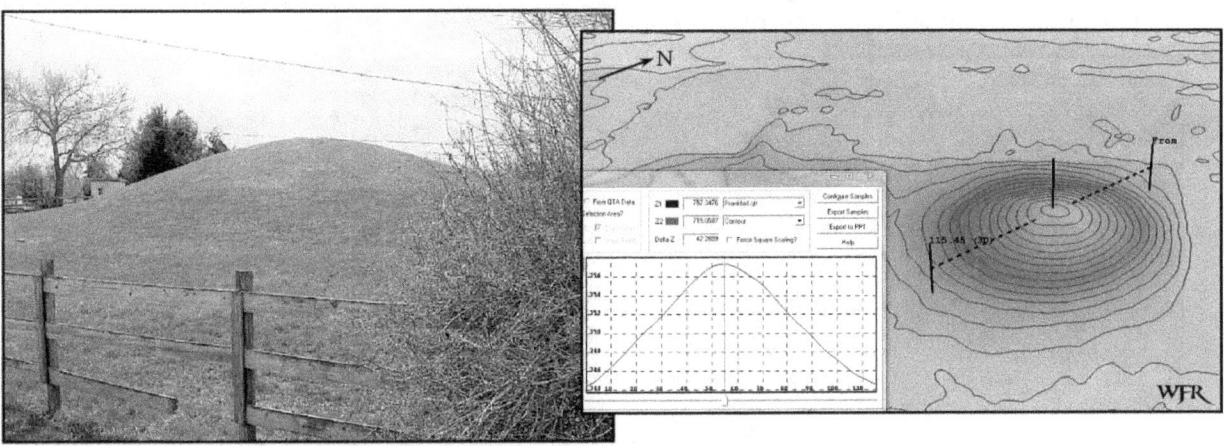

Above left: view of the Henneberger Mound. The Henneberger Mound has not been excavated. Other mounds within the Large Circle, however, were found to contain multiple burials and grave goods (note 9) (Moorehead 1892:113-143).
Above right: LiDAR profile of Henneberger Mound. The mound is about 120 feet in diameter and 12 feet high.

Figure 6.13. Paint Creek North Fork: Frankfort Works Alignments

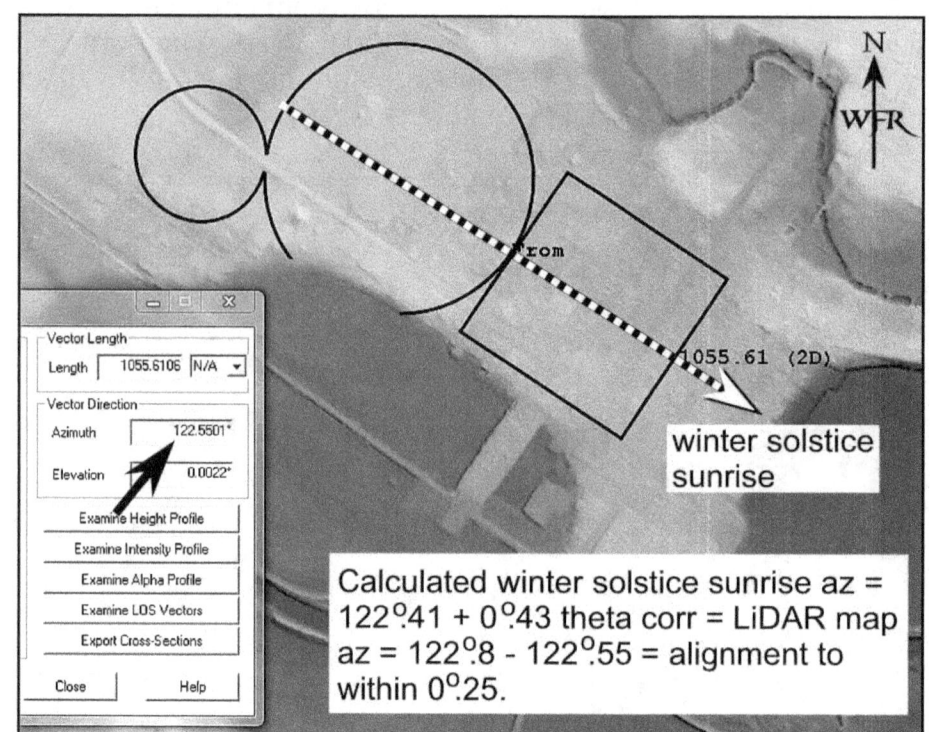

Above: sketch of Frankfort Works superimposed over LiDAR image. If I have plotted the location of the Frankfort Works correctly, then the major axis of the Large Circle and Square is aligned to the winter solstice sunrise (note 10).

Below: the major axis of Frankfort is not only aligned to the winter solstice sunrise, but also to the Hopewell Mound Group - about 5 1/2 miles distant (note 11). Map provided by MyTopo, annotation added, earthworks to scale.

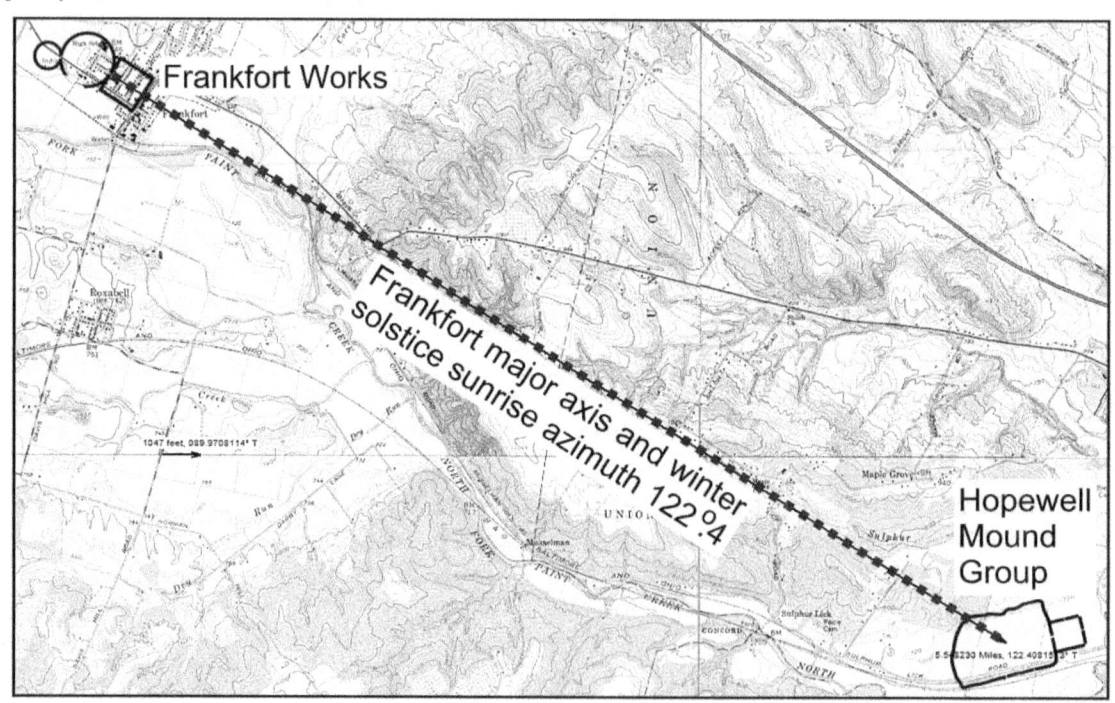

Figure 6.14. Paint Valley Sites: Bourneville Circle

The Bourneville Circle is located in Paint Valley, across from Spruce Hill. Originally, its walls were 8 - 10 feet high. The earthwork has been considerably reduced in height by farming. However, it is still visible in LiDAR imagery.
Below: Squier and Davis (1848:Pl. III, no. 1) map showing location of Bourneville Circle.

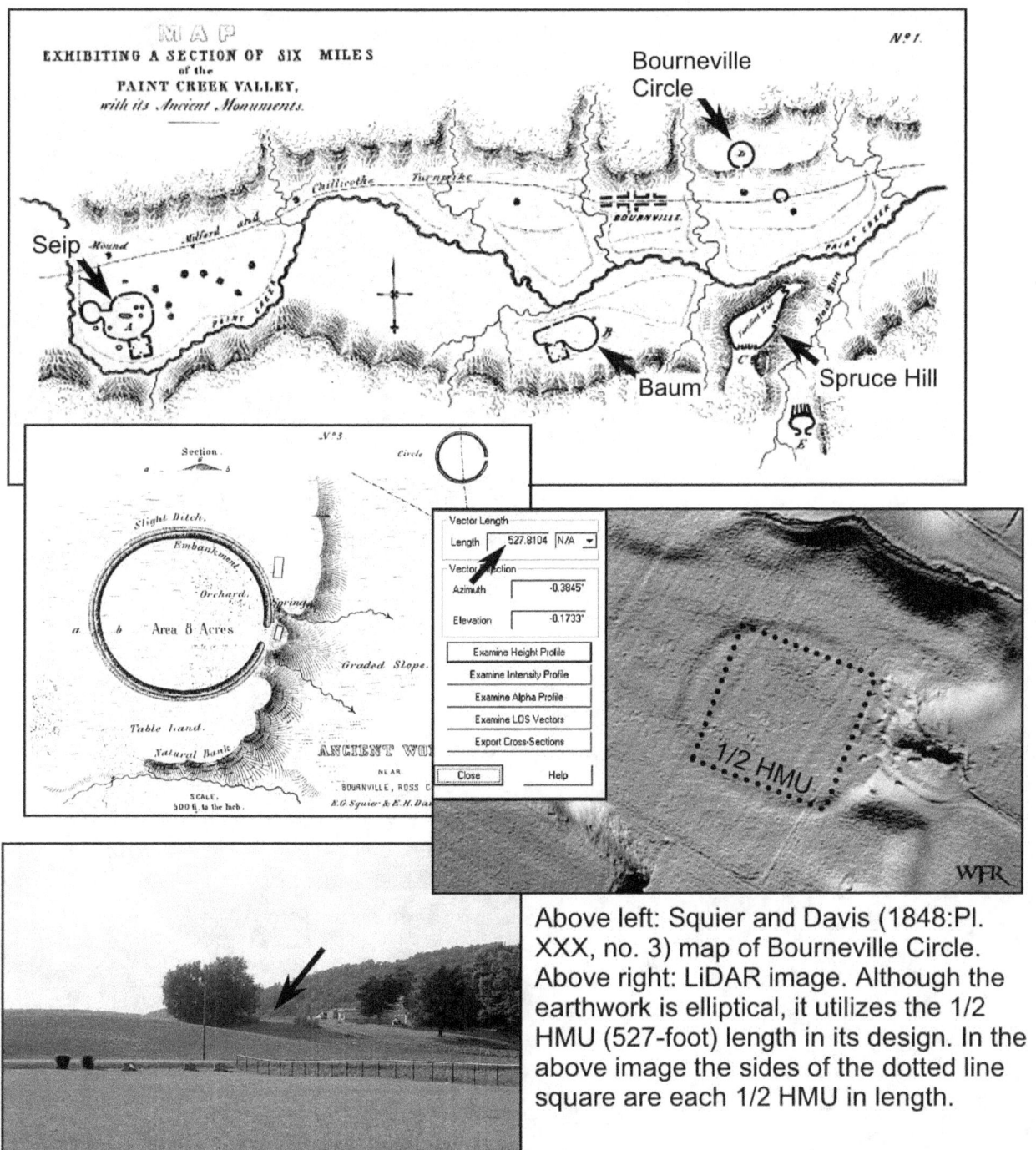

Above left: Squier and Davis (1848:Pl. XXX, no. 3) map of Bourneville Circle. Above right: LiDAR image. Although the earthwork is elliptical, it utilizes the 1/2 HMU (527-foot) length in its design. In the above image the sides of the dotted line square are each 1/2 HMU in length.

Above: Bourneville Circle is situated on a high terrace overlooking Paint Creek.

Figure 6.15. Paint Valley Sites: Spruce Hill

Spruce Hill is the largest hilltop enclosure in Ohio. A stone wall extends across the 'isthmus' of the enclosure and along sections of the hill's perimeter. Diagnostic bladelets and other artifacts indicate that the site was "built and used by Ohio Hopewell" (Ruby 2009:61).

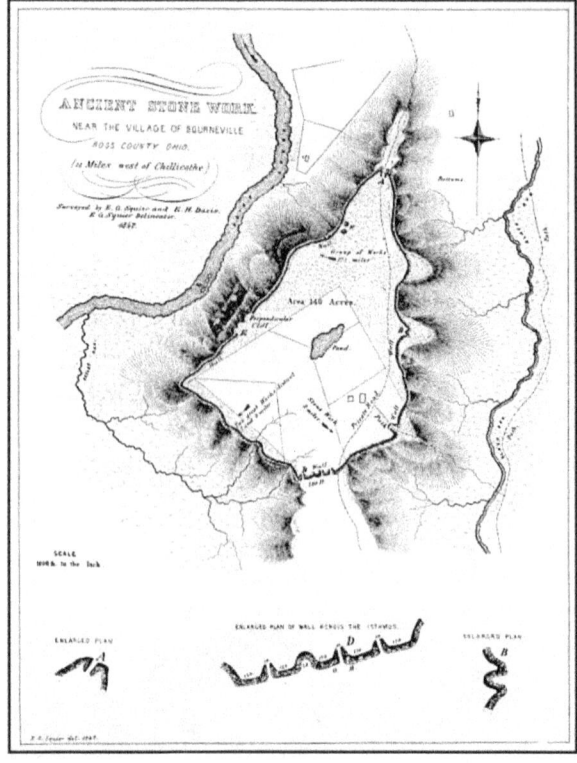

Above: aerial view of Spruce Hill
Left: Squier and Davis (1848:Pl. IV) map of Spruce Hill.
Below: LiDAR image of Spruce Hill.

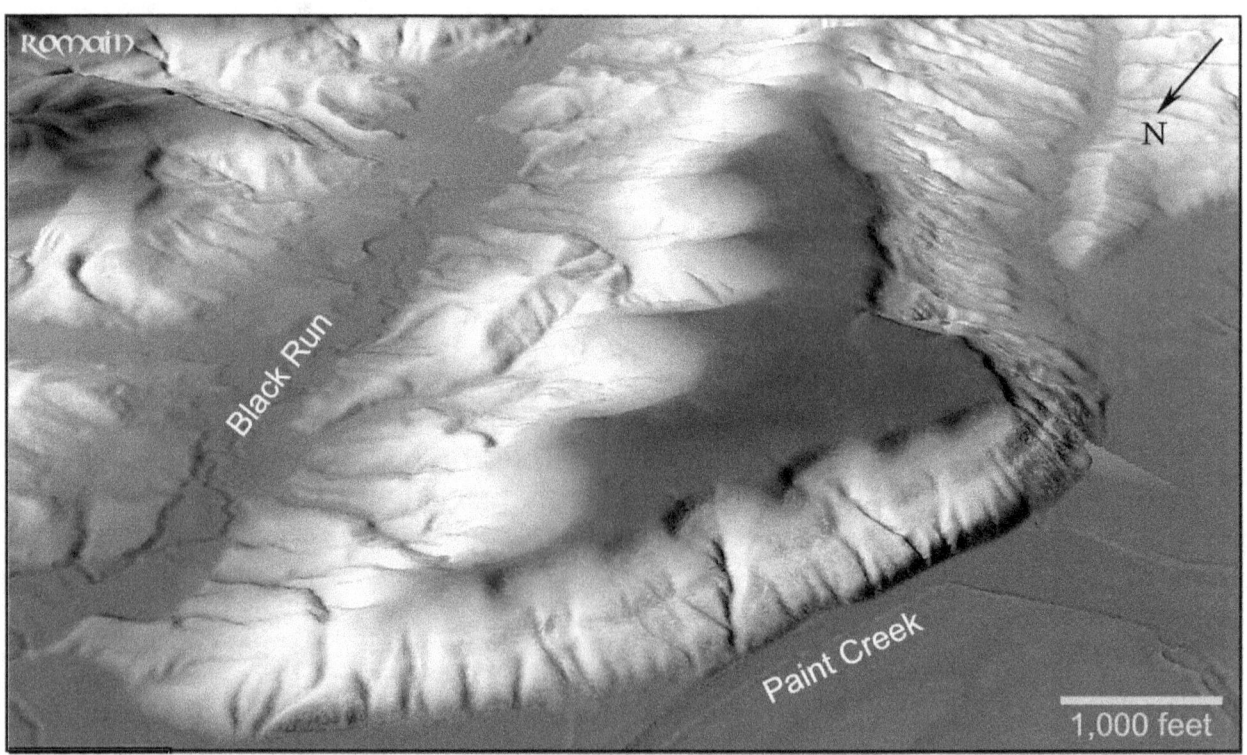

Figure 6.16. Paint Valley Sites: Spruce Hill Geology

Among the interesting features of Spruce Hill are the concretions discussed by Squier and Davis (1848:13) and shown on their map. Presumably, these concretions tumbled down from the Spruce Hill cliff. Unfortunately they are no longer present in Paint Creek. Large concretions are, however, still found in the cliffs of Copperas Mountain - about 4 miles upstream, across from the Seip earthwork.

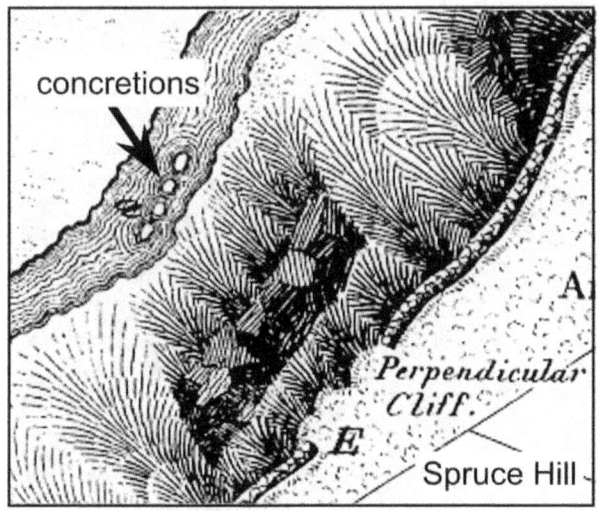

Above: enlarged detail of Squier and Davis (1848:Pl. IV) map showing perpendicular cliff and concretions (annotation added).
Left: view of cliff shown by Squier and Davis. The cliff is comprised of loose shale.

Above left: concretion in shale cliff at nearby Copperas Mountain.
Above right: group of 55 concretions found in a burial mound at the Snake Den Group in adjacent Pickaway County (Moorehead 1899:fig. 3) - suggesting that concretions were considered special by the Moundbuilders.

Figure 6.17. Paint Valley Sites: Baum

Baum is one of five earthwork complexes that share similarities in design attributes to include size and component shapes. The site is located on the south side of Paint Creek, immediately west of Spruce Hill. Much of the site has been destroyed by plowing. Sections of the earthwork walls, however, are still visible in LiDAR imagery.

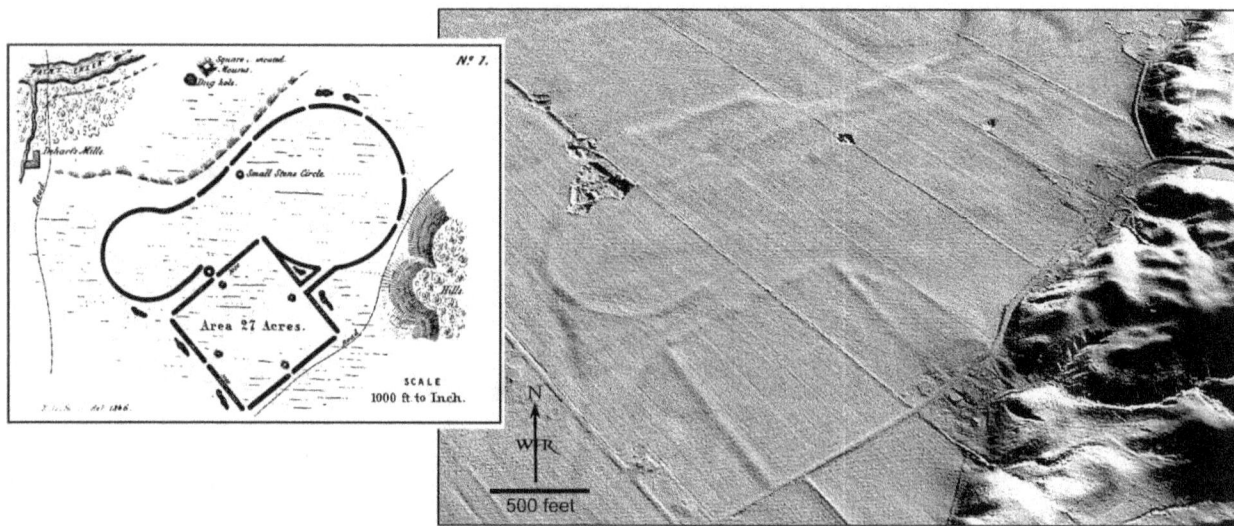

Above left: Squier and Davis (1848:Pl. XXI, no. 1) map of the Baum Earthworks.
Above right: LiDAR image showing Baum Earthwork.

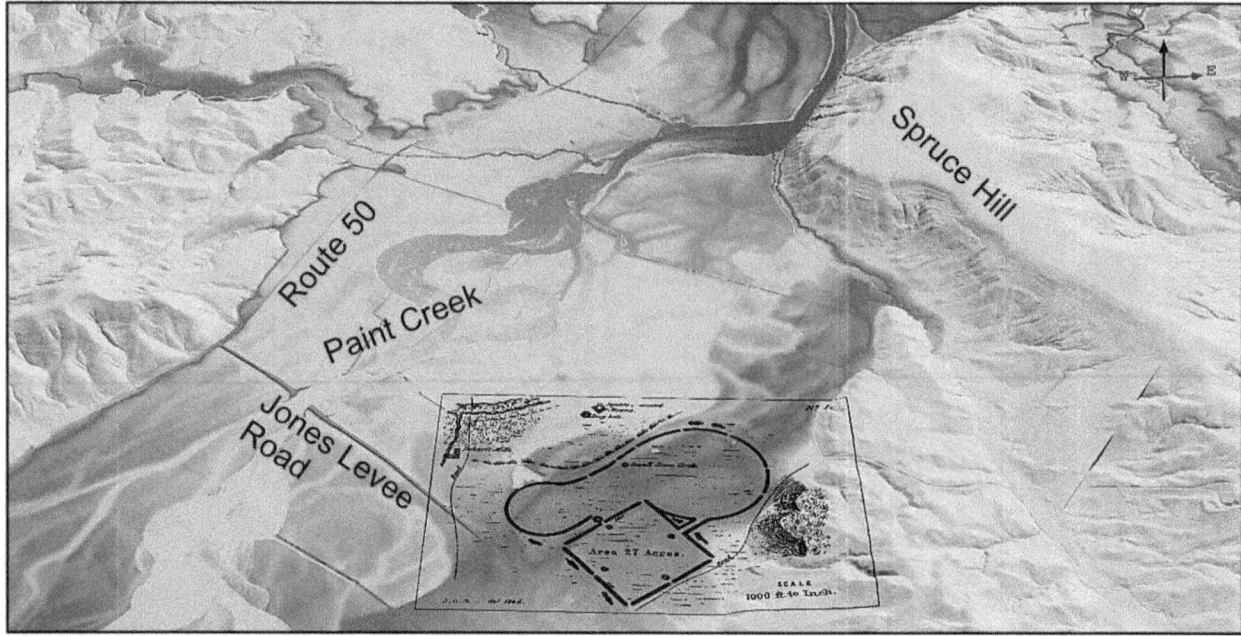

Above: Squier and Davis (1848:Pl. XXI, no. 1) map of the Baum Earthworks over LiDAR image. Notably, Baum is nestled on a narrow strip of high ground between the hills to the east and the Paint Creek floodplain to the west. Presumably, this location afforded some protection from flooding. Flooding is a problem in this area, with nearby Route 50 at Jones Levee Road sometimes closed due to high water (e.g., The Highland County Press 12-April-2011).

Figure 6.18. Paint Valley Sites: Baum Alignment and HMU

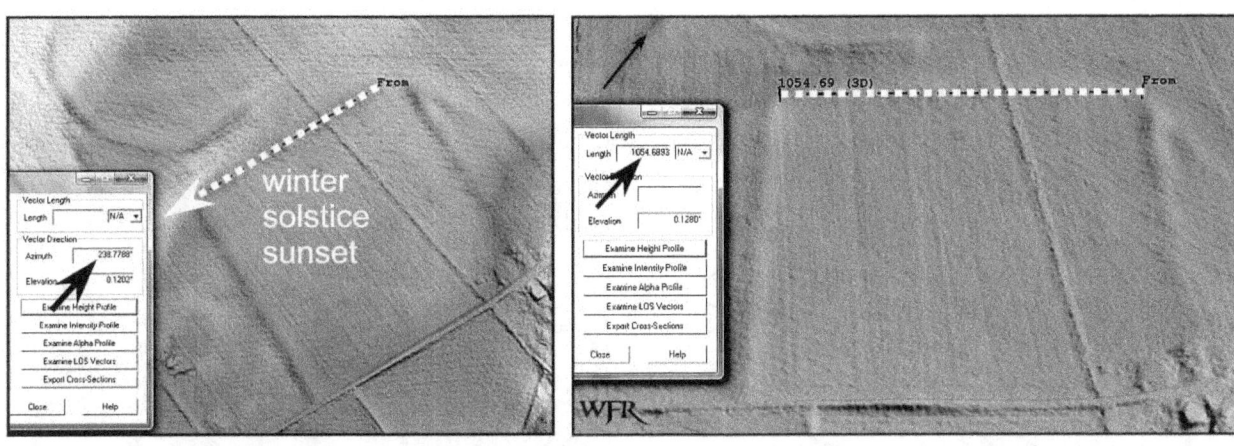

Above left: the Baum Square is aligned to the winter solstice sunset to within 1 degree. (Calculated winter solstice sunset az = 237.57 + 0.41 theta corr = LiDAR az = 237.98 - 238.78 = 0.8; note 12).
Above right: the inside to inside dimensions of the Baum Square are equal to 1 HMU (or 1,054 feet) in length.
Below: the Baum Large Circle, Small Circle, and Square all incorporate the 1 HMU (1,054-foot) length in their designs.

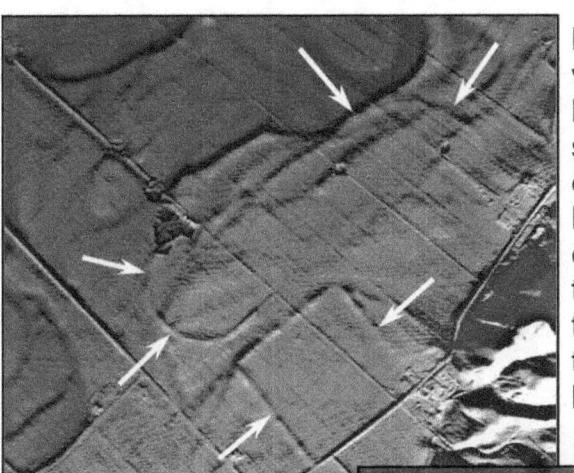

Left: LiDAR image of Baum. Arrows mark visible remnants of earthwork walls.
Below center: superimposed circles and square show the size and locations of earthwork components.
Below right: The Baum Large Circle circumcribes a 1 HMU Square. The Square, in turn, defines the size of the Small Circle. Since the sides of the Square are 1 HMU, it follows that the size of the Small and Large Circles are based on the 1 HMU length. Figures to scale.

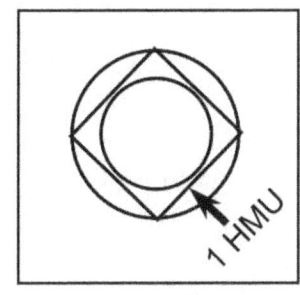

Figure 6.19. Paint Valley Sites: Seip

Seip is located on the north side of Paint Creek, about 3 1/2 miles upstream from Baum. Much of the site has been destroyed by plowing. However, sections of the earthwork walls are still visible in LiDAR imagery. Several excavations have been carried-out at Seip (note 13).

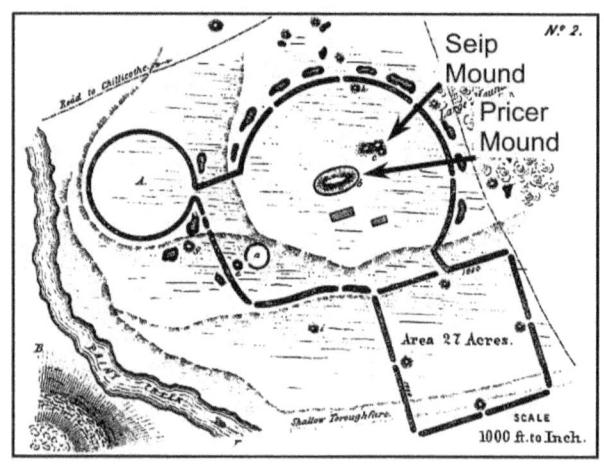 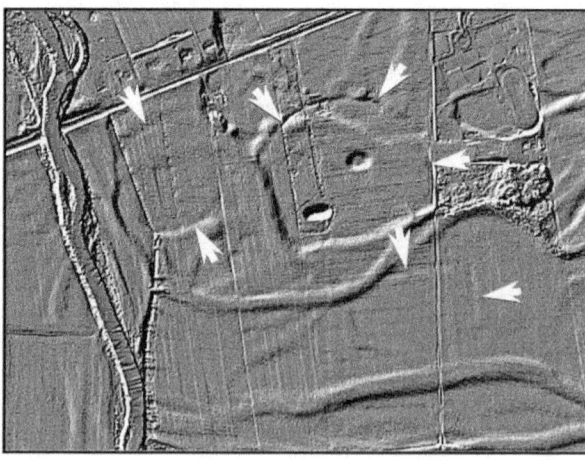

Above left: Squier and Davis (1848:Pl. XXI, no. 2) map of the Seip Earthworks.
Above right: LiDAR image showing remnants of the Seip Earthworks.
Below: Squier and Davis (1848:Pl. XXI, no. 2) map of Seip over LiDAR image.

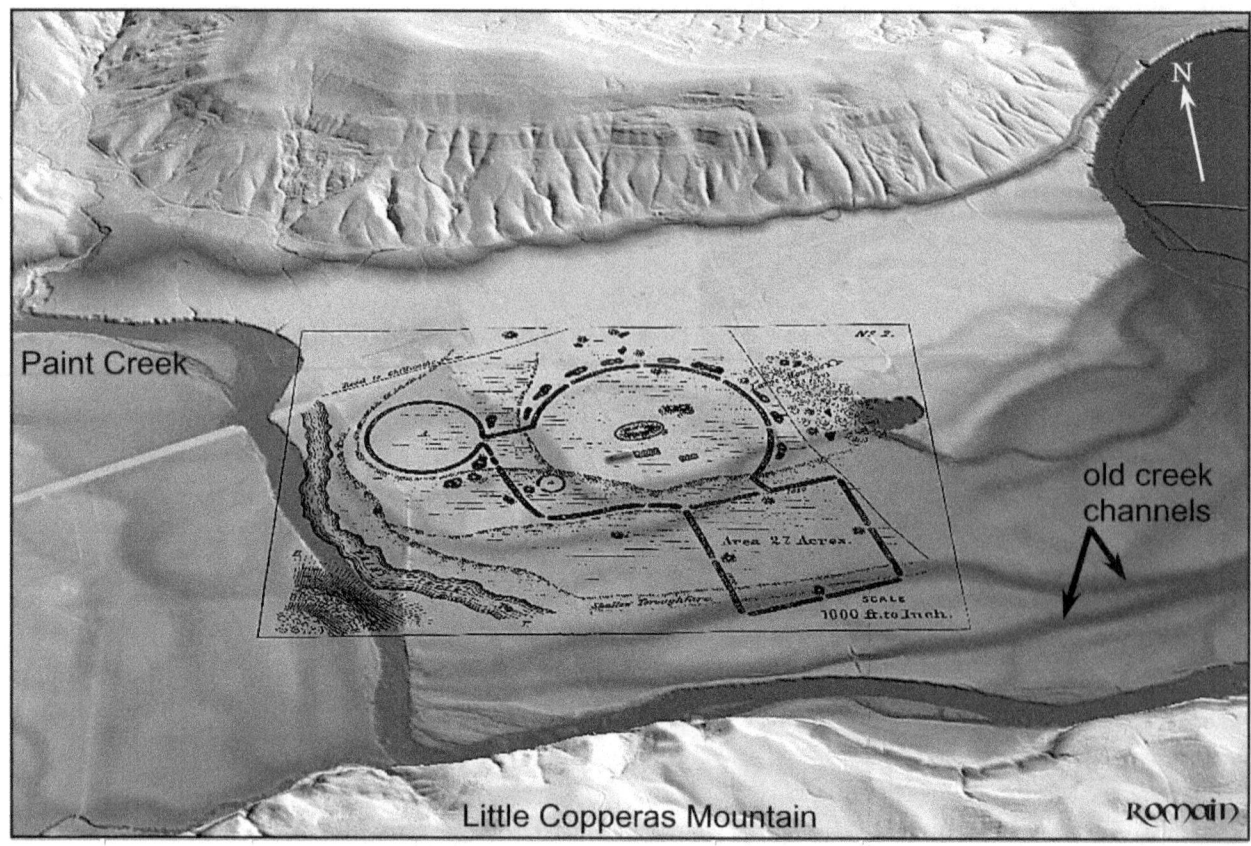

Figure 6.20. Paint Valley Sites: Seip HMU

The dimensions of the Seip Earthworks appear based on the 1 HMU (or 1,054-foot) length (note 14).

Above: using LiDAR-visible wall remnants as a guide, the configuration of the original earthwork can be established. The Seip Square seems intended as a 1 HMU square, the Small Circle nests inside a 1 HMU square, and the Large Circle circumscribes a 1 HMU square.

Below: the Seip Square is oriented to the winter solstice sunrise (note 15).

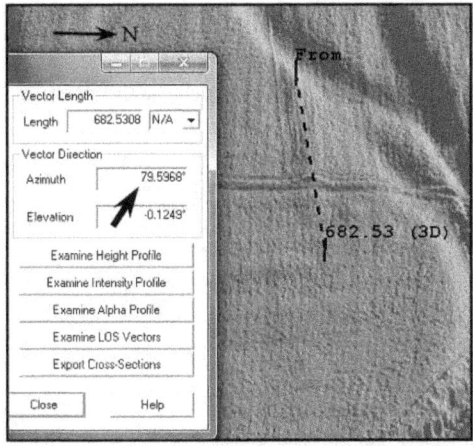 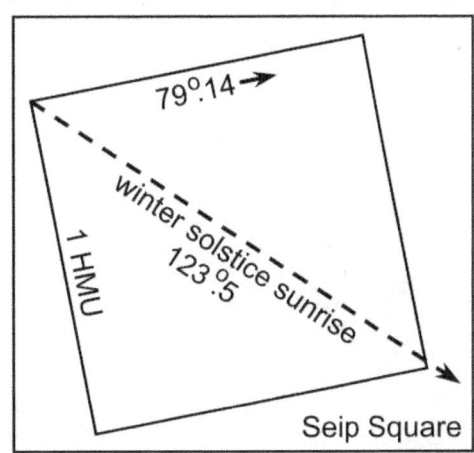

Above left: only a few sections of the Seip Square walls are visible in LiDAR imagery. One of these sections is part of the north wall. The theta corrected true azimuth for that wall section is 79.14 degrees (79.59 - 0.45 theta corr = 79.14). This closely matches the azimuth of 79.22 provided by Thomas (1894:488-489). Using the known LiDAR azimuth as a reference, a 1 HMU square is constructed. The resulting diagonal of this square extends along an azimuth of 124.14. The calculated azimuth for the winter solstice sunrise is 123.45. Thus the alignment of the posited Seip Square is to within about 0.7 degrees.

Figure 6.21. Paint Valley Sites: Seip Pricer Mound

At least 18 mounds have been found in and around the Seip Works. The largest was the Pricer Mound (note 16). Of interest is the strong visual resemblance between the Pricer Mound and the outline of Spruce Hill as seen from the Bourneville Circle, about 5 miles distant.

Above left: view of Spruce Hill from the Bourneville Circle.
Above right: view of Pricer Mound at Seip.

Above: viewed from the Bourneville Circle, the winter solstice sun will rise over Spruce Hill (note 17). Map courtesy of MyTopo, annotation added.

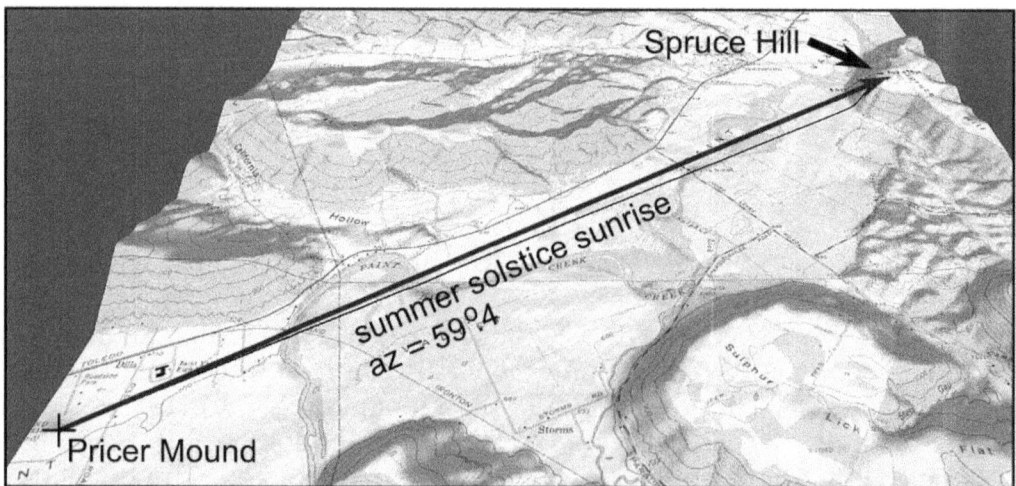

Above: viewed from the Pricer Mound, the summer solstice sun will rise over Spruce Hill and along the Paint Creek Valley (note 18). Map courtesy of MyTopo, annotation added.

Figure 6.22. The Paint Creek and Chillicothe Group of Five

There are five earthworks in the Chillicothe area that share remarkable similarities. Two are in the Scioto River Valley, two are in the Paint Creek River Valley, and one is in the North Fork Paint Creek River Valley. All five square enclosures are the same size. Four of the small circles are the same size and four of the large circles are the same size. The dimensions of the circles and squares that comprise these works are based on the HMU (1,054 ft.) length. Map figures below from Squier and Davis (1848:Pls. XX and XXI), figures to scale, annotation added.

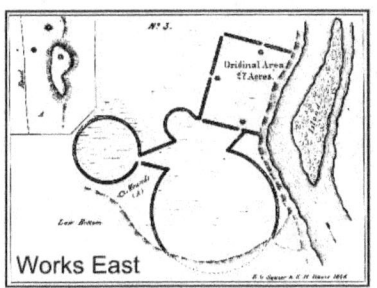 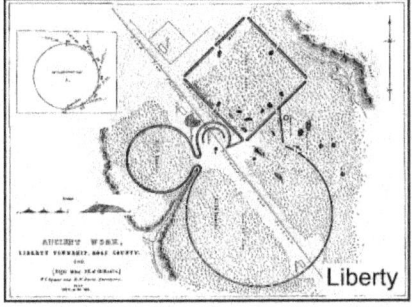 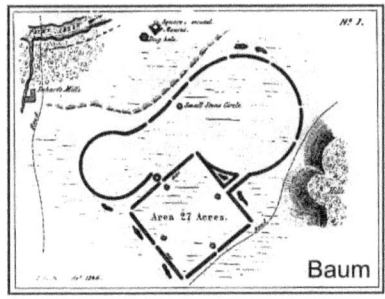

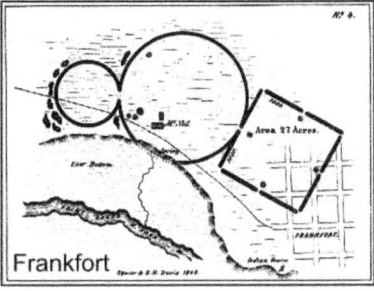 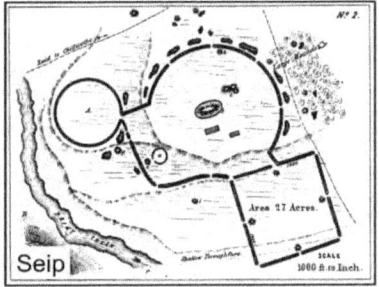

Below: schematic drawings showing how the earthwork components comprising the group of five are all related to the HMU (1,054 feet) length. (Also see Romain 1992a, 1996, 2000). With reference to Liberty, recall from earlier discussions how the 1,178-foot length is derived from the HMU length.

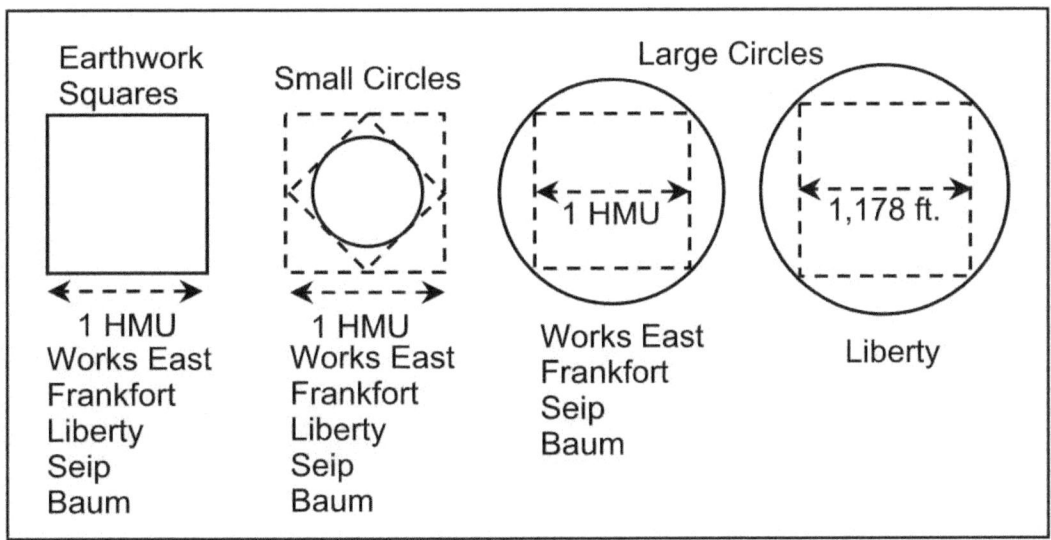

Figure 6.23. Inter-site Relationships
Several Chillicothe area earthworks are situated on inter-site lunar azimuths.

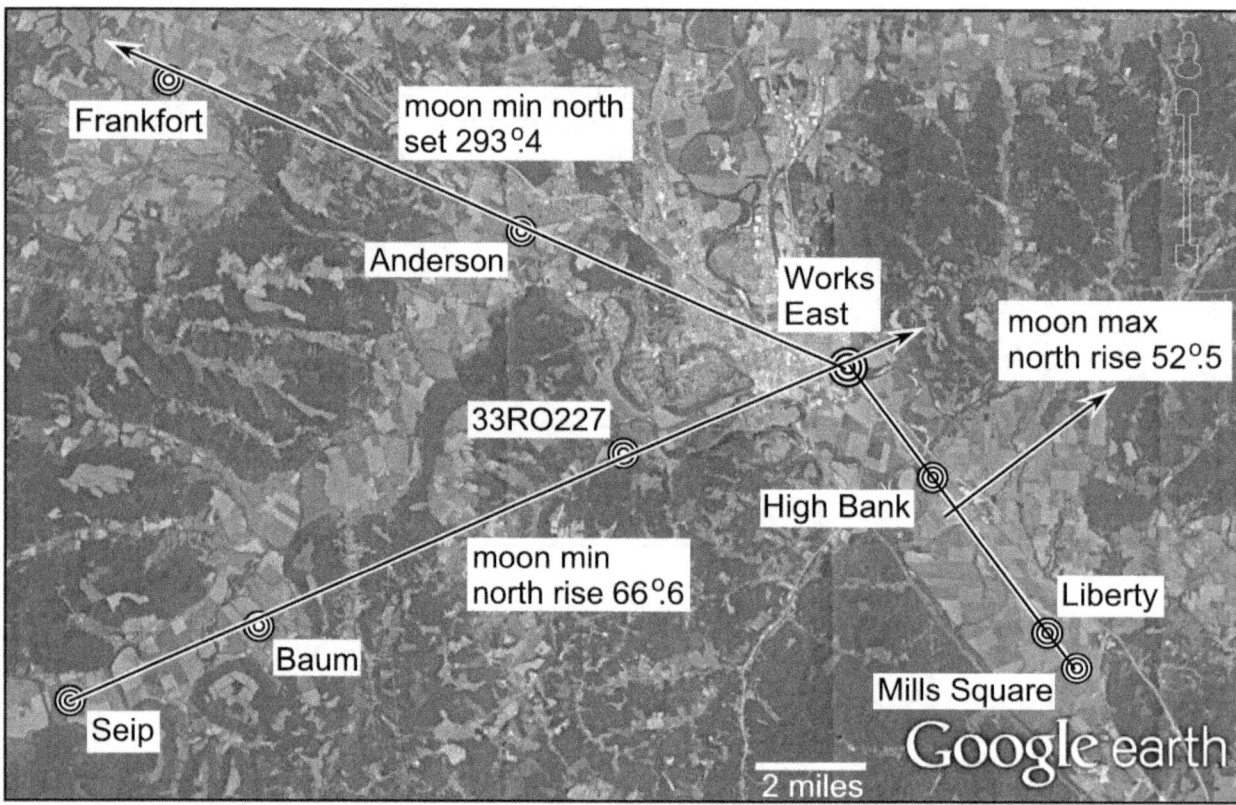

Above: a considerable number of celestial inter-site alignments have been suggested for the Chillicothe area earthworks (see e.g., Romain 1992b, DeBoer 2010, Hively and Horn 2010). The image presented here shows the inter-site alignments that I believe were either intentional or recognized by the Adena-Hopewell (note 19).

CHAPTER 7
BRUSH CREEK WATERSHED

Ohio Brush Creek is a 60-mile long tributary of the Ohio River. It drains about 435 square miles (Figure 7.1). The watercourse originates in southeastern Highland County and flows generally southward to its confluence with the Ohio River. Several tributaries feed into Ohio Brush Creek, including Baker Fork. Baker Fork originates in the northwest corner of Pike County, in a place called Whiskey Hollow. From Whiskey Hollow, Baker Fork passes through an area known as Beech Flats, past Fort Hill, and south to where it meets Ohio Brush Creek. Serpent Mound is located very near to the confluence of Baker Fork and Ohio Brush Creek.

Trefoil

Little is known about the Trefoil Complex. Based on Squier and Davis's map, it appears to have been a series of conjoined circle earthworks having interior ditches and several nearby mounds. Technically, the Trefoil Complex is located in the Paint Creek River Valley. I have included it in this chapter, however, because it is at the entrance to a unique geological area known as Beech Flats (Tight 1895). Beech Flats is a glacially-formed valley. Its significance is that it connects the Paint Creek Valley to the Brush Creek Watershed and Fort Hill and Serpent Mound. This valley passageway is important, as the hills south of Paint Creek extending into Highland County are among the most rugged in Ohio. For those who wanted to transition from Paint Creek to the Fort Hill and Serpent Mound areas, the Beech Flats passageway was basically the only way in - and it was Trefoil that marked the way.

Fort Hill

Once in the Beech Flats Zone, the most prominent watercourse is Baker Fork. From its origin in Whiskey Hollow, the creek flows more or less along the longitudinal center of Beech Flats and along the west side of Fort Hill. The place where Baker Fork meets Fort Hill is quite spectacular, with stone arch bridges and deep gorges. No doubt the impressive scenery at this location contributed to the appreciation of Fort Hill as a special place. Adding to that, however, is that, viewed from the north lookout promontory of Fort Hill, the summer solstice sun can be seen to rise along the length of the Beech Flats Valley. As Figure 7.5 shows, the summer solstice sightline cuts right through the heart of Whiskey Hollow - the point of origin for the Baker Fork Creek. Fort Hill is in effect, a connecting point for earth, sky, and water through Baker Fork, Whiskey Hollow, and the summer solstice sunrise.

Serpent Mound Meteorite Impact Crater

When I said that Trefoil marks the way into a special area, I meant just that. About 2 miles south of Fort Hill, another unusual feature is found. Sometime between 250 - 330 million years ago, a meteorite slammed into the area, resulting in a 7-mile diameter crater known as the Serpent Mound Impact Crater (Milam 2010). Although it is doubtful that the Moundbuilders surmised that a meteorite hit the area, they certainly would have been aware of the unusual results. The rock formations within the crater have been visibly uplifted, vertically-tilted, and down-dropped rock. Indeed the peculiar nature of the area led geologist John Locke in 1838 to observe:

> "...it became evident that a region of no small extent had sunken down several hundred feet, producing faults, dislocations and upturning of the layers of the rocks.... On traveling from Locust Grove to Sinking Spring, I found that a tract large enough for a township, reaching within a mile of Sinking Spring and extending several miles up Straight Creek, was in the same manner dislocated and sunken about four hundred feet....it is evident that this mountain at some remote period of time, has sunk down from its original place, and I venture to call it the Sunken Mountain"(Locke 1838:266-267).

It may be that Serpent Mound was located where it is due to the unusual nature of the area - suitable as a place of emergence from the Lowerworld for the Great Serpent of Indian myth and legend. Further contributing to the special character of the Serpent Mound area are the occurrence of large sinkhole fields. As a matter of fact, three sinkholes are found at Serpent Mound park itself. Recent electrical resistivity ground imaging (Zaleha and Romain 2012) shows that one of these sinkholes leads to a silt-filled void, or cave.

Serpent Mound

As to Serpent Mound, there is in the profile view of the Serpent Mound promontory, looking from Brush Creek up to the Serpent, an uncanny resemblance of that landform to the profile of a real serpent. William Holmes's (1886:627) remarks describing this visual heirophany deserve full quotation:

> "Having the idea of a great serpent in the mind, one is at once struck with the remarkable contour of the bluff, and especially of the exposure of rock, which readily assumes the appearance of a colossal reptile lifting its front from the bed of the stream. The head is the point of rock, the dark lip-like edge is the muzzle, the light-colored under side is the white neck, the caves are the eyes, and the projecting masses to the right are the protruding coils of the body. The varying effects of light must greatly increase the vividness of the impressions, and nothing would be more natural than [to]...recognize this likeness and...regard the promontory as a great manito."

Figure 7.8 shows the resemblance of the Serpent Mound promontory to the head of a snake.

Figure 7.1. Ohio Brush Creek Watershed and Beech Flats.

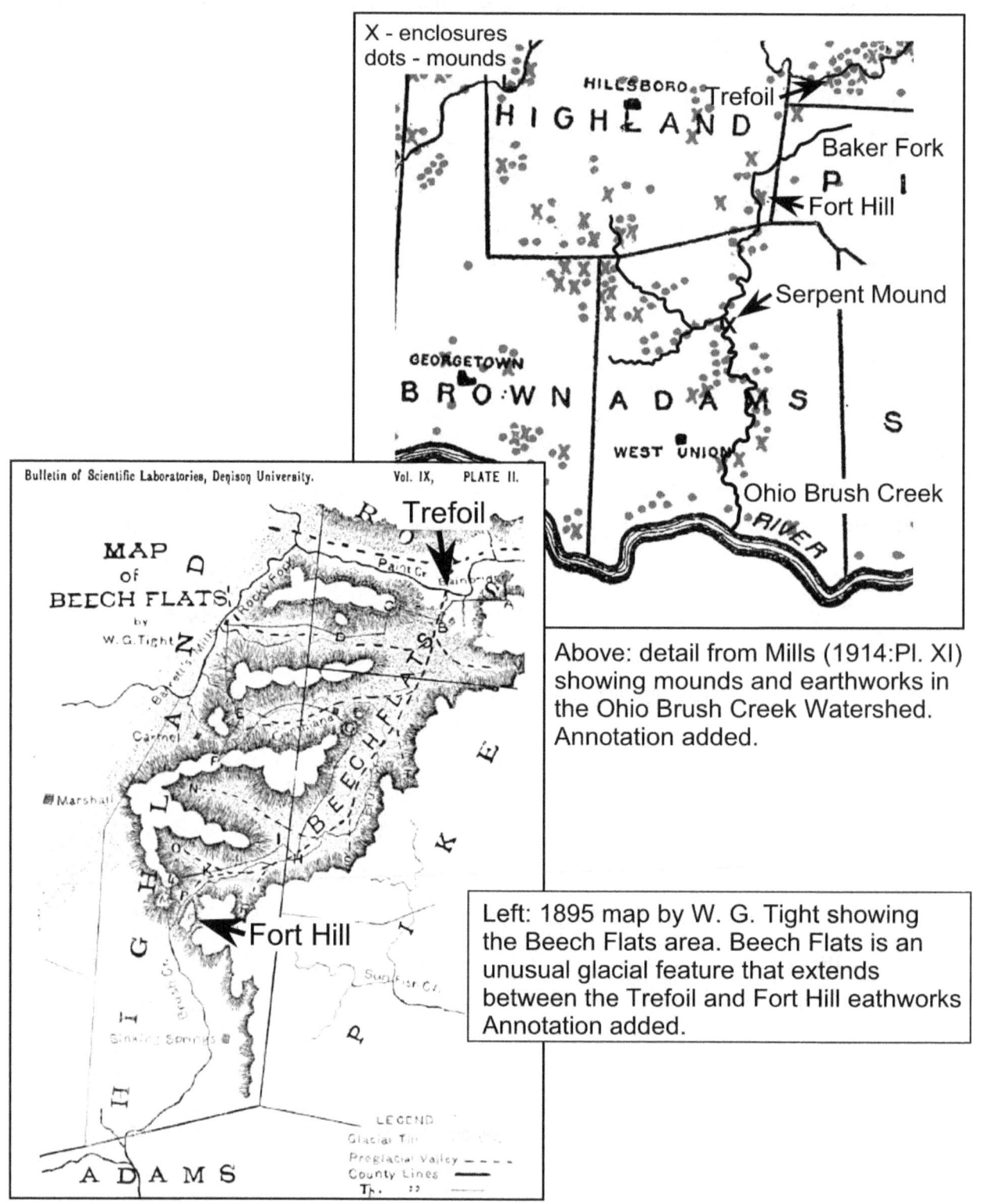

Above: detail from Mills (1914:Pl. XI) showing mounds and earthworks in the Ohio Brush Creek Watershed. Annotation added.

Left: 1895 map by W. G. Tight showing the Beech Flats area. Beech Flats is an unusual glacial feature that extends between the Trefoil and Fort Hill eathworks Annotation added.

Figure 7.2. Trefoil

Trefoil is located about one mile west of Bainbridge, on the south side of Paint Creek. The site appears to have been a series of conjoined circle earthworks with a set of parallel walls leading to another circle. To the east of the circles were seven mounds. None of the circles are visible today. However, the largest of the mounds - i.e., the Campbell Mound is still mostly intact. It is not known to have been excavated.

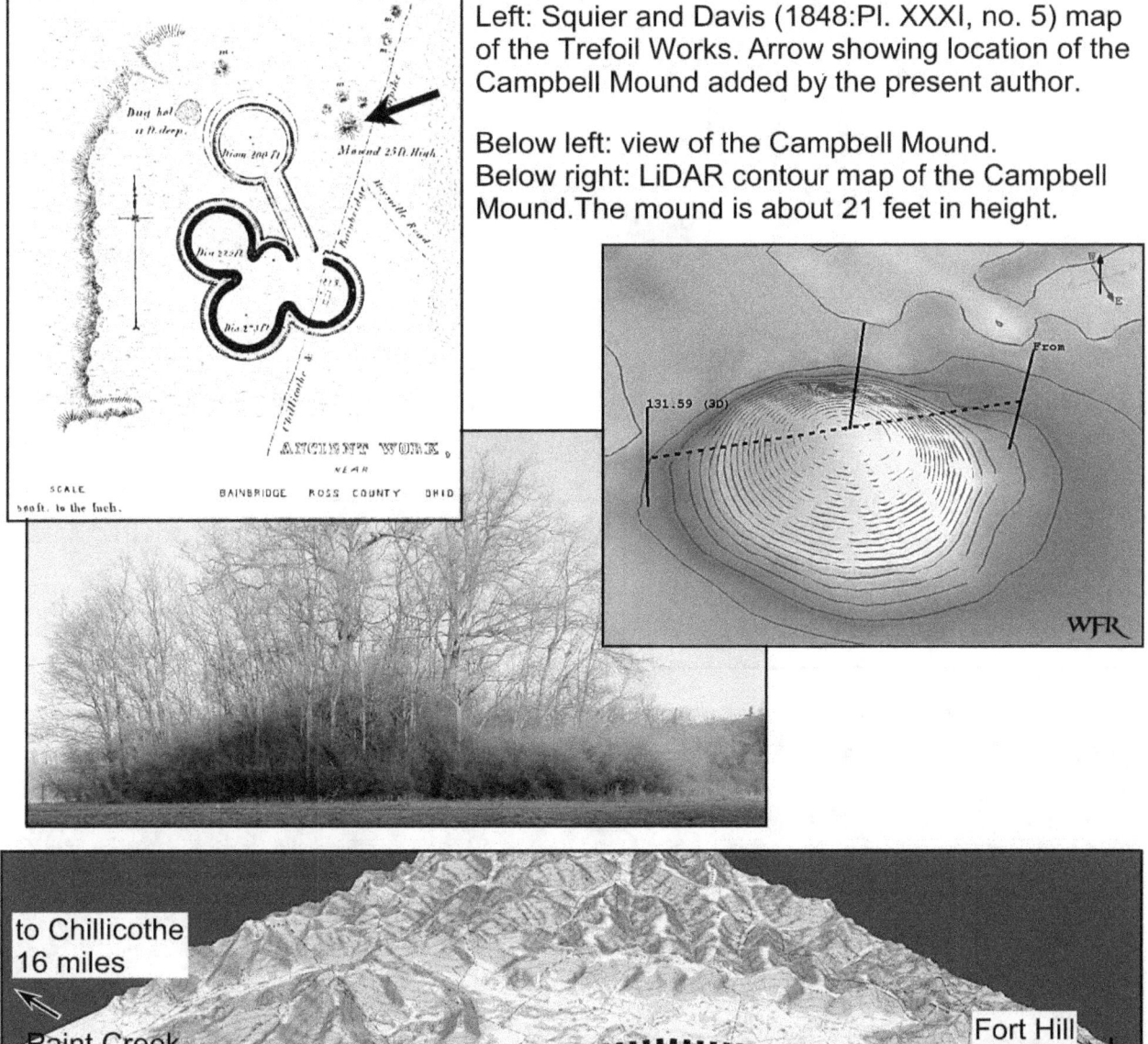

Left: Squier and Davis (1848:Pl. XXXI, no. 5) map of the Trefoil Works. Arrow showing location of the Campbell Mound added by the present author.

Below left: view of the Campbell Mound.
Below right: LiDAR contour map of the Campbell Mound. The mound is about 21 feet in height.

Above: Trefoil marks the easiest approach from Paint Creek Valley into the Fort Hill area - about 10 miles distant. Map provided by MyTopo, annotation added.

Figure 7.3. Brush Creek - Baker Fork: Fort Hill.

Fort Hill is a hilltop enclosure. The site is situated in Highland County, about 26 miles southwest of Chillicothe. An earthen wall with interior ditch extends around the near top of the hill (note 1). A unique promontory extends to the north and may have played an important role in how this site was used.

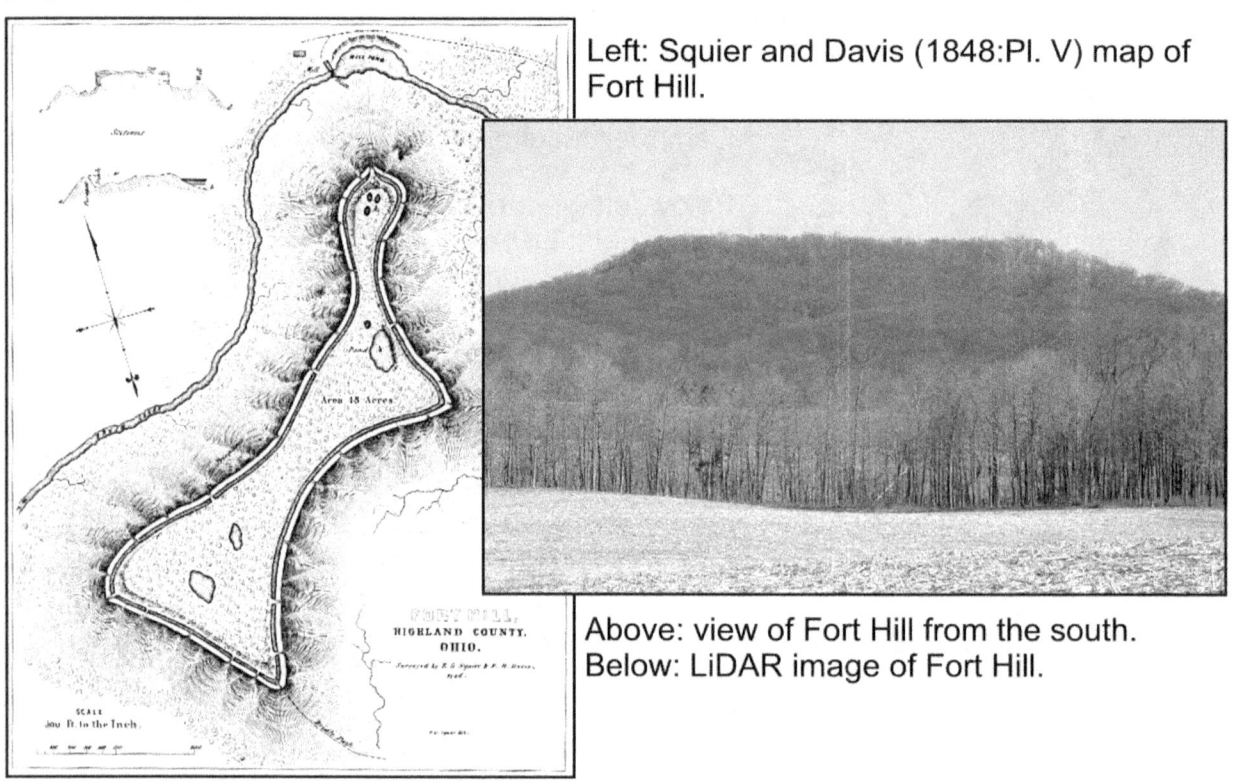

Left: Squier and Davis (1848:Pl. V) map of Fort Hill.

Above: view of Fort Hill from the south.
Below: LiDAR image of Fort Hill.

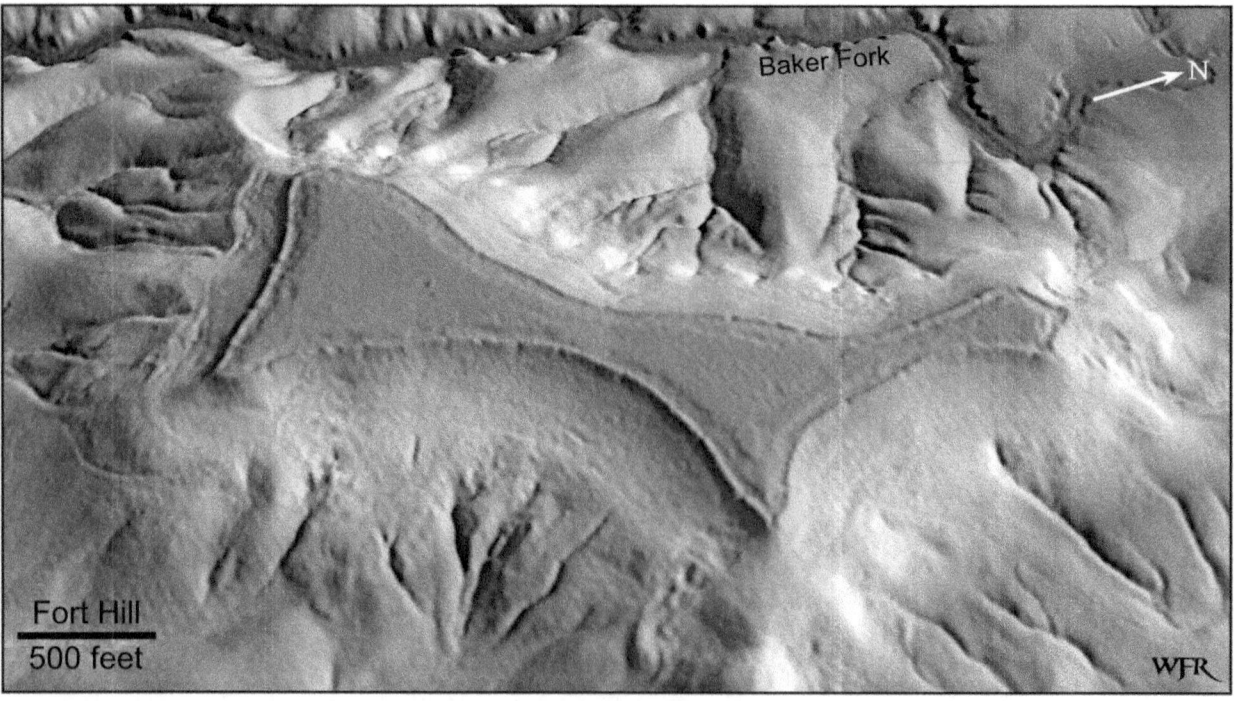

196

Figure 7.4. Brush Creek - Baker Fork: Fort Hill Alignment

Fort Hill is situated at the southwest terminus of an unusual geological region known as Beech Flats. Formed by glacial still water deposits (Wright 1890:92), the area is quite flat - hence its name. Flanking the Beech Flats formation are rugged mountains.

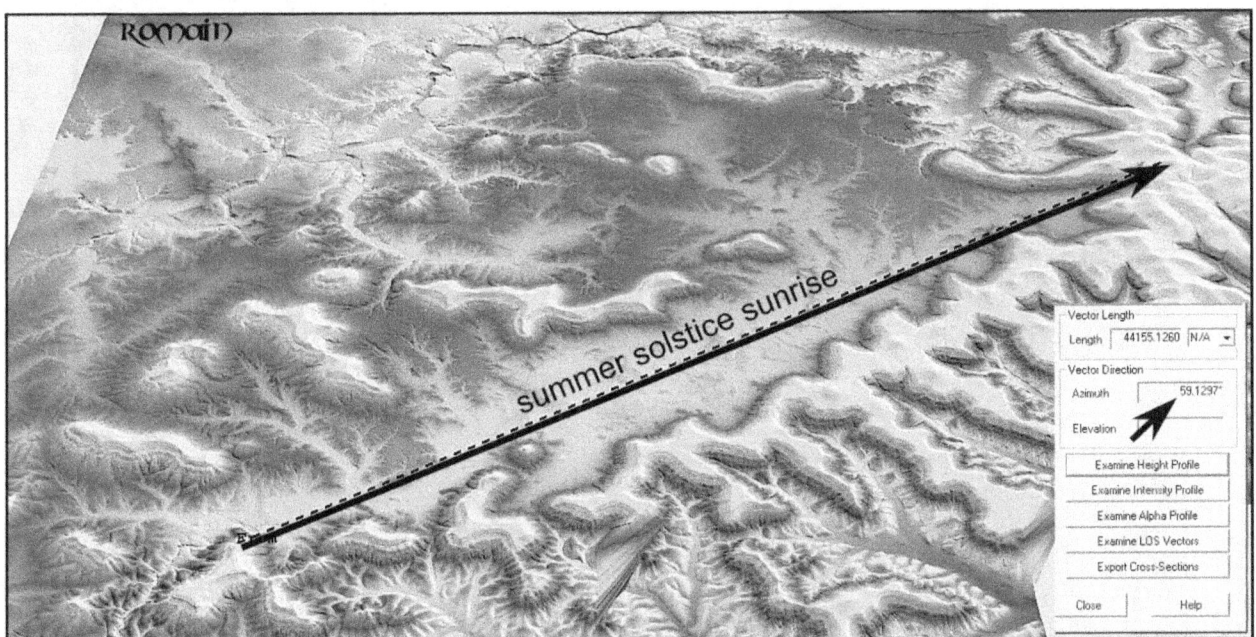

Above: LiDAR image. Viewed from the north end of Fort Hill, from the 'lookout promontory,' the summer solstice sun will rise in alignment with the Beech Flats Valley. (Calculated summer solstice sunrise az = 58.5 + 0.57 theta corr = LiDAR az = 59.1; note 2).

Below: LiDAR view of the Fort Hill 'lookout promontory.'

Above: view from the Fort Hill 'lookout promontory' in the direction of the summer solstice sunrise. Black dot shows where sunrise will occur.

Figure 7.5. Brush Creek - Baker Fork: Fort Hill Circle Alignment II

Below: enlarged details showing the Fort Hill north observation point and distant features for the summer solstice sunrise. Maps provided by MyTopo, annotation added.

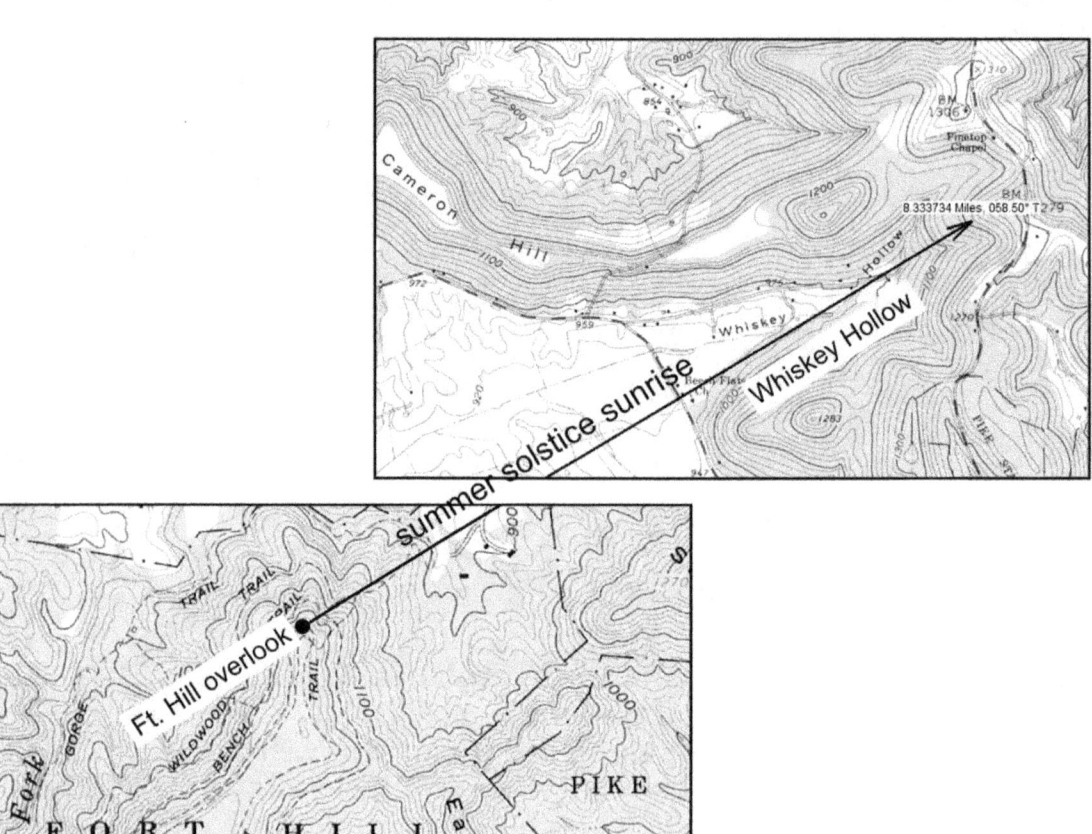

Above: Whiskey Hollow is at the northeast terminus of the Beech Flats Valley. Its significance is that it marks the direction of the summer solstice sunrise; and it is the place of origin for Baker Fork Creek. Baker Fork Creek leads past Fort Hill, to Serpent Mound.

Figure 7.6. Brush Creek - Baker Fork: Fort Hill Circle Earthworks

There are two small circle earthworks and a Hopewell habitation area located about three-quarters of a mile south of Fort Hill (note 3).

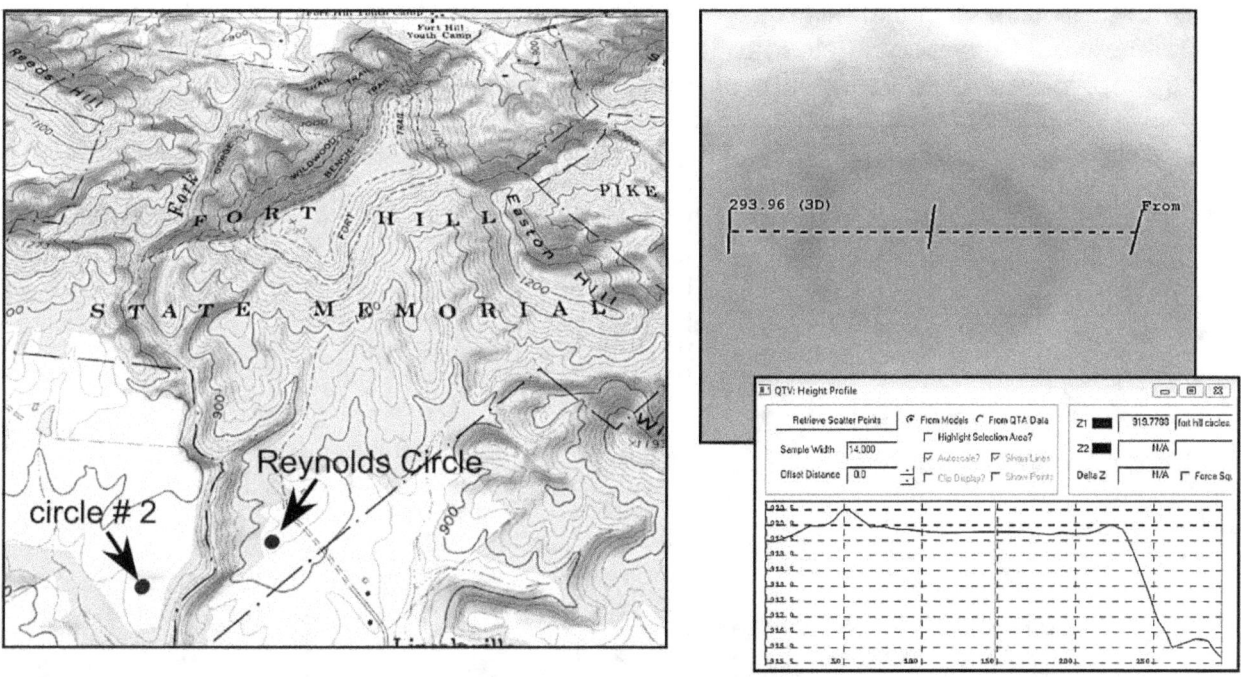

Above left: map showing location of Fort Hill circle enclosures. Map provided by MyTopo, annotation added.
Above right: LiDAR image and profile of the William Reynolds Earthwork. The walls of the circle are slight - i.e., only about 6 inches in height.
Below: the size of the Reynolds Circle is based on an iteration of the 1 HMU length - i.e., 186 feet (note 4).

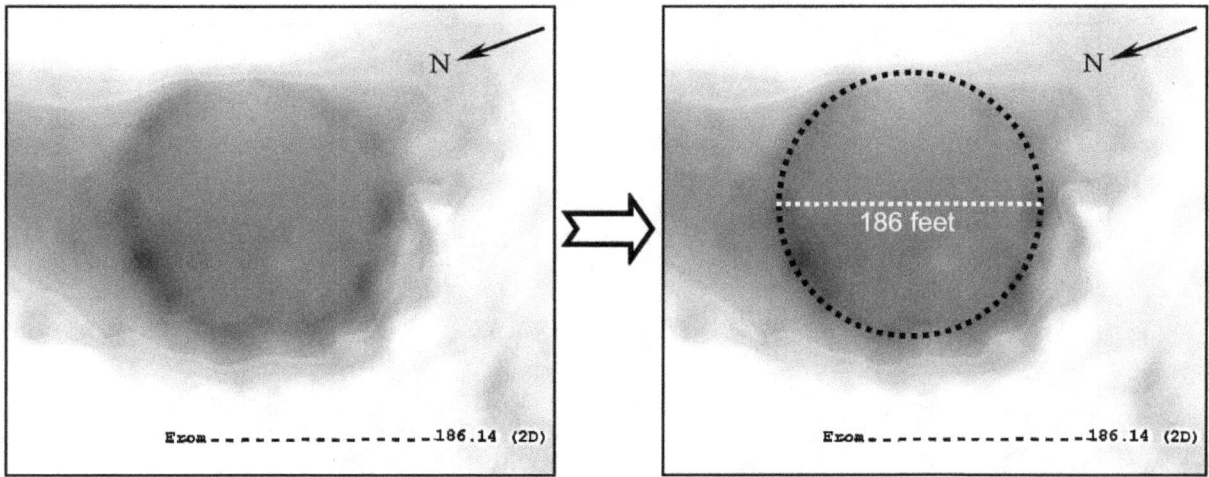

Figure 7.7. Baker Fork: Fort Hill to Serpent Mound

The nearest major site to Fort Hill is Serpent Mound - about six miles distant. Navigation between Fort Hill and Serpent Mound might have been accomplished by following the course of Baker Fork - a tributary of Brush Creek. Following this route, travellers would have come close to the center of the Serpent Mound Impact Crater central uplift area. Below left: view of Baker Fork south of Fort Hill.

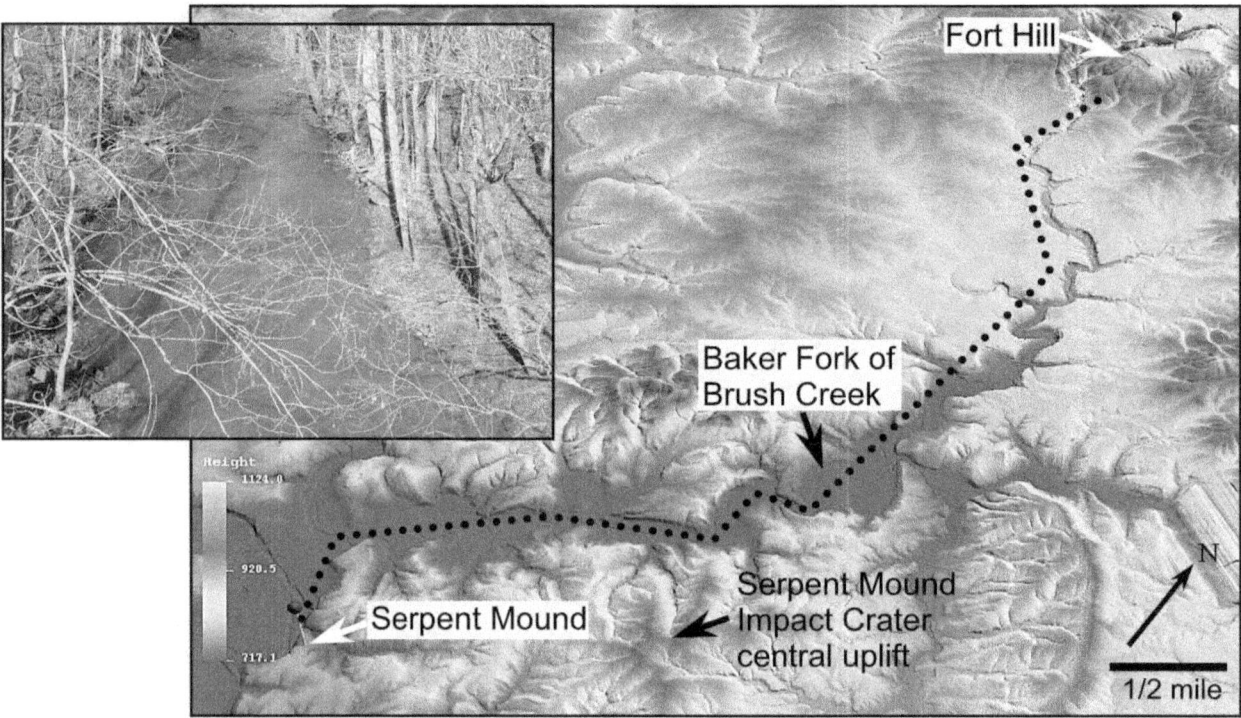

Above: LiDAR view of the area between Fort Hill and Serpent Mound.
Below: LiDAR view of the Serpent Mound Impact Crater (note 5).

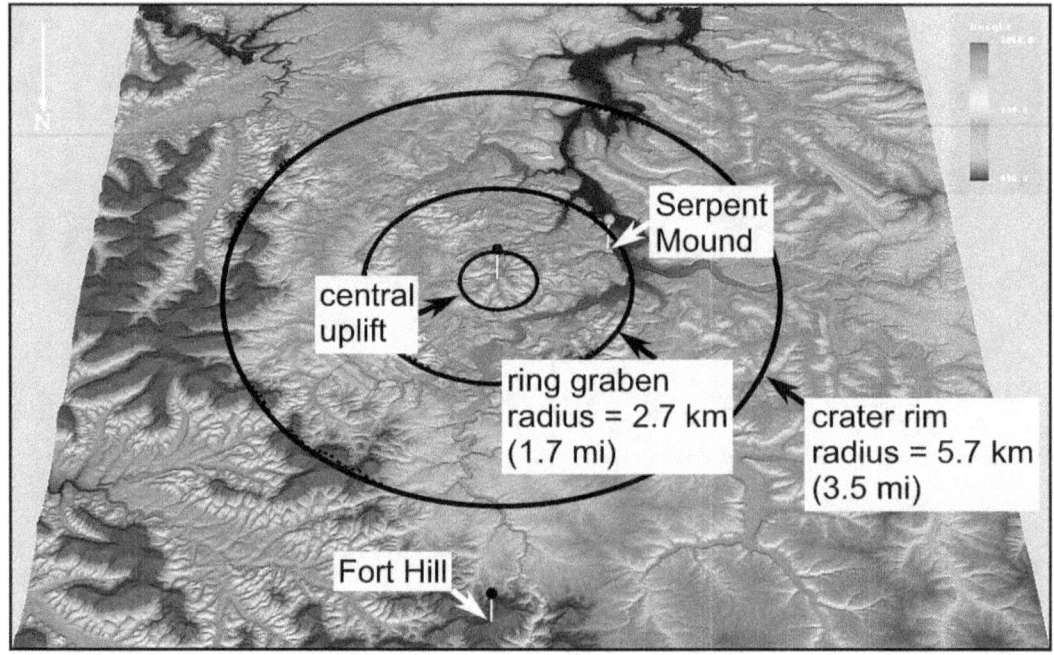

Figure 7.8. Brush Creek: Serpent Mound

Serpent Mound is situated on a promontory overlooking Brush Creek. Radiocarbon dating indicates the effigy was built during the Early Woodland period - ca. 300 B.C. (note 6).

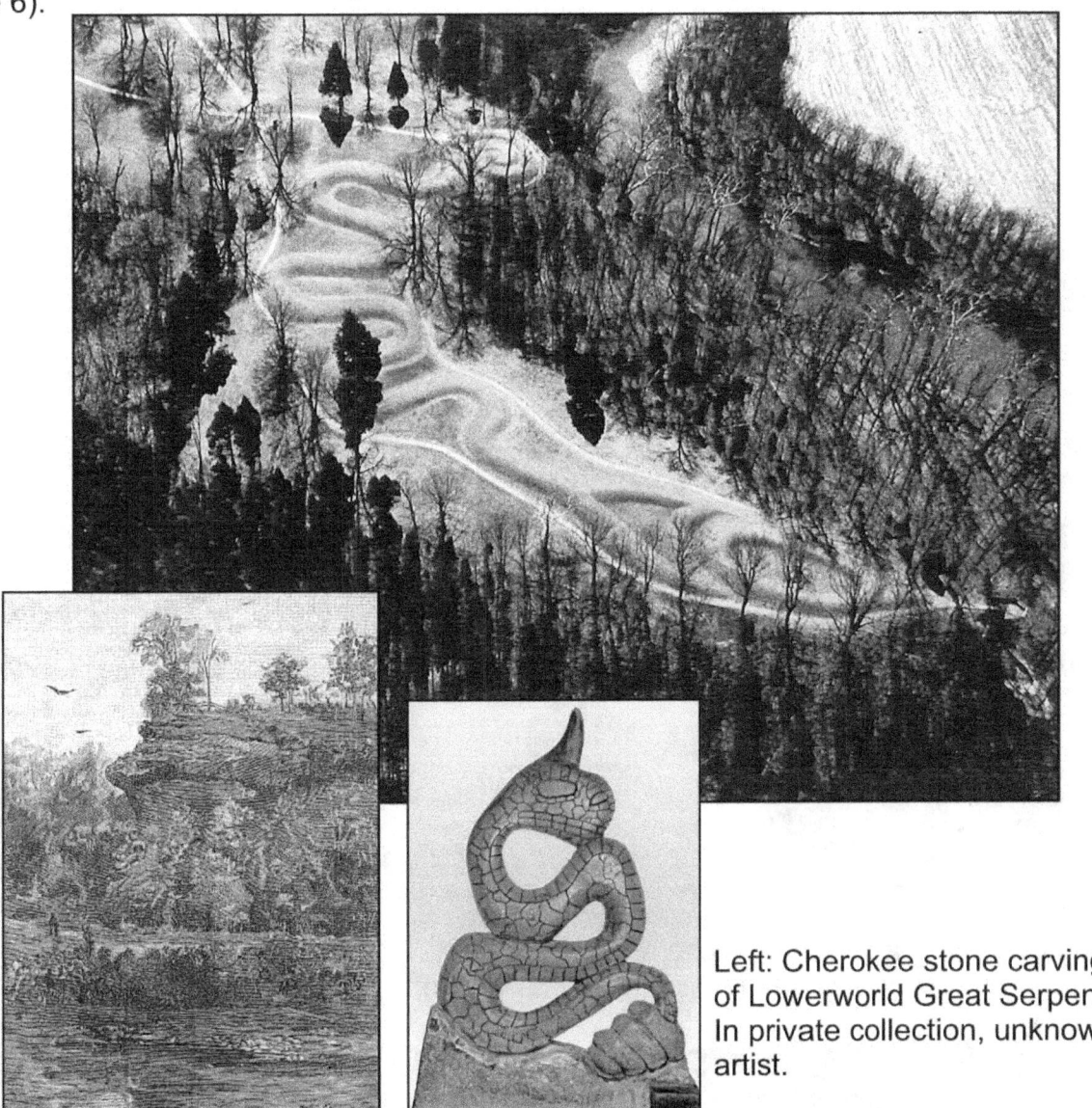

Left: Cherokee stone carving of Lowerworld Great Serpent. In private collection, unknown artist.

Many Eastern Woodlands cosmologies tell of a monstrous serpent that rules the Lowerworld (e.g., Hudson 1976, Lankford 2007a, Reilly 2011). This Serpent is an antagonist of the Upperworld Sun and Thunderbirds. It has long been suggested that Serpent Mound represents the Lowerworld Great Serpent (note 7). Building upon this, ethnologist George Lankford (2007a, 2007b) suggests that the star constellation Scorpius represents the celestial aspect of the Great Serpent. I agree with the Scorpius interpretation and would add that the oval earthwork within the jaws of the Serpent was likely meant as a sun symbol. Thus the mythic struggle between the two entities is memorialized.

Above left: sketch by Frederic Putnam (1890:871) showing resemblance of Serpent Mound cliff to the head of a monstrous serpent emerging from the earth.

Figure 7.9. Brush Creek: Serpent Mound Oval Embankment

Several lines of evidence suggest that the oval embankment is a sun symbol. As shown below, the oval is solar-aligned. Specifically the oval is aligned to the summer solstice sunset. Analysis of other posited alignments through the Serpent's body shows most other solar alignments are in error by several degrees. Posited lunar alignments are a better fit, but fail to account for all the convolutions (note 8).

Above: summer solstice sunset in alignment with the oval (note 9).

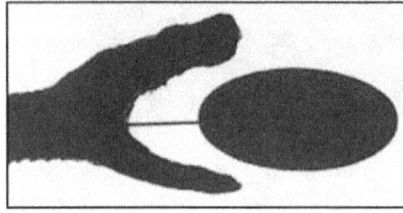

Above left: the Serpent Mound resembles the side view of a serpent attempting to bite a circle disk that has been rotated in a similar fashion as the serpent's head (note 10).

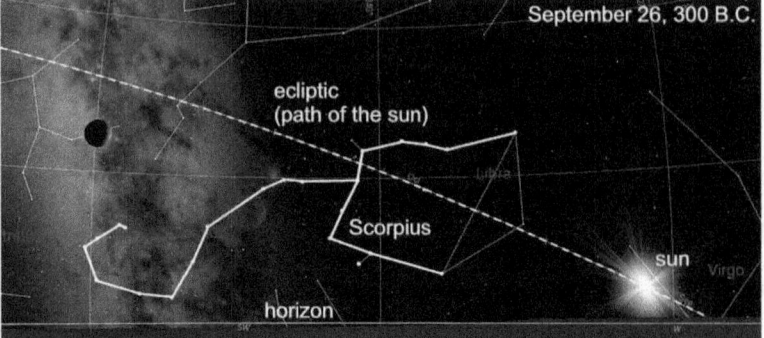

Above: the ecliptic is an imaginary line that marks the annual path of the sun. It appears higher in the summer and lower in the winter.

Above: in the summer months the constellation Scorpius chases after the sun along the ecliptic (note 11). Note that the Serpent gets closer to the sun toward the time of the autumnal equinox. Map by Starry Night, annotation added.

Figure 7.10. Brush Creek: Serpent Mound and Scorpius Rotation
The body convolutions of the Serpent Mound can be explained by reference to how Scorpius rises from the southern horizon and simultaneously rotates clockwise over the course of the summer months. Each convolution references an incremental clockwise turn. (Also note the path of the ecliptic through the Serpent's jaws.)

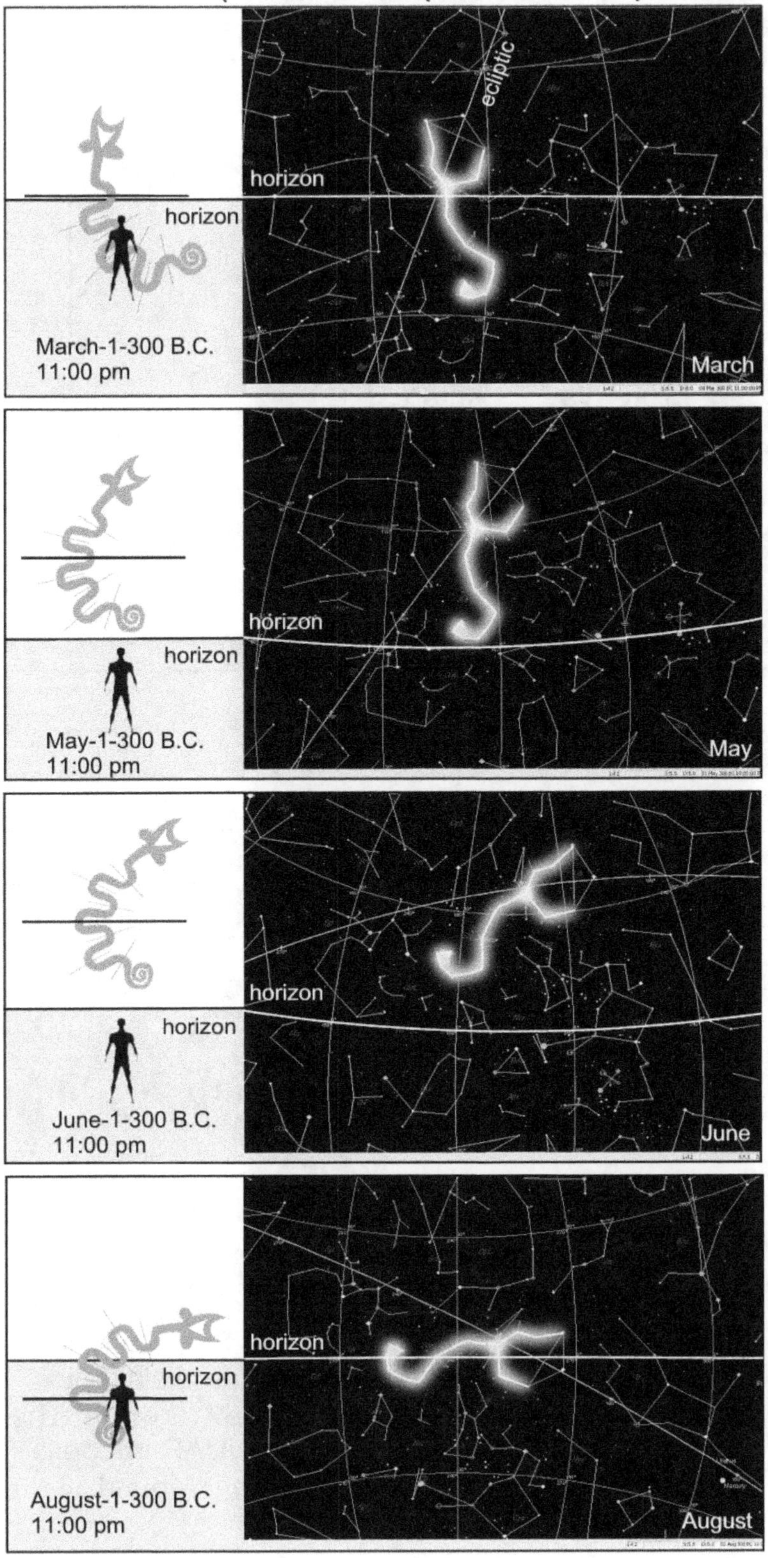

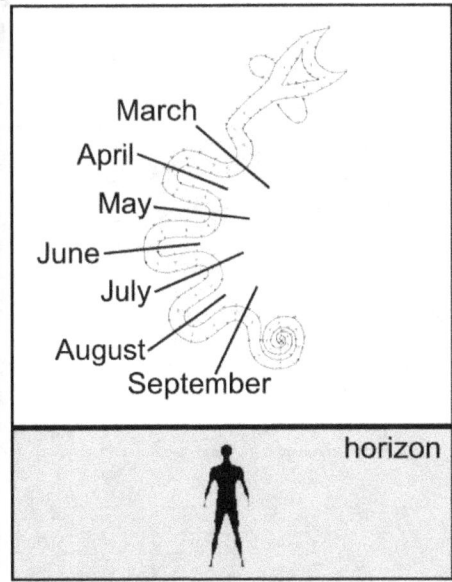

Above: in the interpretation provided here, the first step is to lift the Serpent off the flat earth surface and place it in a three dimensional sky. In the above figure the little person is looking out across the southern horizon to Scorpius (note 12). Each line that bisects a body convolution represents one month of rotation of the Serpent. The rising of Scorpius begins in March as the head of the Serpent rises from the Lowerworld. The Serpent continues its movement in an upward and clockwise manner until the end of summer - when it descends back into the Lowerworld, below the horizon. Due to space limitations not all summer months shown. Maps by SkyMap Pro, annotation added.

Figure 7.11. Brush Creek: Serpent Mound and Scorpius

Lankford's (2007a, 2007b) interpretation of Serpent Mound as a cognate for Scorpius comports well with another set of astronomical data. As shown below, at about midnight, on the night of the summer solstice, as viewed from Sugarloaf Mountain (the Hopewell axis mundi), Scorpius (and its bright star Antares) are situated at the end of the Milky Way on an azimuth of 225.9 degrees. Likewise, as plotted from Sugarloaf, the Serpent Mound effigy itself is located on the same 225.9 degrees, 37 miles to the southwest (note 13). Map by Starry Night, annotation added.

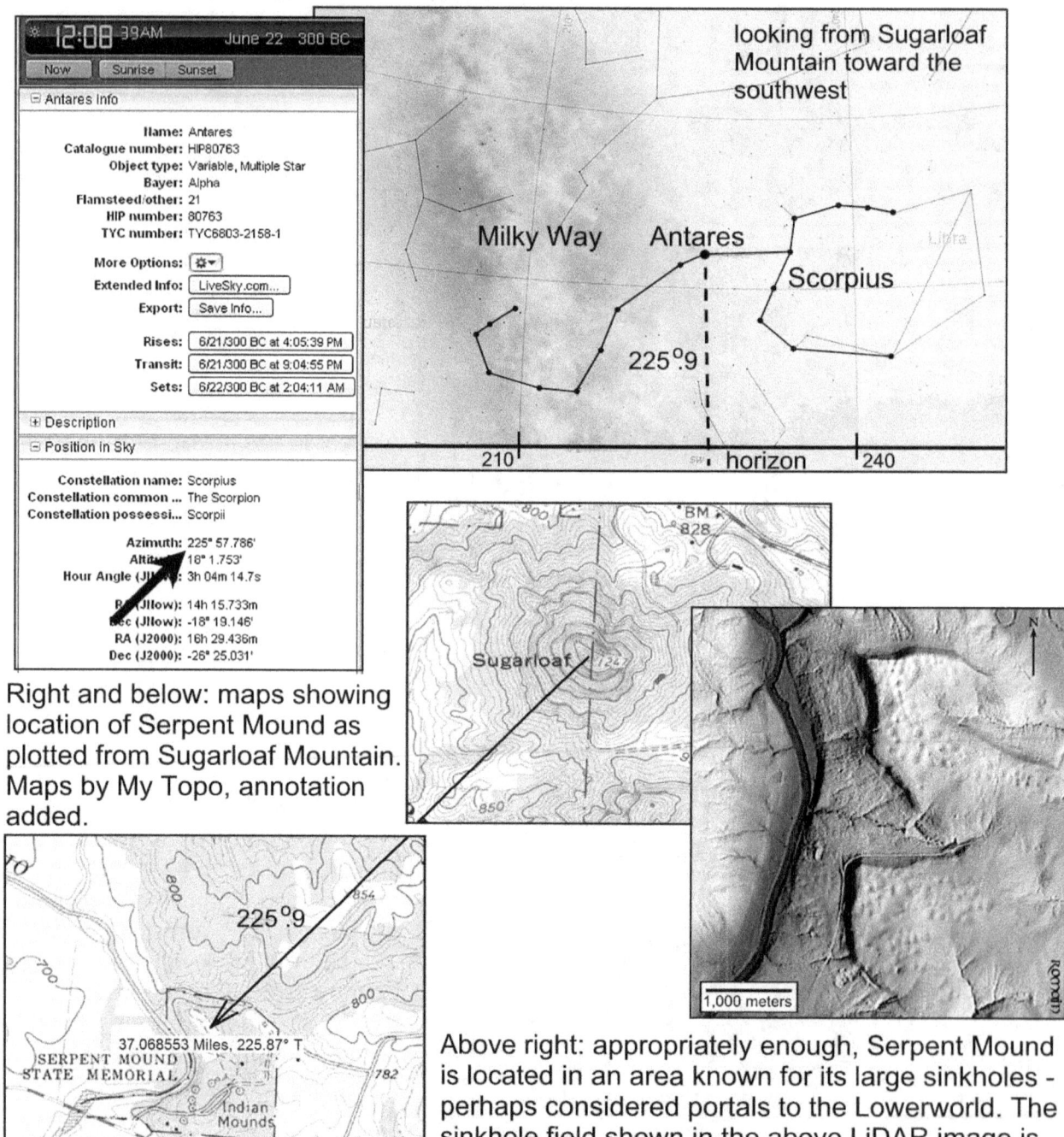

Right and below: maps showing location of Serpent Mound as plotted from Sugarloaf Mountain. Maps by My Topo, annotation added.

Above right: appropriately enough, Serpent Mound is located in an area known for its large sinkholes - perhaps considered portals to the Lowerworld. The sinkhole field shown in the above LiDAR image is about two miles southeast of Serpent Mound.

CHAPTER 8
LITTLE MIAMI RIVER WATERSHED

The Little Miami River is located in southwestern Ohio. The river flows through the glaciated Till Plains physiographic region, from the southwest-central part of the state, south to the Ohio River. The region is characterized by gentle, rolling till plains. River terraces occur in some areas. Hillsides along major rivers are steep and in some areas the terrain is rugged. Among the Hopewell earthworks found in this river valley are Fort Ancient, Milford, Turner, and Stubbs.

Cedarville Earthworks

The most northern of the earthworks in the Little Miami Watershed considered here are the Cedarville Earthworks. Comprised of at least two geometric enclosures and an impressive mound - i.e., the Williamson Mound, these earthworks are situated along Massie's Creek, a tributary of the Little Miami River. Several things combine to make this a special location. First is the Massie Creek Gorge. This is quite an impressive gorge and as viewed from the Williamson Mound the winter solstice sun rises in alignment with the gorge. Across Massie's Creek from the Williamson Mound, is the Pollock Earthwork. Of special interest is that an old channel of Massie's Creek surrounds three sides of the earthwork. The river and gorge-cut sides of Pollock are quite steep. Thus the visual result is that Pollock resembles an island surrounded by watercourses. The effect is further accentuated by perimeter walls and an intricate walled entrance into the earthwork.

Fort Ancient

Although it bears the name 'Fort Ancient' the site was built by people of the Hopewell culture. The Fort Ancient site is the most visually impressive of all the hilltop-promontory enclosures. As with the Pollock site, Fort Ancient seems to form something of an earth island with its surrounding watercourses, deep gullies, ravines, and very steep sides. Multiple numbers of small ponds ring the inside perimeter of the earthwork walls. One can imagine that during Hopewell times these water features supported a variety of wildlife such as frogs, turtles, salamanders, and small fish.

Two things stand-out about the location of Fort Ancient as special. First is the spectacular view of the Miami River Valley afforded from the South Fort area. Most of the burials found at Fort Ancient have been found in the South Fort area, especially on the terraces facing the river, thus providing the dead with incredible views of the valley.

Next, it is of considerable interest to find that the longitudinal axis of the site extends in an orthogonal fashion to the summer solstice sunset. In this the Fort Ancient promontory bears a natural geometric relationship to the solstice sunset. The likelihood that the Hopewell recognized this relationship is suggested by a second summer solstice sunset alignment through a square formed by four mounds situated in the North Fort area. Thus the naturally occurring solstice orthogonal alignment of the site is duplicated by the man-made solstice alignment provided by the North Fort mounds and square. (Posited solstice alignments from the North Fort Square through gateways to the east are likely fortuitous given that the are more than 60 gateways through the perimeter walls.)

The long parallel walls leading from the North Fort area are also of interest. Not only are these walls aligned to the summer solstice sunrise, they also appear to have terminated at a shrine-like feature (see Figure 8.7). The implication is that processional movements timed to the solstice originated at the main entrance to Fort Ancient and from there, proceeded to the shrine along a celestial azimuth.

Milford-Milford West-Turner Complex

The Milford, Milford West, and Turner earthworks are situated at the confluence of the Little Miami River and the East Fork of the Little Miami River. The Milford and Milford West earthworks take their name from a ford across the Little Miami River at that location. Today, except for a few traces, all three earthworks are mostly destroyed.

One of the interesting things about these earthworks is that each has a parallel-walled walkway that leads to an elevation that is either higher or lower than the main enclosure. Further, each of these walled passageways ends at a circle enclosure. In the case of the Milford Earthwork, the walkway extends from the ellipse earthwork, to a circle at the top of a nearby hill. From there, a second set of walls extends from the hilltop circle, down to a lower terrace. The West Milford Earthwork has a similar set of parallel walls extending from the square, to the terrace below, opposite the terminus for the Milford walkway. The Turner Earthwork also has a parallel-walled walkway, in this case extending from the ellipse to a small circle situated at a slightly higher elevation than the ellipse. All three sets of linear walls are generally oriented parallel to a section of the Little Miami River and all terminate downstream from their associated main earthwork.

Interestingly, two of the complexes – i.e., Milford and Turner include large ellipse-shaped earthworks. Recall that an ellipse earthwork is also found at Newark. Notably, both the Turner and Newark ellipses are oriented to the summer solstice sunset – albeit along different axes, and both ellipse earthworks have burials within.

East Fork Works

The East Fork Works is a 'lost' earthwork in the sense that its location is not known with certainty. No trace of the earthwork is visible on the ground or in aerial photographs. Of historic interest is that the East Fork Earthwork has been variously called the General Lytle Works, Gridiron Works, Hannukkah Mound, and Candelabra Earthworks, reflective more of Western interpretations than Native American design intentions. It is doubtful that the earthwork was made to resemble a football field gridiron, or Jewish menorah and oil lamp.

My best estimate for the location of the earthwork is shown in Figure 8.10 and is based on correspondences between topographic features identified on early maps and LiDAR imagery. To verify the earthworks' location, ground-truthing is needed, using either geophysical survey methods such as magnetic gradiometer survey, or actual excavation.

Figure 8.1 Map of sites in the Little Miami River Watershed

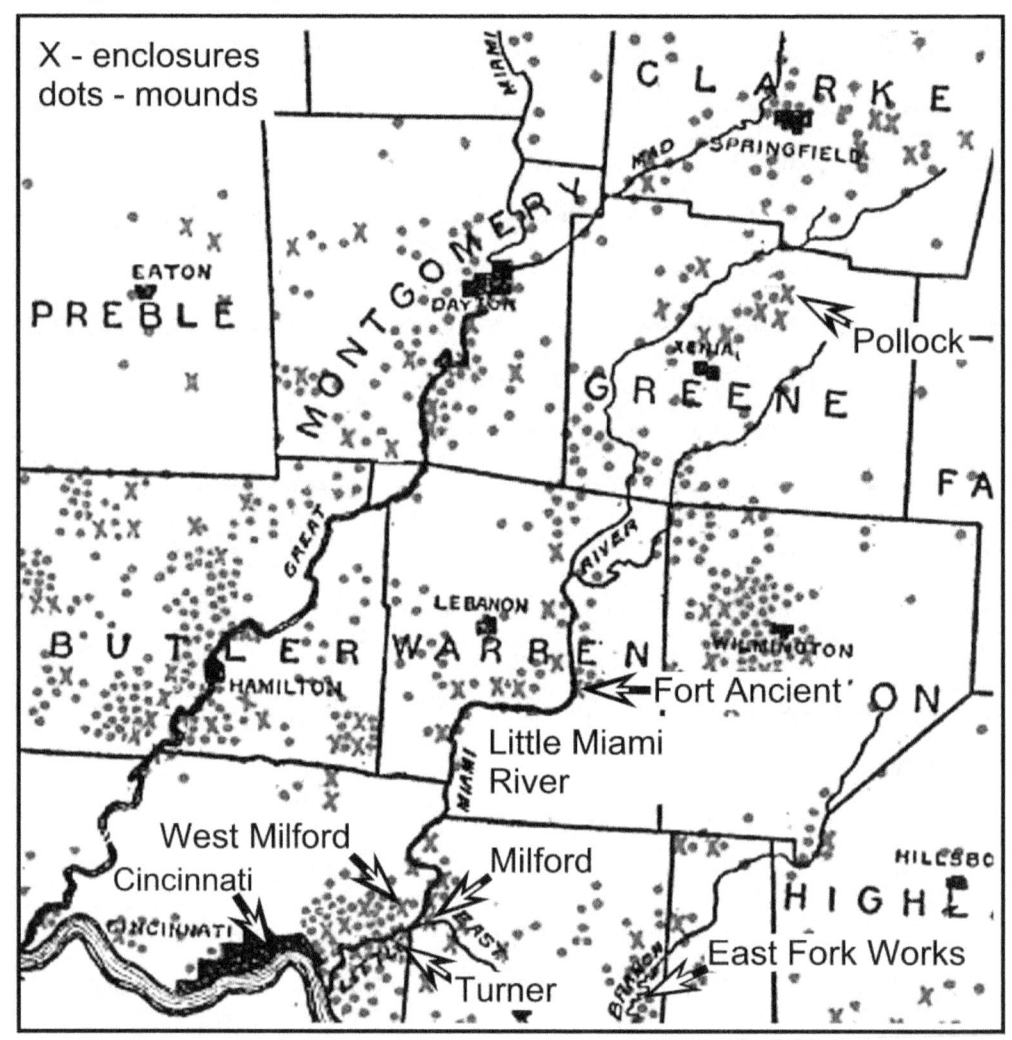

Above: detail of Mills (1914:Pl. XI) map showing earthworks and mounds in the Little Miami River Watershed (annotation added).

Figure 8.2. Cedarville Earthworks

The Cedarville Earthworks are located in Greene County, on Massie's Creek, a tributary of the Little Miami River. The earthworks include the Pollock Works, Bull Earthwork, Williamson Mound, several small mounds, circle, and cresent earthworks. Archaeologist Robert Riordan (2010) proposes that the Pollock and Bull earthworks comprise a set of symbolically linked rectilinear and circular forms analogous to the linked circle and square earthworks of south-central Ohio.

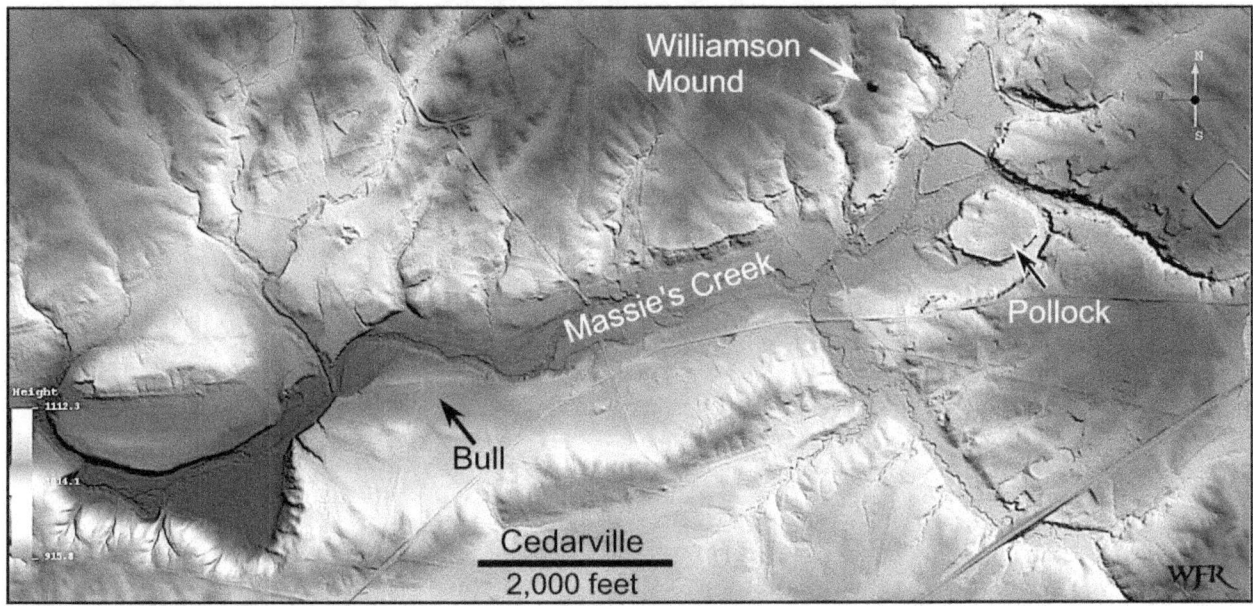

Above: LiDAR view of the Cedarville Earthworks.

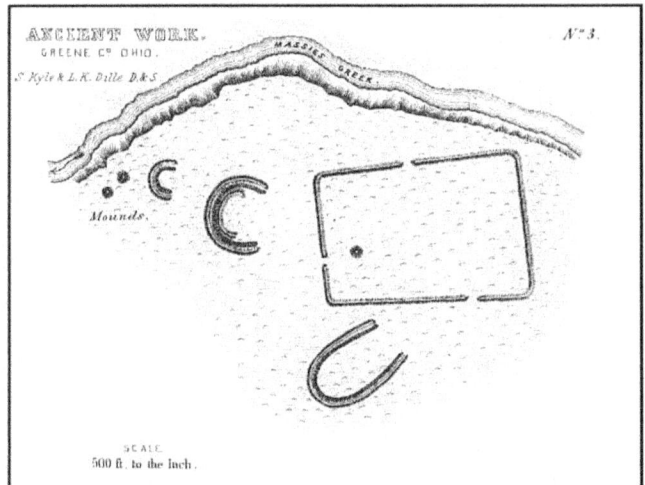

Above: Squier and Davis (1848:Pl. XXXIV, no. 3) map of the Bull Enclosure.
Right: Squier and Davis (1848:Pl. XII, no. 3) map of Pollock.

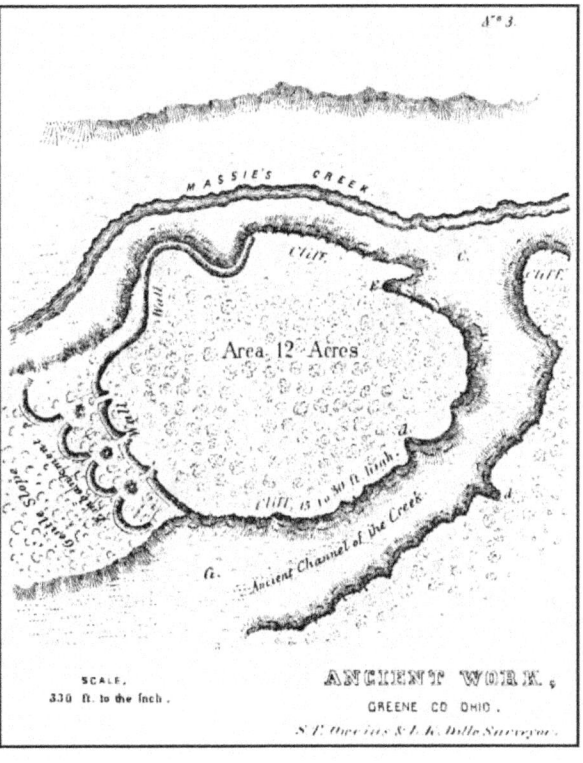

Figure 8.3. Pollock Earthworks

The Pollock Earthwork is a low, mesa-like promontory, bounded on three sides by a steep cliff - 15 to 30 feet high, and an earth and stone wall on the isthmus side. Three gateways cut through the isthmus wall (note 1).

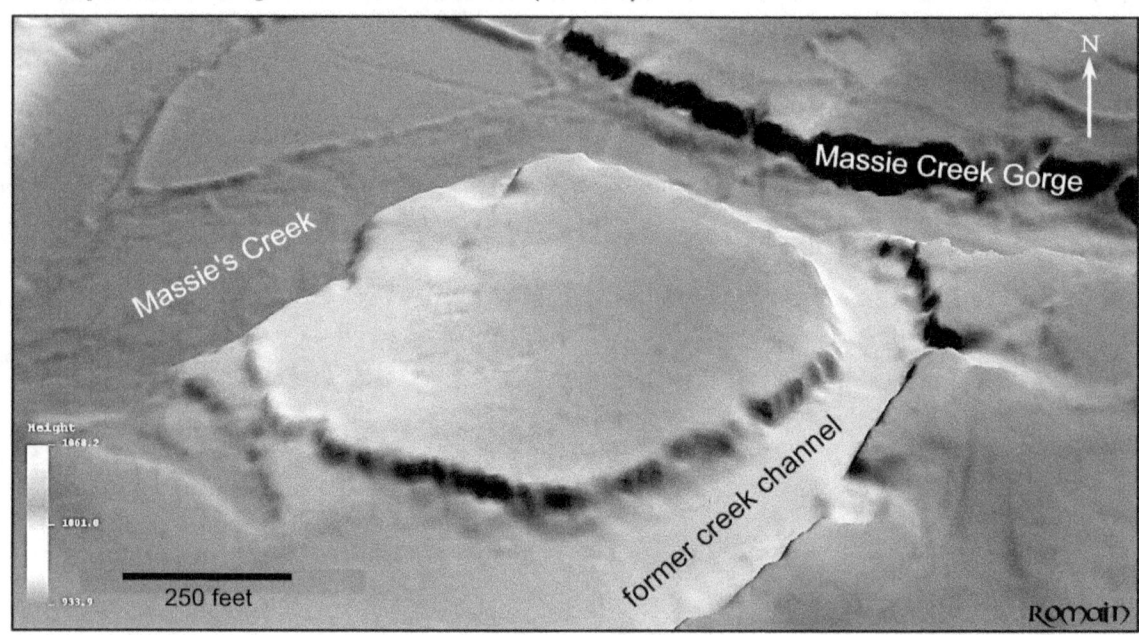

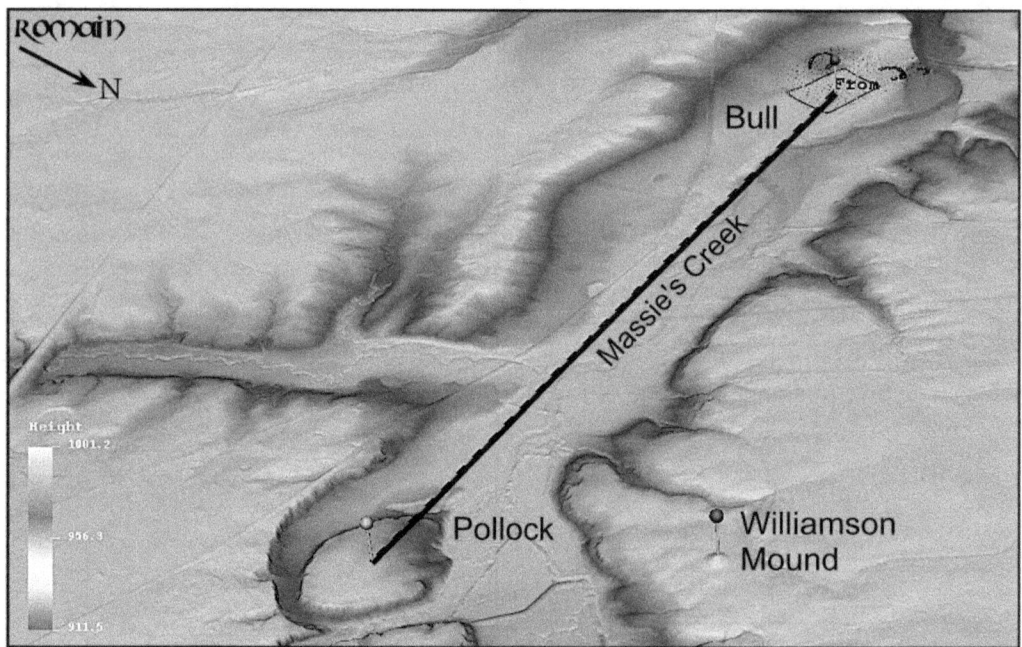

Above: the sightline between the Pollock and Bull earthworks extends parallel to the Massie Creek Valley - reminiscent of the situation in the Chillicothe area involving the High Bank, Liberty, and Works East sites relative to the Scioto River Valley. A distance of about one mile separates the Pollock and Bull eartworks.

Figure 8.4. Williamson Mound

The Williamson Mound is believed to be an Adena mound. It has not been excavated. The mound is about 25 feet high. It overlooks the entrance to the Massie Creek Gorge - one of the most spectacular geological formations in the area.

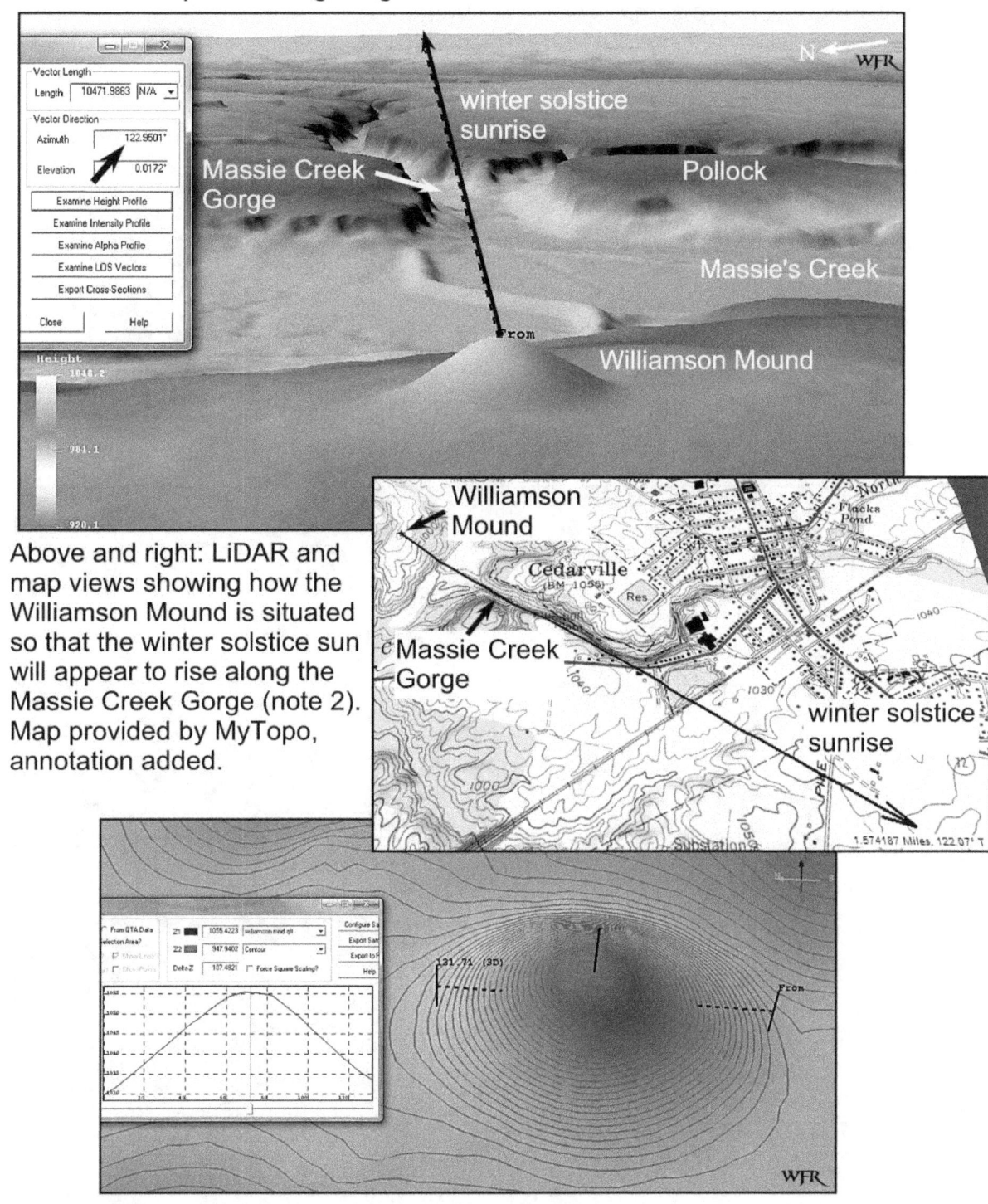

Above and right: LiDAR and map views showing how the Williamson Mound is situated so that the winter solstice sun will appear to rise along the Massie Creek Gorge (note 2). Map provided by MyTopo, annotation added.

Above: the Williamson Mound has suffered from erosion. However, LiDAR analysis suggests the diameter of the mound was based on the 1/8 HMU length (131.8 ft.).

Figure 8.5. Little Miami River Valley: Fort Ancient

Fort Ancient is a very large, flat-topped promontory enclosed by perimeter walls (note 3). The site is located in Warren County, overlooking the Little Miami River. What may have made this location special is that the promontory and resulting site axis are oriented orthogonal to the summer solstice sunset.

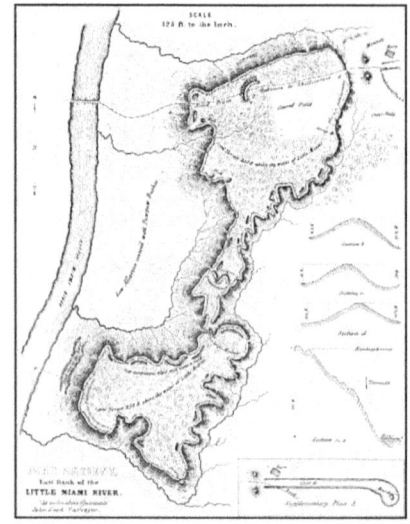

Above: aerial view of Fort Ancient.
Left: Squier and Davis (1848:Pl. VII) map of Fort Ancient.

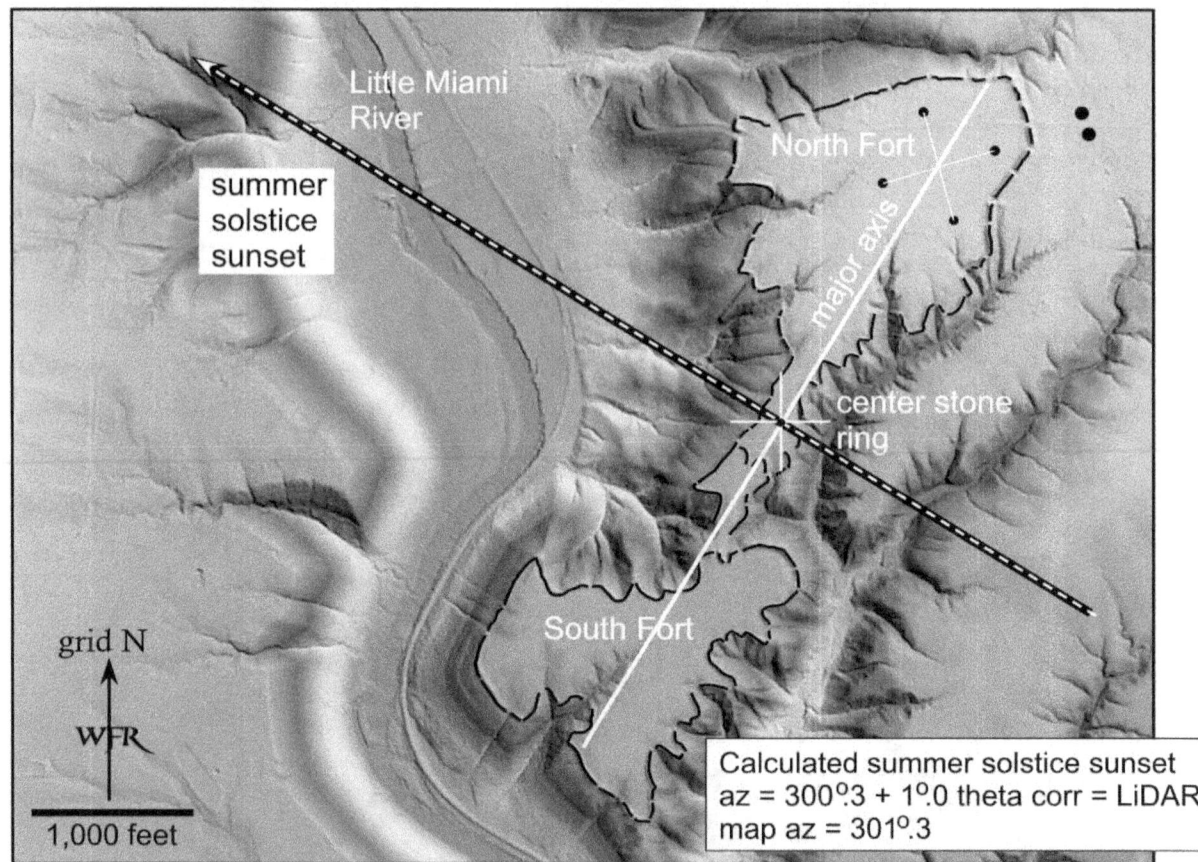

Calculated summer solstice sunset
az = 300°.3 + 1°.0 theta corr = LiDAR
map az = 301°.3

Above: LiDAR image showing the Fort Ancient major axis and summer solstice sunset azimuth through the center stone ring (note 4). (Perimeter walls highlighted.) The cross, or square in the North Fort is comprised of four stone mounds.

Figure 8.6. Fort Ancient Features

The major axis for Fort Ancient is not arbitrary. It is established by a line extending from the center of the North Fort Square through the center stone circle (note 5).

Above left: view showing the four stone covered mounds that comprise the North Fort Square.
Above right: view of the Fort Ancient center stone ring.
Right: one of the four mounds that comprise the North Fort Square.

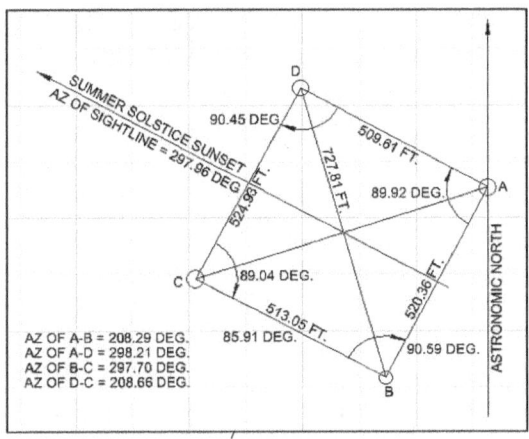

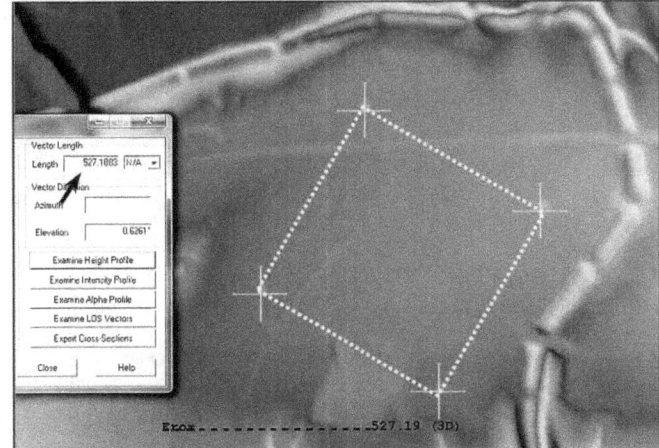

Above left: as if to emphasize the othogonal alignment of the overall Fort Ancient site to the summer solstice sunset, the Fort Ancient Square is also oriented to the summer solstice sunset (note 6). Noteworthy too is that the corners of the Square are accurate right angles to within one degree.
Above right: the sides of the North Fort Square are each equal to 1/2 HMU to less than 2 percent. The above right LiDAR image shows how a 1/2 HMU square matches the Square.

Figure 8.7. Fort Ancient Parallel Walls

Early maps (e.g. Moorehead 1890) show a set of low parallel walls extending from the north gateway area of Fort Ancient to the northeast for about 2,700 feet. The walls were 130-140 feet apart. Today the walls are not visible on the ground, or in LiDAR imagery. Faint traces of the walls, however, are visible in a 1930s USDA aerial photo.

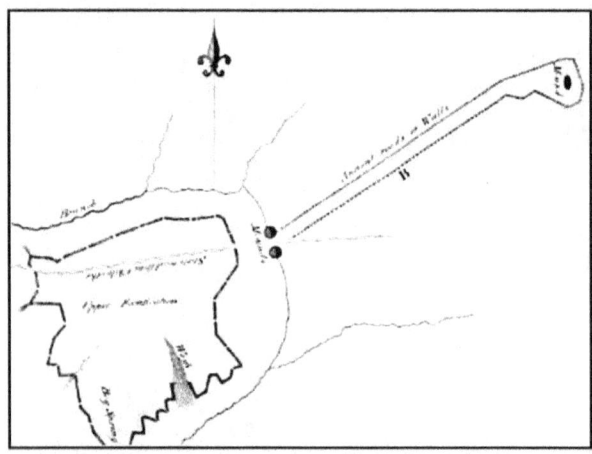
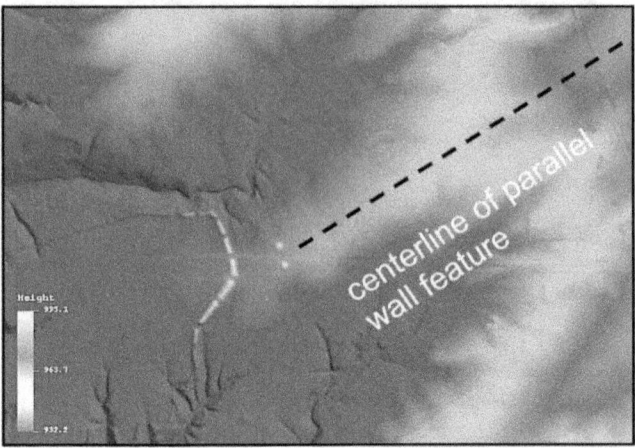

Above left: enlarged detail of Atwater's (1820:Pl. IX) map of Fort Ancient (note 7). Above right: one of the interesting features of the parallel walls is that they extend along a stretch of elevated ground (lighter color in the above LiDAR image). At ground level, the changes in elevation are very subtle and difficult to detect with the naked eye.

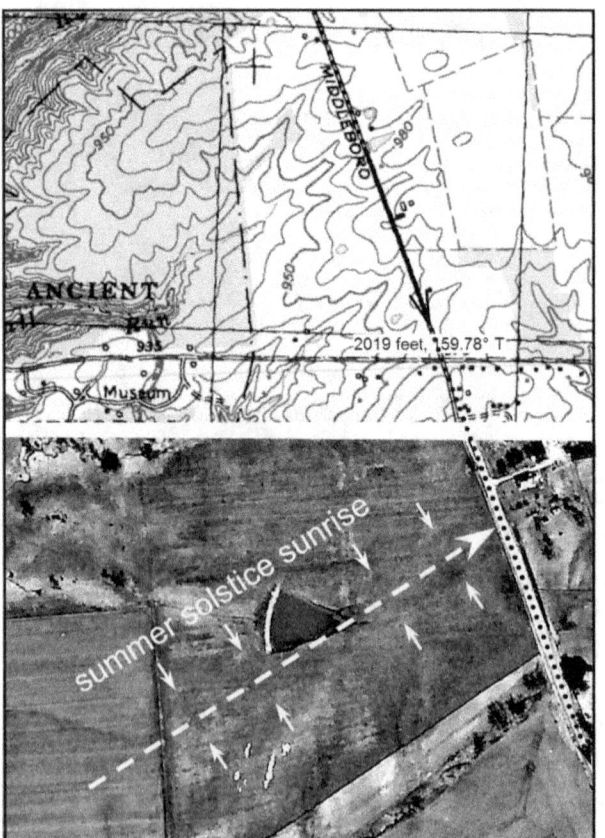

Left: in the 1930s aerial photo shown, faint traces of the parallel walls are shown by the white arrows. The dashed line shows the centerline trajectory of the parallel wall feature.

Analysis of the USGS topographic map and correctly oriented aerial photo shows the centerline trajectory of the parallel walls extends along an azimuth of about 58.4 degrees. The calculated summer solstice sunrise was at 58.7 degrees - hence the alignment is accurate to within about 0.3 degrees (note 8).

Both Atwater (1820:Pl. IX) and Squier and Davis (1848:Pl. VII) show a mound at the closed northeast end of the parallel wall feature. Presumably, the idea was for an observer to witness the summer solstice sunrise from the entrance to Fort Ancient - marked by the two gateway mounds, over the third mound at the northeast end of the parallel walls.

Figure 8.8. Little Miami River Valley: Milford-Turner Group

About 19 miles southwest of Fort Ancient, on the banks of the Little Miami River, there were three major earthworks - Milford, West Milford, and Turner. Today, the earthworks are mostly destoyed.

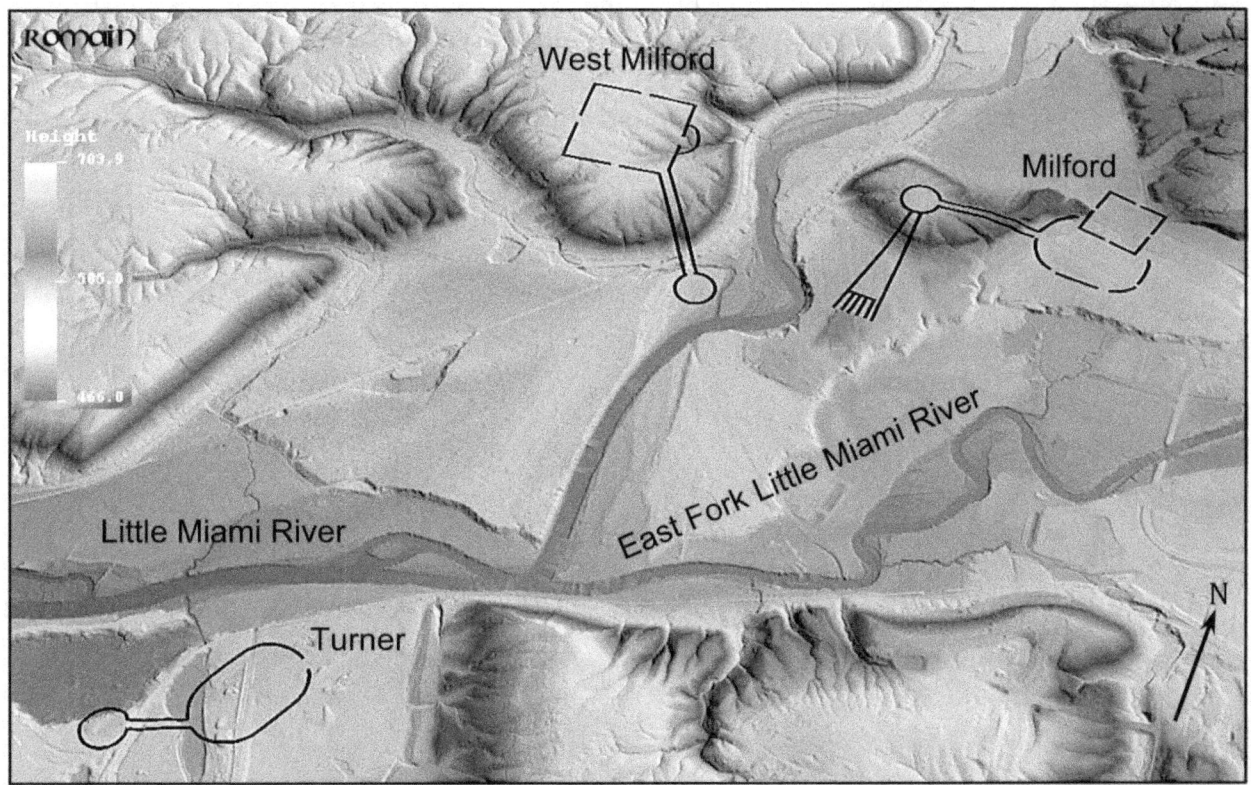

Above: LiDAR image with Turner, Milford, and West Milford earthworks superimposed at correct scale. Locations for Milford and West Milford estimated (note 9).

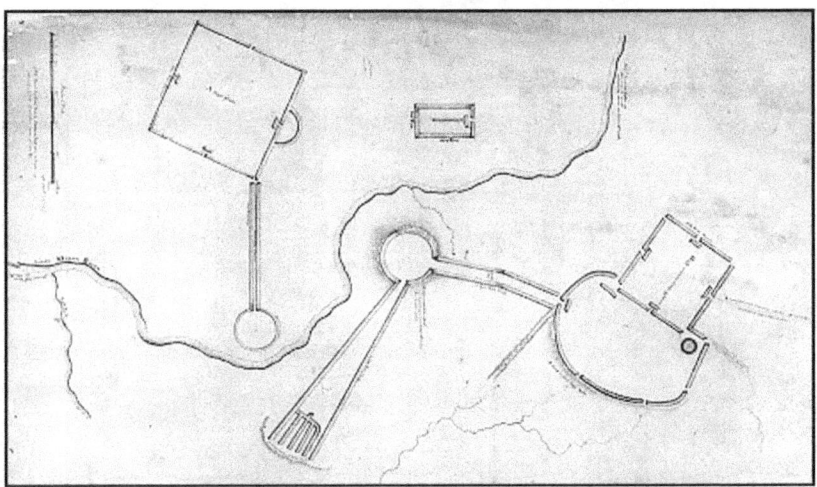

Above left: Enlarged detail of Roberdeau map from 1823 (from McCullough 1991 [2010]).
Above right: map of the Turner Works from survey made in 1887 (Willoughby 1922:Pl. 1).

Figure 8.9. Little Miami River Valley: Turner Alignment

Without ground-truthed data it is difficult to know what the West Milford and Milford earthworks were oriented to. Possibly West Milford was aligned to the cardinal directions. For the Turner Earthworks, the situation is better. In the 1930s aerial photo below, sections of the embankment walls are visible. Using the visible wall sections and the Willoughby (1922) map as a guide, the location for the Turner Ellipse can be established. From analysis of the USGS topo map for the area, the azimuth for the railroad (dashed line) is established and a true north arrow is added to the figure.

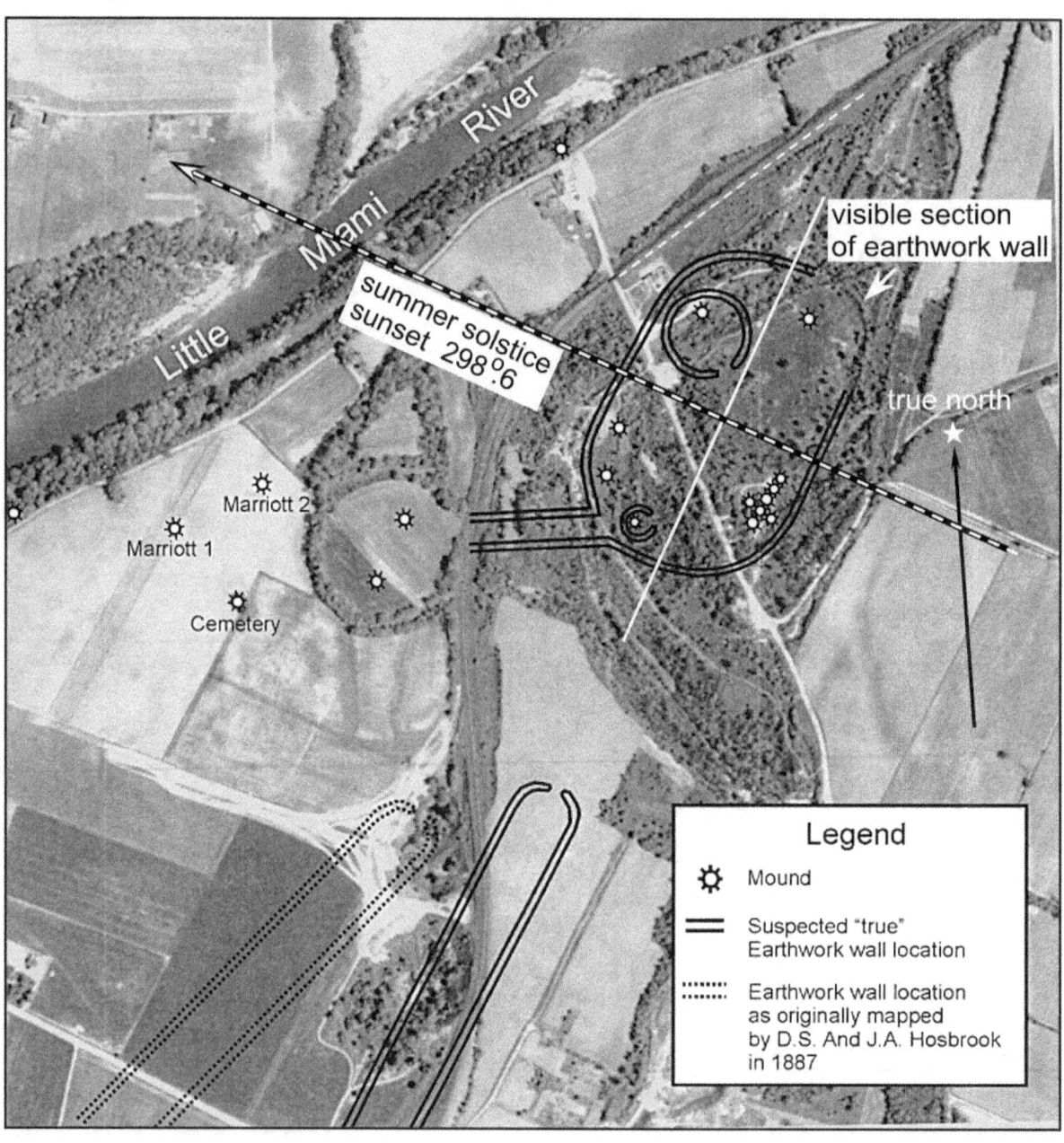

Above: the minor axis of the Turner Ellipse was oriented to the sumer solstice sunset (note 10). Aerial photo courtesy of Matthew Purtill, earthwork annotation by Purtill, directional annotation by Romain.

Figure 8.10. Little Miami River Valley: East Fork Works

The East Fork Works is a 'lost earthwork.' Although mentioned and illustrated by Squier and Davis (1848:95, Pl. XXXIV, no. 2B), the exact location for the earthwork has never been definitely established (note 11). Early references describe the site as located on the 'waters of the East Fork of the Little Miami River,' about 20 miles upstream from the confluence of the Little Miami River and its East Fork. Depending on how it is viewed, the design could represent a thunderbird. Others think it looks like a menorah.

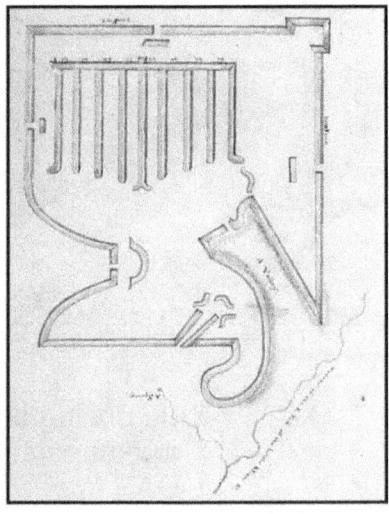
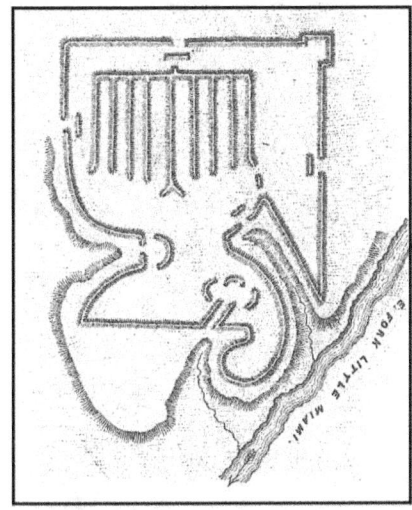

Above left: Squier and Davis (1848:Pl. XXXIV, no. 2B) map of the East Fork Works. Above right: probable ultimate source for Squier and Davis map - i.e., map made in 1823 by Major Isaac Roberdeau (from McCullough 2010). Unfortunately, the Roberdeau map does not include a north arrow. The Roberdeau map does, however, show several topographic features that can help identify its location.

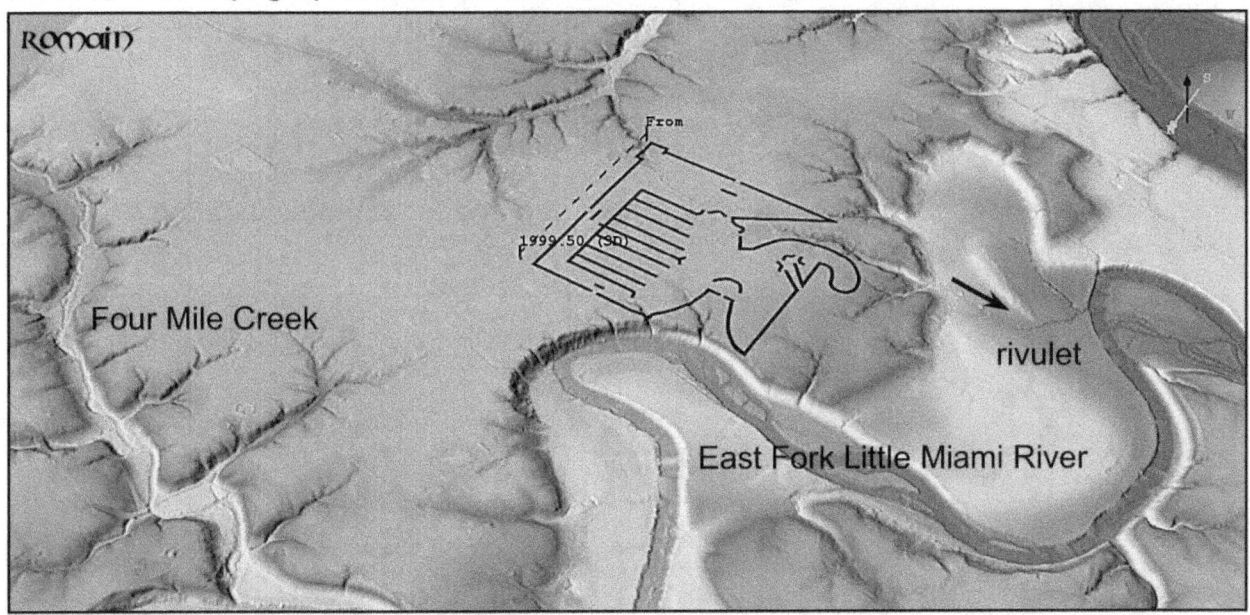

Above: LiDAR image showing possible location for the East Fork Works (note 12). The posited location is roughly 20 miles upstream from the confluence of the Little Miami River and its East Fork. In the above figure, the earthwork is 2,000 feet on its side, matching the dimensions given by Roberdeau 1923 (in McCullough 1996).

Figure 8.11. Little Miami River Valley: East Fork Works II

There are several correspondences between the Roberdeau map and local terrain supportive of the posited location for the East Fork Earthwork.

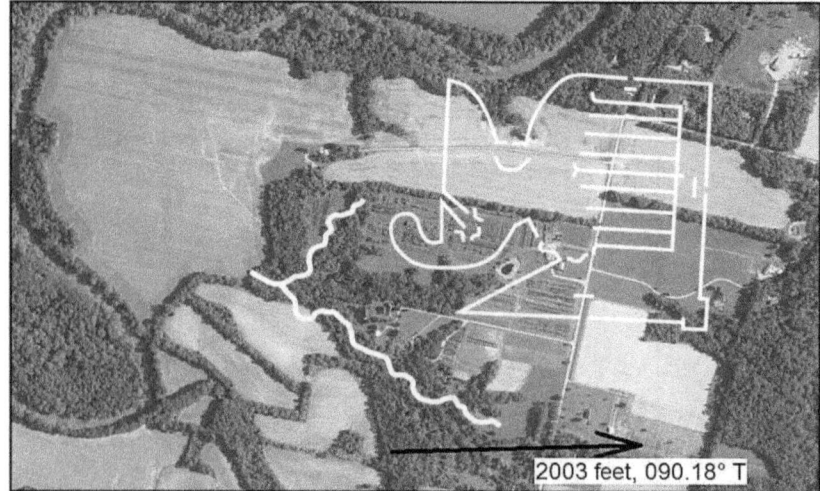

Left: outline of Roberdeau figure superimposed over aerial photo. Aerial photo provided by My Topo.

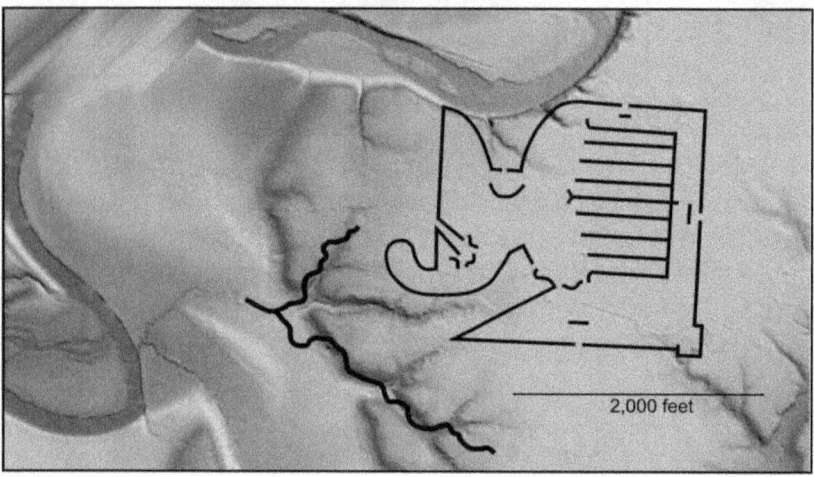

Left: outline of Roberdeau figure superimposed over LiDAR image.

Below: correspondences between Roberdeau map and topographic features revealed by LiDAR imagery.

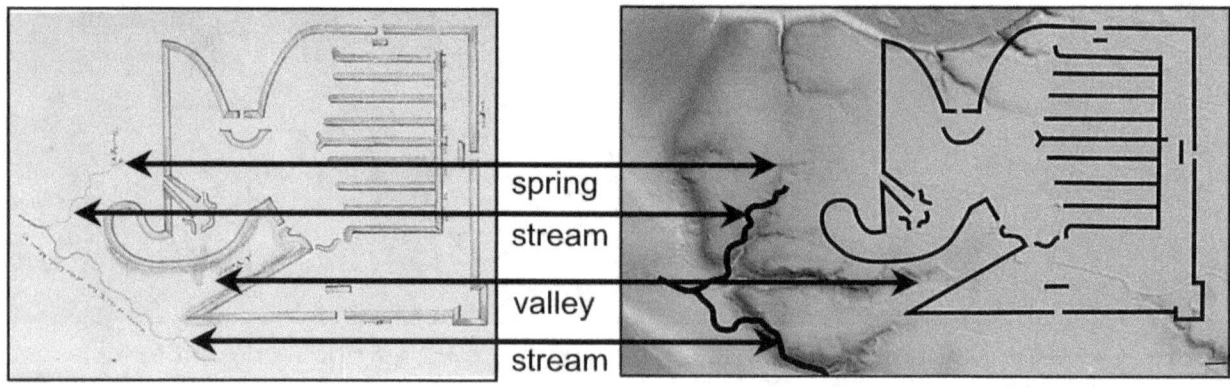

CHAPTER 9
GREAT MIAMI RIVER WATERSHED

The Great Miami River is a major tributary of the Ohio River. The river is located west of Cincinnati and from there, extends northeast for about 160 miles. A considerable number of large mounds and promontory enclosures are located along its length.

Relatively few large flat terraces suitable for the building large geometric enclosures are found in this river valley. What the river valley lacks in flat terraces it makes-up in its abundance of promontories that overlook the Great Miami River. As a group, the promontory enclosures typically have one or more walls that extend across the promontories, so that the space within is physically distinct from and separated from the tableland. The effect is that each promontory enclosure is basically its own little world, balanced between water below and sky above. I suspect, but cannot prove (indeed how would someone prove such a thing), that this effect was intentional, with each enclosure representing a miniature world. For those who entered these areas, the effect was to situate the individual at the center of the world (albeit a man-made one). From the center of the world, or cosmic center, the individual would presumably have been able to exercise some control or effect on the world through ritual activities.

The Question of Signal Mounds

The notion that mounds were used as signal stations for communication by fire, smoke, or mirrors - whether up and down the Scioto River Valley, or along the Great Miami River Valley has been around for some (e.g., MacLean 1879:59; Squier and Davis 1848:141). As remarked by early observers, many mounds that are located on high vantage points along the Great Miami River exhibit traces of fire.

In 1990, a small group tested the mound-signal hypothesis for a section of the Great Miami River Valley (Archaeological Conservancy 1990:4). Volunteers positioned themselves on the Great Mound, Hill-Kinder Mound, and Miamisburg Mound. Using mirrors, flashes of light sent from the Great Mound were seen by the group on the Hill-Kinder Mound. Likewise, mirror signals send from the Hill-Kinder Mound were seen by those at the Miamisburg Mound.

LiDAR analysis corroborates these experimental findings. Further, by applying line-of-sight analyses using modern topographic map computer programs, it can be demonstrated that line of sight communications were possible for a distance of about 40 miles along the Great Miami River Valley - from the Two Mounds site in the south, to the Calvary Enclosure at Dayton. Of course just because a mound or series of mounds 'could' have been used for signaling does not mean that they necessarily were. Nevertheless, the idea remains an interesting and even likely possibility.

Figure 9.1. Map of sites in the Great Miami River Watershed

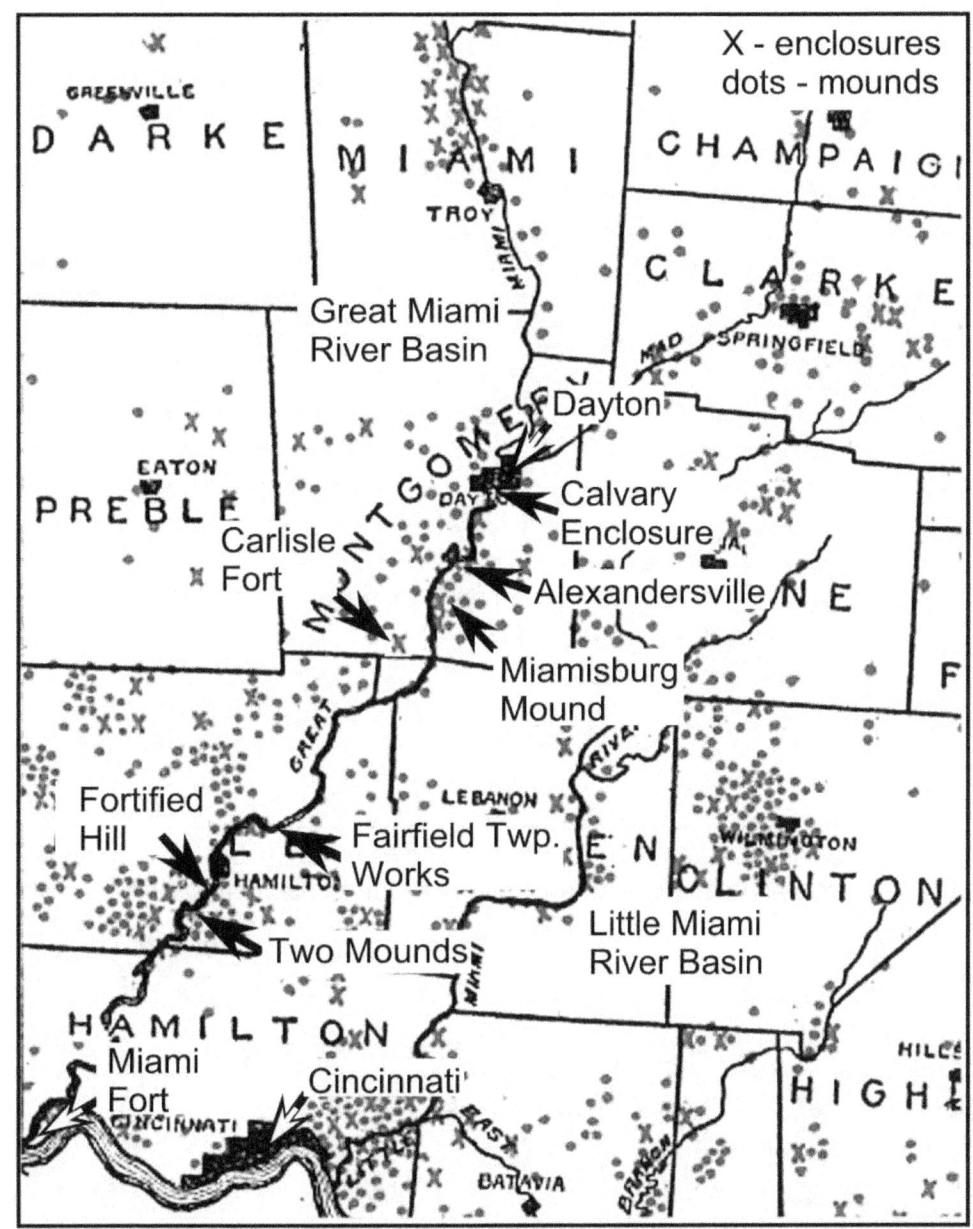

Figure 9.1. Detail of Mills (1914:Pl. XI) map showing earthworks and mounds in the Great Miami River Watershed (annotation added).

Figure 9.2. Great Miami River Valley: Calvary Cemetery Enclosure

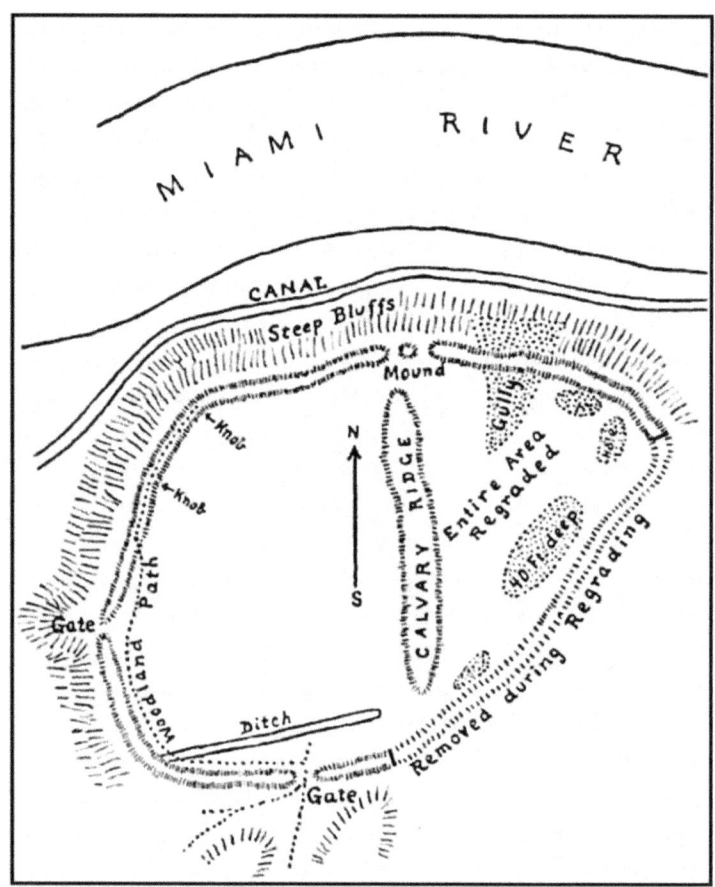

Above: map of the Calvary Cemetery Enclosure (Foerste 1915:38). The earthwork enclosed about 24 acres, making it one of the largest hilltop enclosures in Ohio.

Below: the Calvary Cemetery Enclosure commanded an impressive view of the Miami River Valley and three river confluences. For perspective, the mouth of Stillwater River is four miles distant.

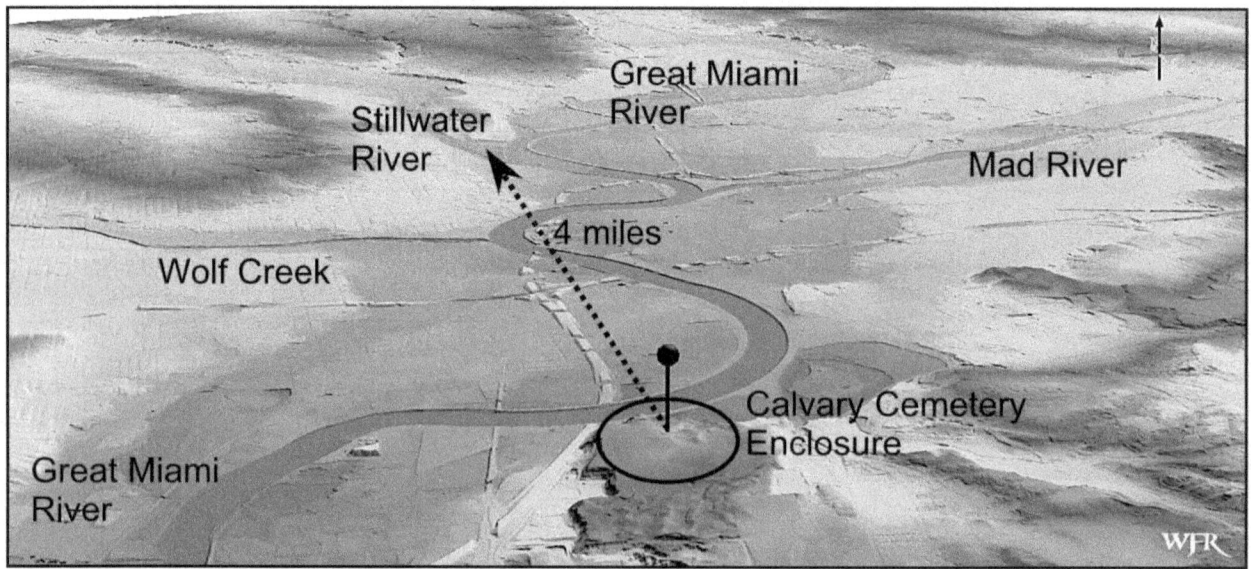

Figure 9.3. Great Miami River Valley: Alexandersville

The Alexandersville Earthworks are located about six miles south of Dayton, on the banks of the Great Miami River. Few remnants of the earthwork are visible today.

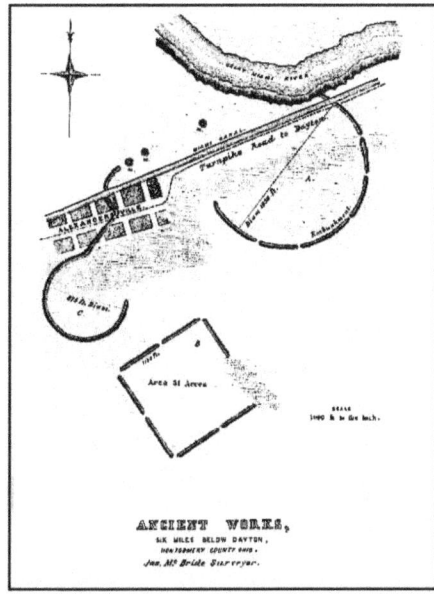 

Above left: map of Alexandersville Works by James McBride (Squier and Davis 1848: Pl. XXIX, no. 1) (note 1).
Above right: 1961 Ohio Department of Transportation aerial photo. In this photo the northwest corner of the Square and part of the Small Circle are visible. Photo courtesy of Stanley Baker, Ohio Department of Transportation, annotation by the present author.
Below left: photo from early 1900s showing cut through one of the walls of the Alexandersville Works (Foerste 1915:143).
Below right: schematic plan of Alexandersville plotted on USGS map. The major axis of the earthwork is oriented to the summer solstice sunrise to within about 1/2 of one degree (note 2).

 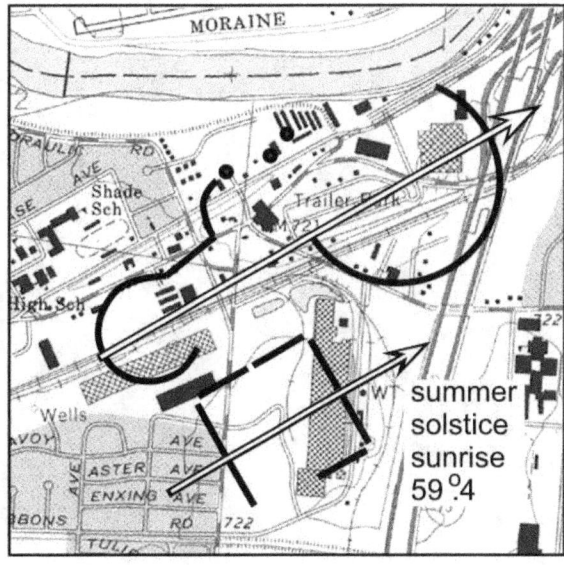

223

Figure 9.4. Great Miami River Valley: Miamisburg Mound
The Miamisburg Mound is the largest conical mound in Ohio. It is 68 feet in height. The mound is situated so that it commands an impressive view of the surrounding area.

Above left: view of the Miamisburg Mound.
Above right and right: views from the top of Miamisburg Mound.
Below: LiDAR image showing location of the Miamisburg Mound overlooking two river confluences.

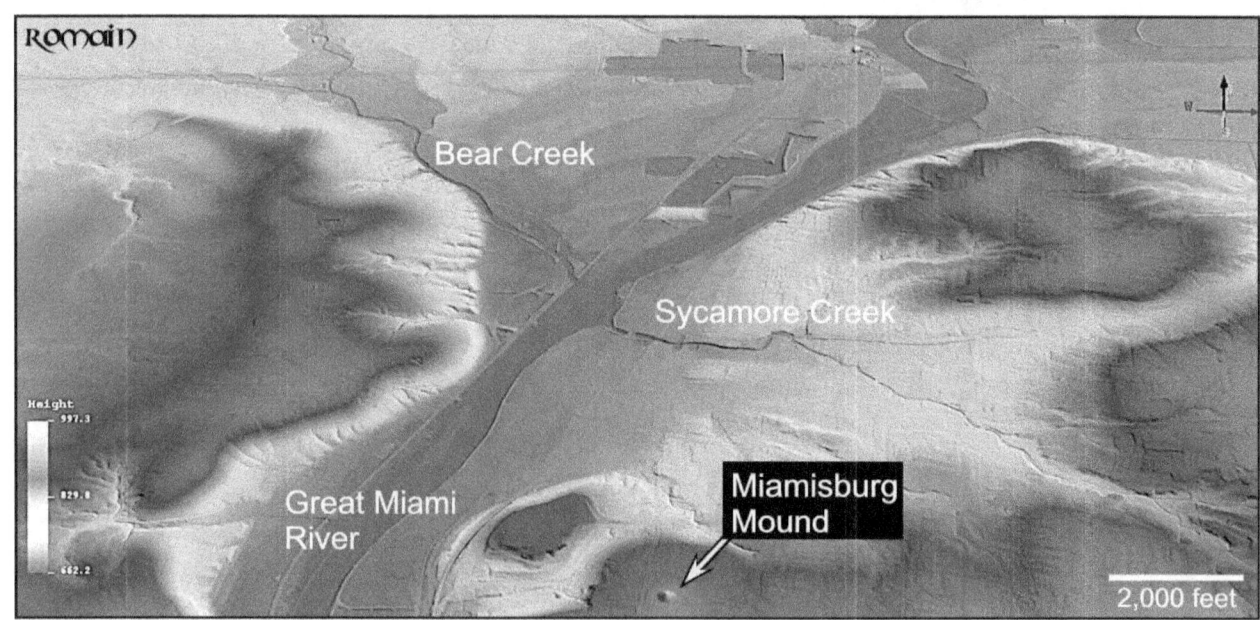

Figure 9.5. Great Miami River Valley: Miamisburg Mound II

The Miamisburg Mound was tunneled into in 1869 (Brine 1894). One bark-covered human skeleton and an empty log tomb were found at different levels. Found at yet another level was what appeared to be a 'stone altar.'

Left: Squier and Davis (1848:fig. 1) sketch of the Miamisburg Mound.

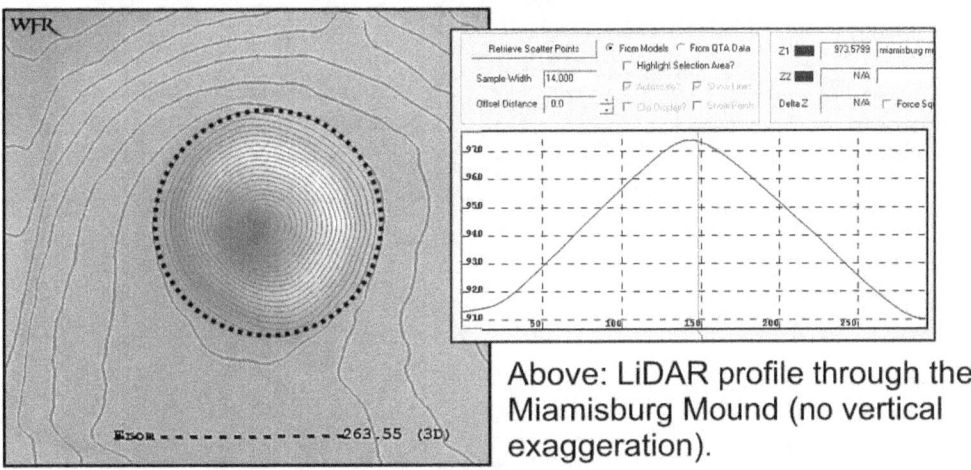

Above: LiDAR profile through the Miamisburg Mound (no vertical exaggeration).

Above: LiDAR image with 3-foot contour intervals. In this image, a 263.5-foot diameter circle (dotted line) has been superimposed to show how the mound incorporates the 1/4 HMU length (263.5 feet) in its design.

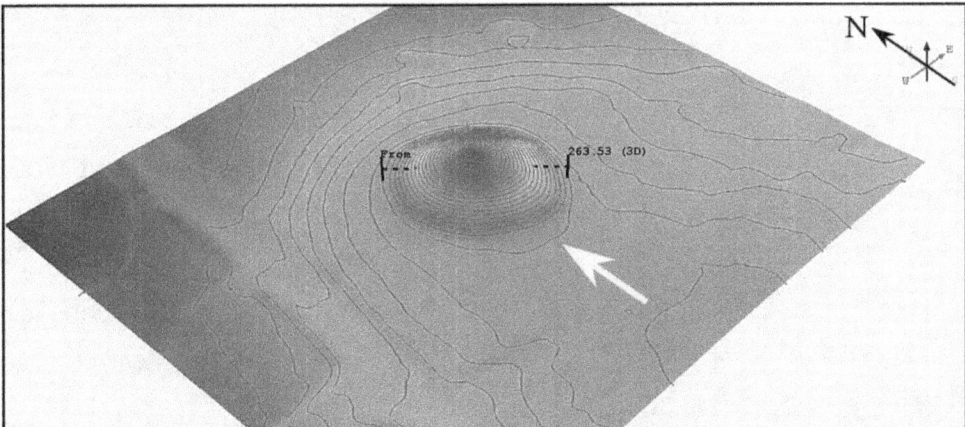

Above: LiDAR shows an elevated area leading due north to the mound. Perhaps this represents an intended avenue of approach.

Figure 9.6. Great Miami River Valley: Hill-Kinder and the Great Mound

The Hill-KInder Mound is a large mound located about 3 1/2 miles south of the Miamisburg Mound and just under 1 mile east of the Great Miami River. In the late 1800s it was reported to be 17 feet in height. Its present height is about 7 feet.

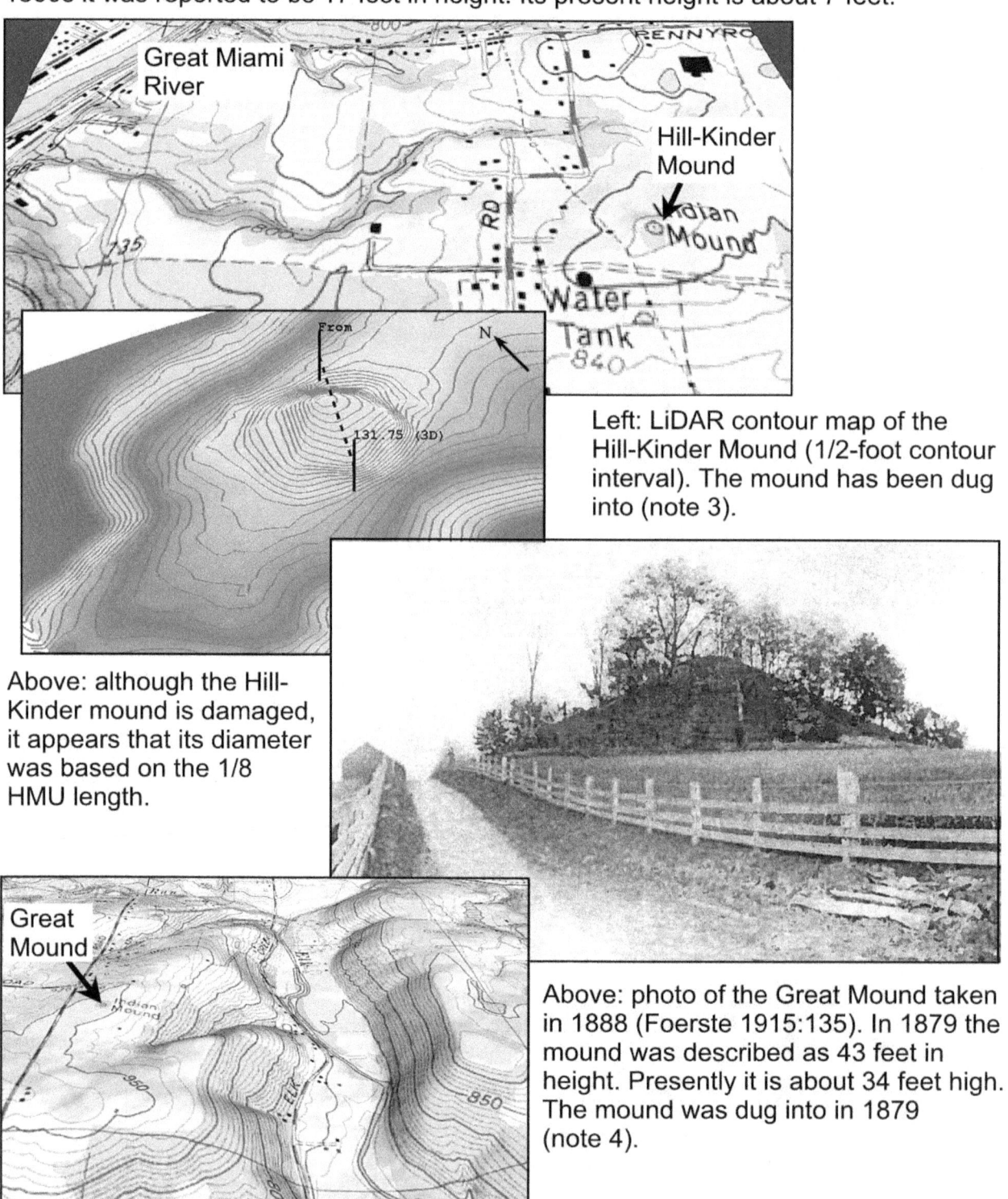

Left: LiDAR contour map of the Hill-Kinder Mound (1/2-foot contour interval). The mound has been dug into (note 3).

Above: although the Hill-Kinder mound is damaged, it appears that its diameter was based on the 1/8 HMU length.

Above: photo of the Great Mound taken in 1888 (Foerste 1915:135). In 1879 the mound was described as 43 feet in height. Presently it is about 34 feet high. The mound was dug into in 1879 (note 4).

Above: the Great Mound is located about 11 miles southwest of Hill-Kinder on the west side of the reat Miami River, overlooking Elk Creek - a tributary of the Great Miami River.

Figure 9.7. Great Miami River Valley: Carlisle Fort

Carlisle Fort is one of the so-called hilltop enclosures in the Great Miami River Valley. It is actually a flat-topped promontory with steep cliffs on three sides and a series of man-made embankments on the fourth side. Carlisle Fort overlooks Twin Creek, a tributary of the Great Miami River.

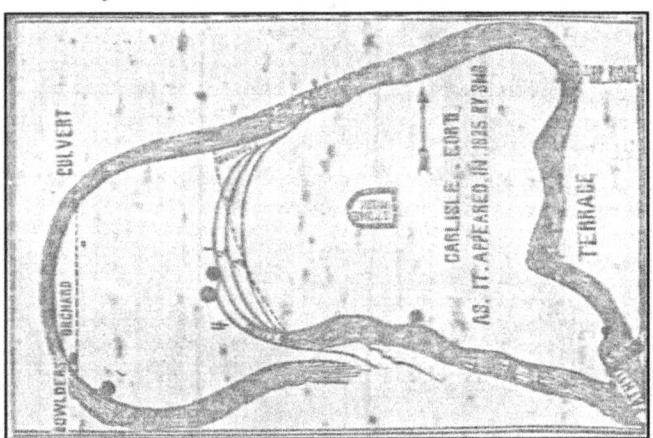
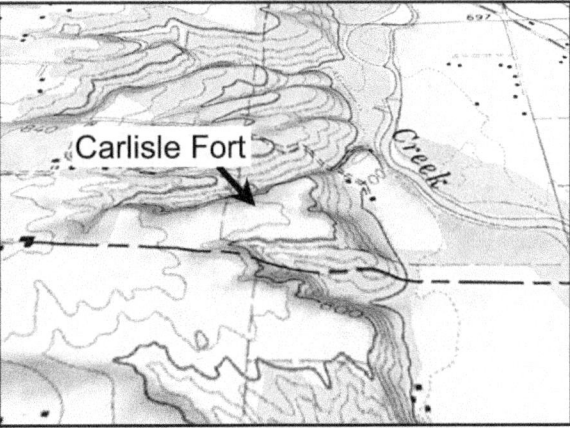

Above left: sketch of Carlisle Fort made in 1835 (Brinkley 1889).
Above right: USGS 7.5-minute series topo map showing location of Carlisle Fort. Map provided by My Topo, annotation added.

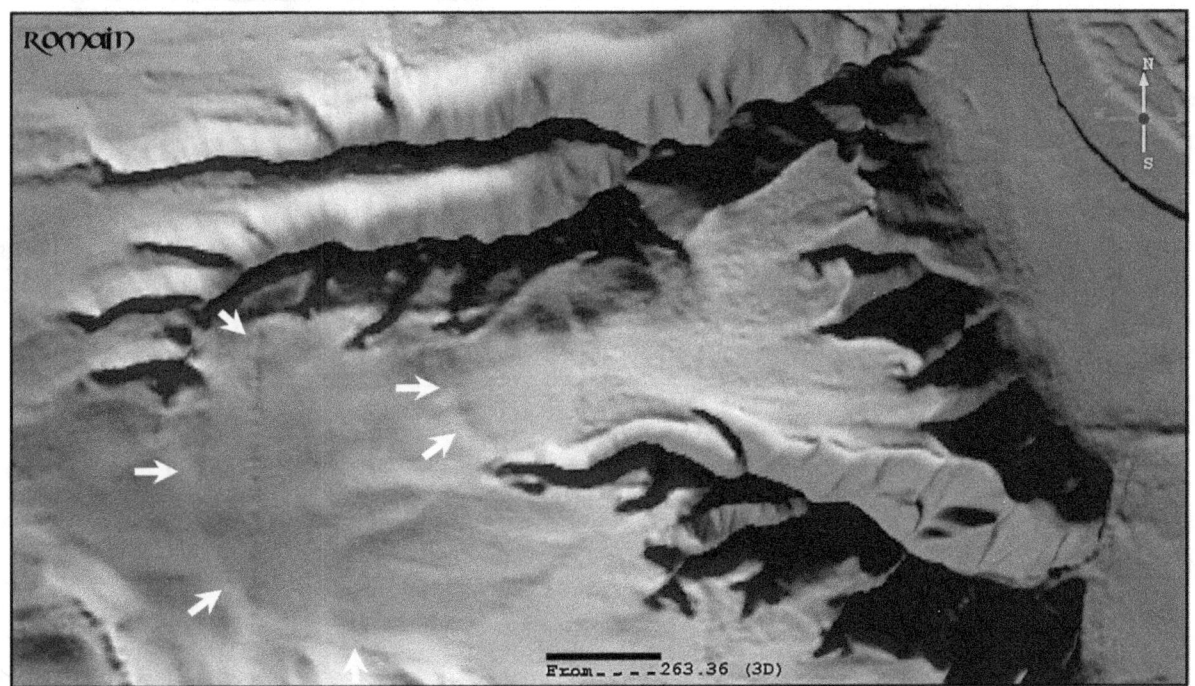

Above: LiDAR image of Carlisle Fort. Two circular embankments are shown by the arrows.

Figure 9.8. Great Miami River Valley: Carlisle Fort Alignment

Carlisle Fort incorporates geometric shapes, a lesser multiple of the HMU, and a solstice alignment in its design.

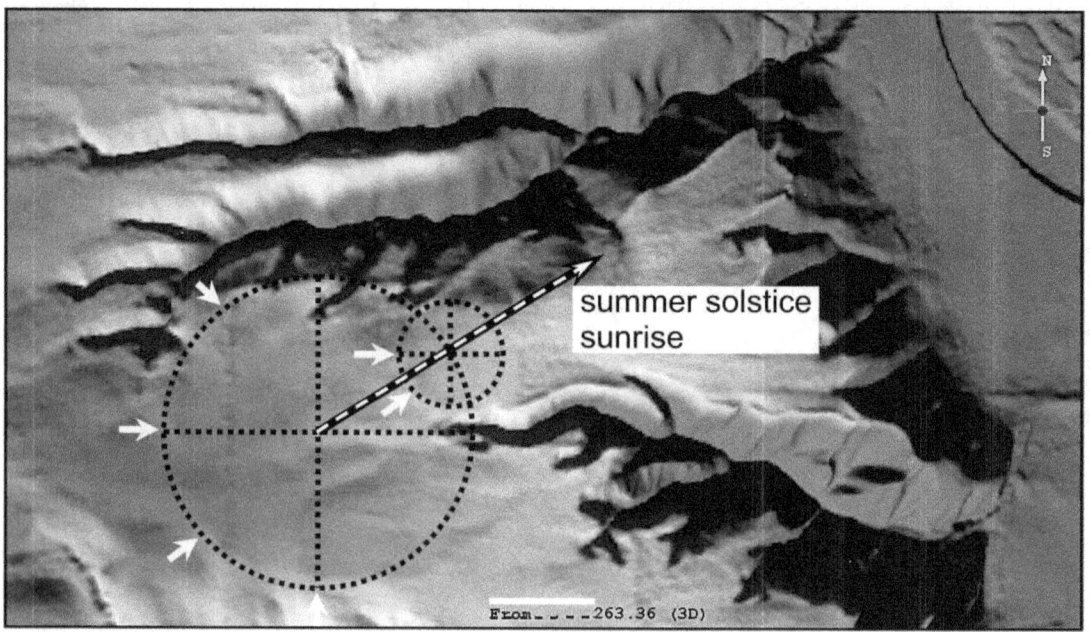

Above: same LiDAR image as presented in Figure 9.7 with the addition of two drawn circles superimposed. Using the LiDAR scale shown, the small circle is made to a diameter of 1/4 HMU (or 263.5 feet). The large circle is 3 X 1/4 HMU (or 790.5 feet). Note how the earthwork arcs closely match the curvature of the drawn circles (note 5). A line drawn from the center of the large circle through the center of the small circle extends along the trajectory of the summer solstice sunrise.

Below: viewed from Carlisle Fort, the summer solstice sun will rise about 500 feet south of the Miamisburg Mound - at a distance of about 4 1/2 miles (note 6). At this distance, the alignment error would not be noticeable. Also shown below, the line-of-sight from Carlisle Fort to the plateau on which the Miamisburg Mound is situated.

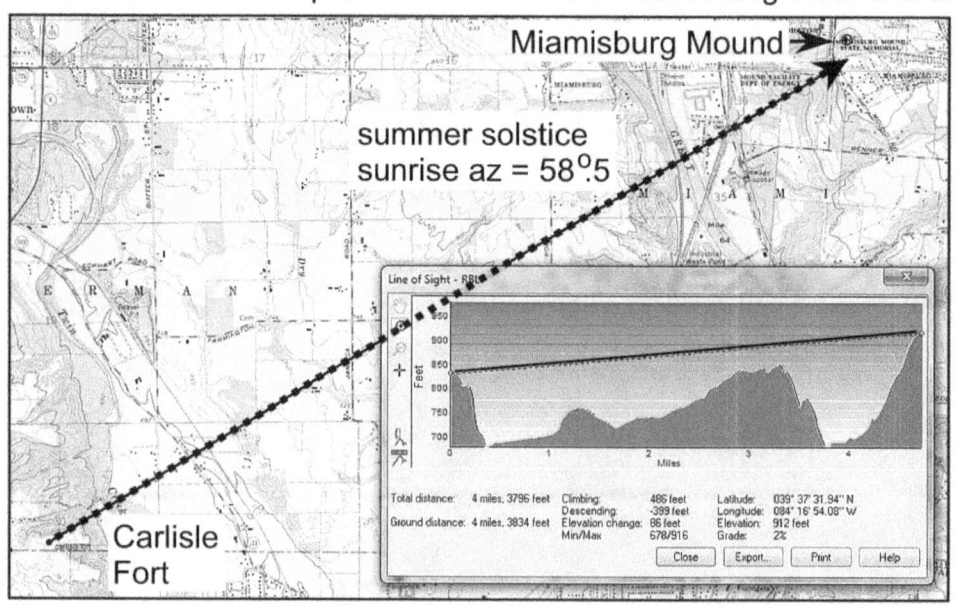

Figure 9.9. Great Miami River Valley: Fairfield Township Works

The Fairfield Township Works is a promontory earthwork located on the south bank of the Great Miami River, about 4 1/2 miles northeast of Hamilton, Ohio. The site is surrounded by deep ravines on two sides and the Great Miami River on its north side. An embankment wall across the fourth side closes the structure.

Right: map of the Fairfield Township Works made by James McBride and John Erwin in 1840 (Murphy 1978:fig.1).

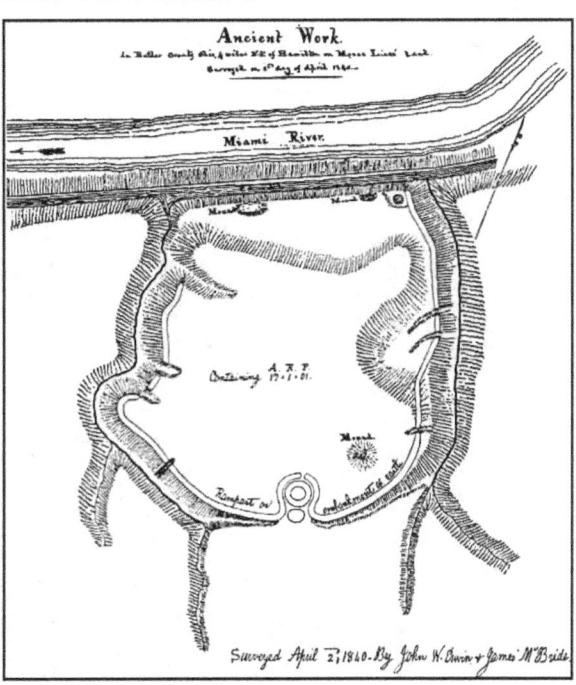

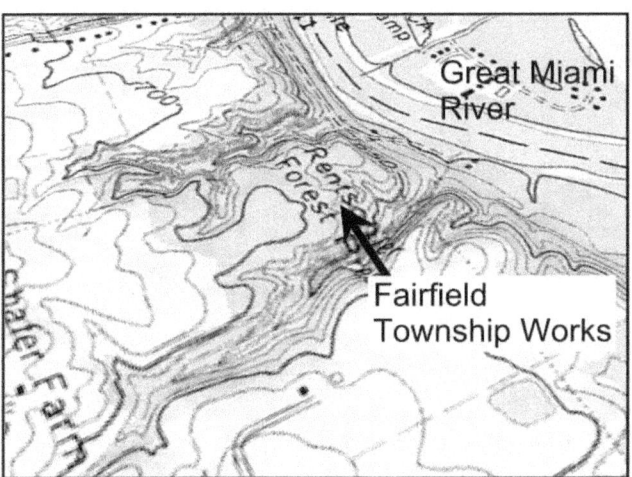

Above left: USGS 7.5-minute series topo map the Fairfield Township Works. Map provided by MyTopo, annotation added.
Below: LiDAR image of the Fairfield Township Works with McBride map superimposed (note 7). Vertical exaggeration = 1.4.

Figure 9.10. Great Miami River Valley: Fairfield Township Works Entrance
One of the interesting things about the Fairfield Township Works is its entrance. The ends of the perimeter wall curve inward forming a circle with an opening to the north. Situated within this configuration is a second circle, with no opening. A small mound "five feet high and forty feet in diameter" is located in the center of the main entranceway (MacLean 1879:182).

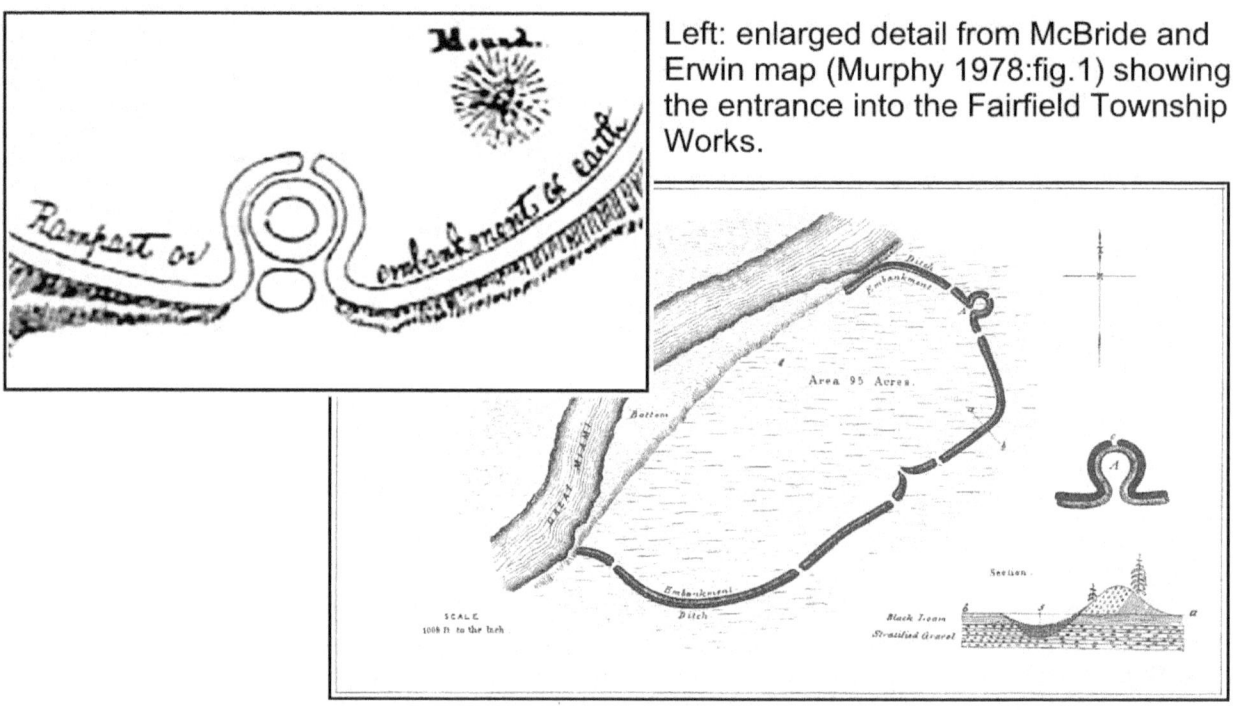

Left: enlarged detail from McBride and Erwin map (Murphy 1978:fig.1) showing the entrance into the Fairfield Township Works.

Above: Colerain Earthwork located about 12 miles south of the Fairfield Township Works (Squier and Davis 1848:Pl. XIII, no. 2). Note the similarity between the entranceways of the Colerain and Fairfield Township Works. (Colerain Works are no longer visible.)

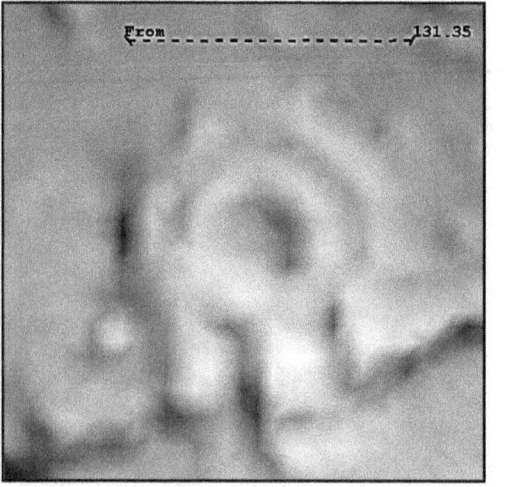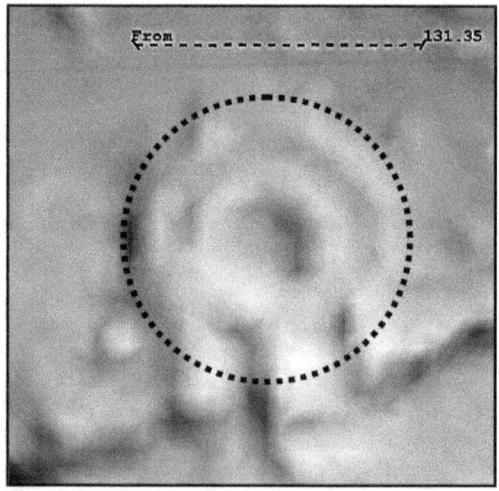

Above: LiDAR analysis suggests that the outside diameter of the circular entranceway into the Fairfield Township Works used the 1/8 HMU length (131.8 feet) in its design. (Vertical exaggeration factor 5.7).

Figure 9.11. Great Miami River Valley: Butler County Fortified Hill
The Butler County Fortified Hill is located about three miles south of Hamilton, Ohio. The site occupies the summit of a high hill more than 280 feet above the Great Miami River.

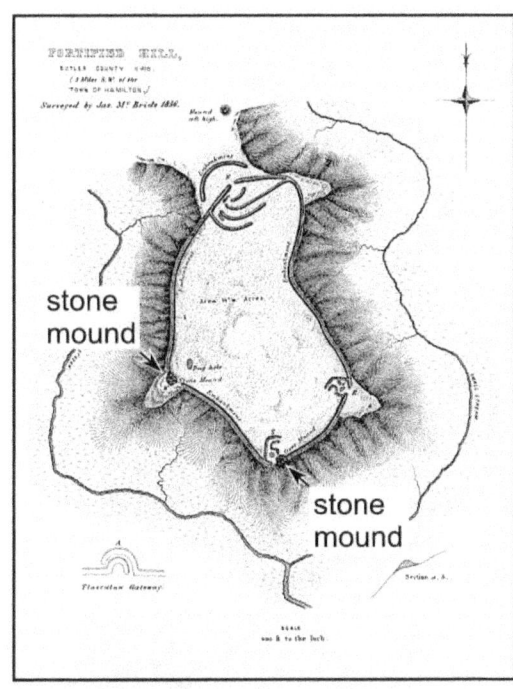

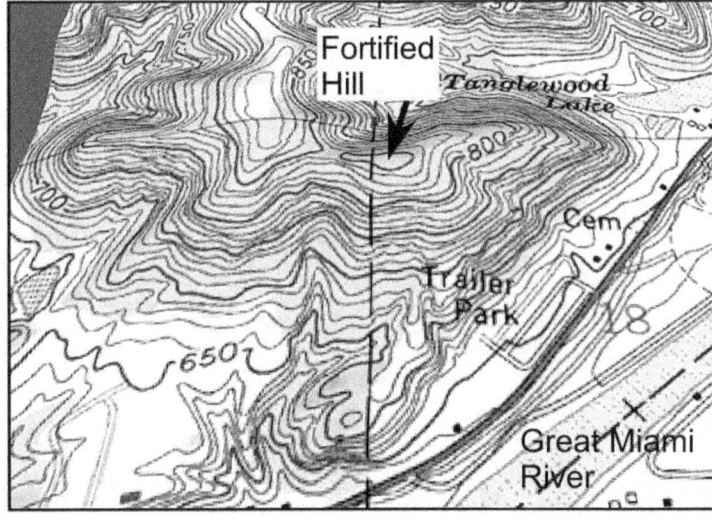

Above: USGS 7.5-minute series topo map detail showing Fortified Hill. Map provided by MyTopo, annotation added.

Above: Map of Fortified Hill made by James McBride in 1836 (Squier and Davis 1848: Pl. VI). A series of overlapping walls cover the north entrance. Eight-foot high stone mounds having indications of fire were at W and E.
Below: LiDAR image of Fortified Hill with McBride map superimposed.

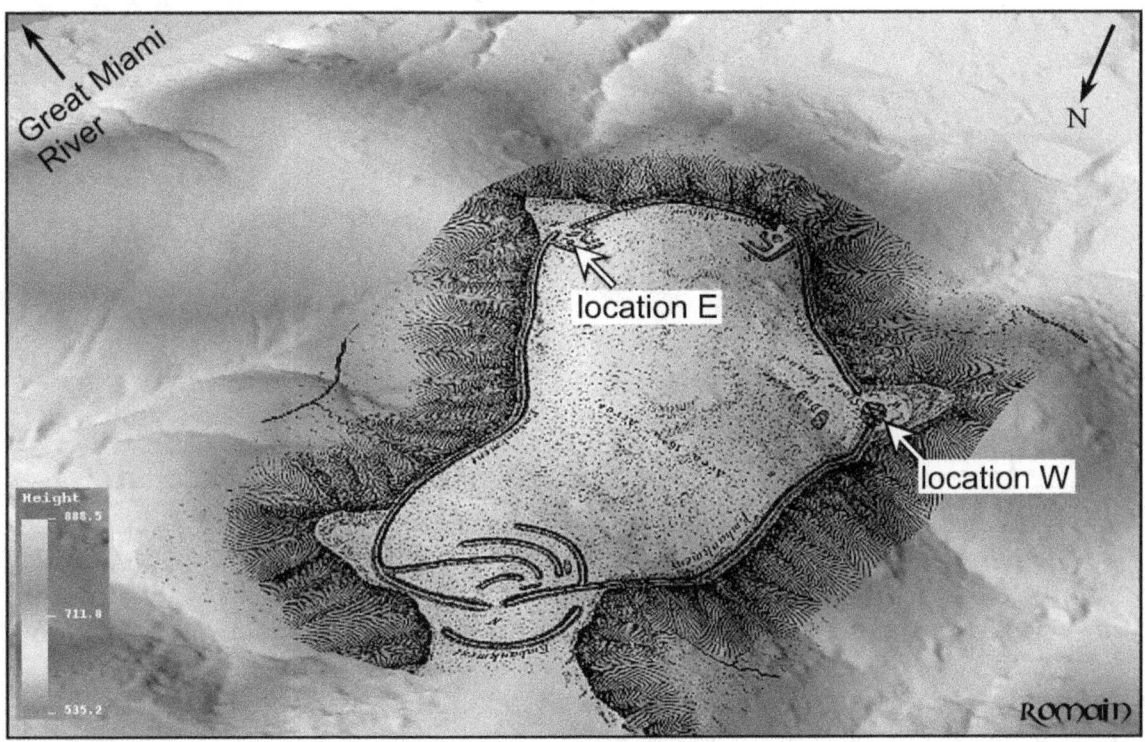

Figure 9.12. Great Miami River Valley: Pollock Wilson and other mounds
Several mounds in the Great Miami River Valley are located on high hills overlooking the river valley.

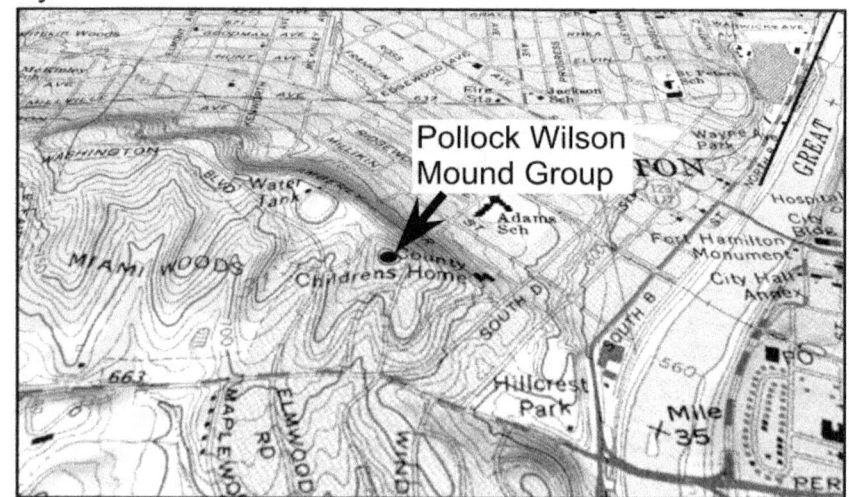

Above: the Pollock Wilson Mound Group consisted of three small mounds located on a high hill overlooking Hamilton, Ohio (MacLean 1879:215) (note 8). The Butler County Children's Home was built on the site in 1875. Map by MyTopo, annotation added.

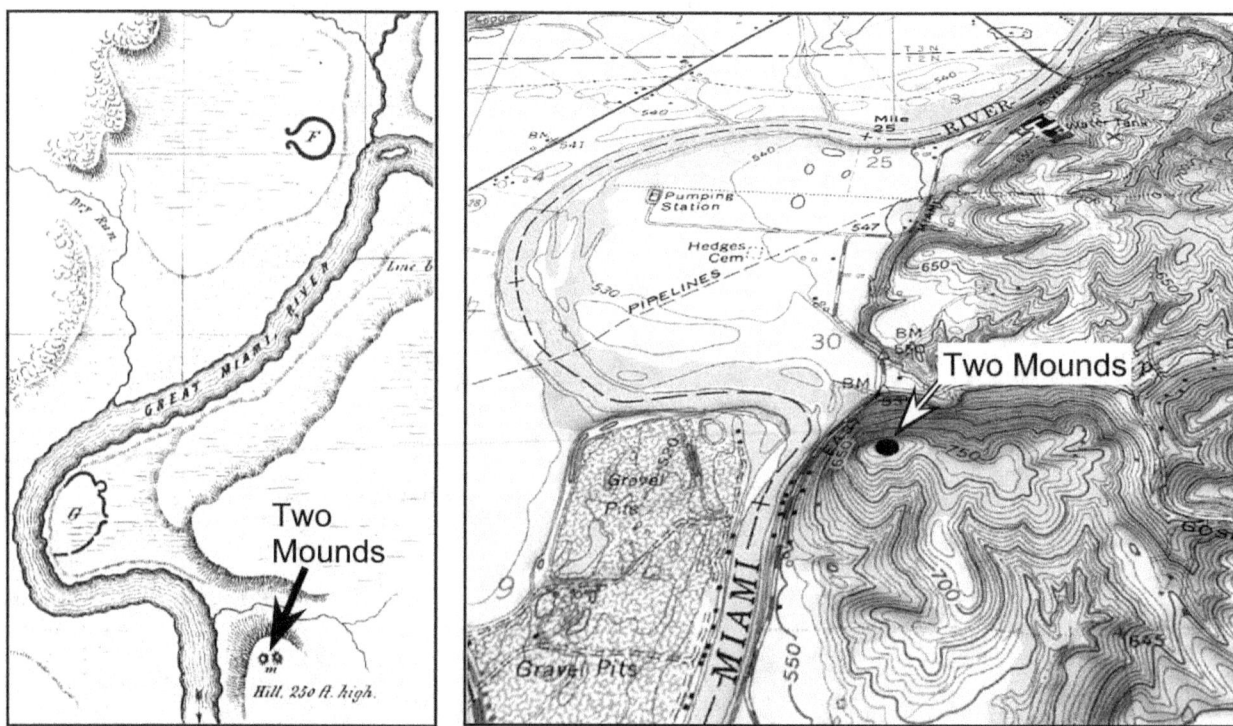

Above left: enlarged detail from Squier and Davis (1848:Pl. III, no. 2). Of interest are the Two Mounds shown on a high hill overlooking the Great Miami River, south of Hamilton. Squier and Davis (1848:36) described these mounds as "five and ten feet high.... composed of earth and stones, considerably burned." The observation that the mounds are "considerably burned" supports their suggested use for signalling.
Above right: USGS 7.5-minute series map with location of the Two Mounds plotted. Map provided by MyTopo, annotation added.

Figure 9.13. Great Miami River Valley: Miami Fort

Miami Fort is situated within the Shawnee Lookout Park at the confluence of the Ohio and Great Miami rivers. A large enclosure, burial mounds, and habitation areas are found within the park (Lee and Vickery 1972). Evidence indicates the area was used by Native Americans for thousands of years (Starr 1960). Miami Fort itself is an earthen enclosure with perimeter walls and an interior ditch.

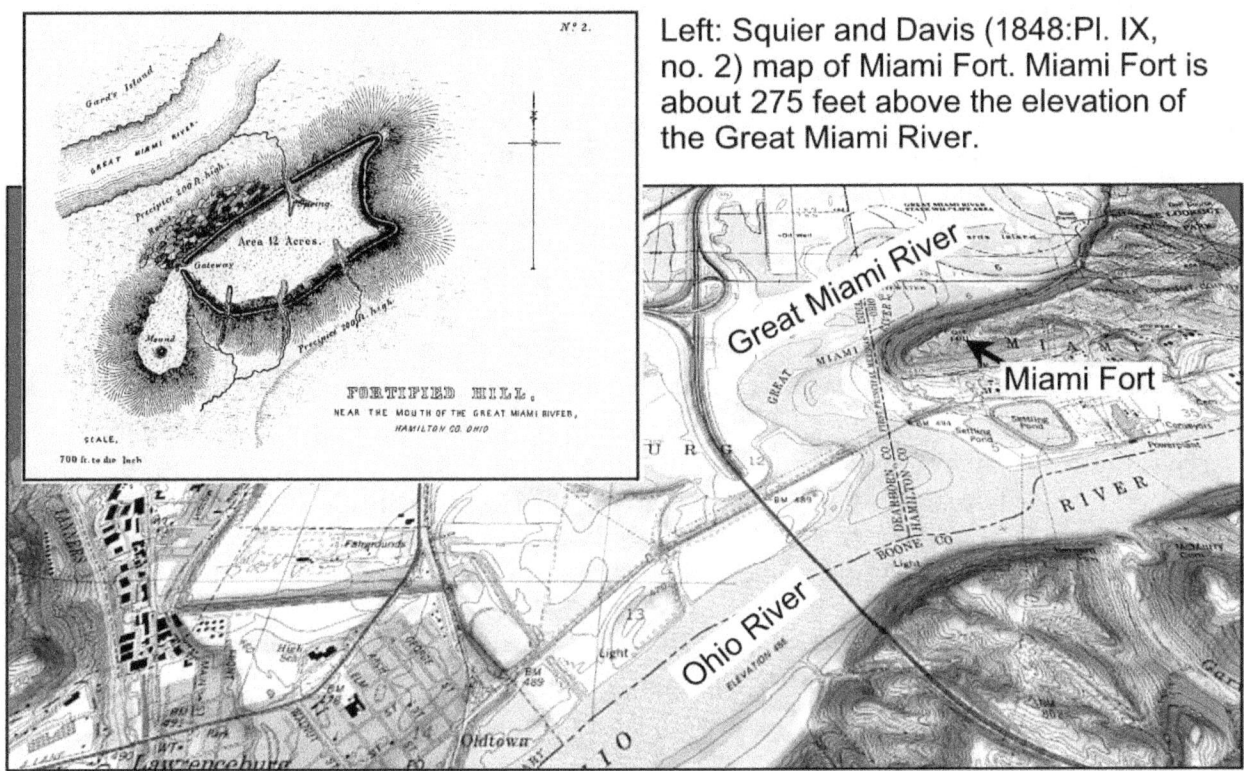

Left: Squier and Davis (1848:Pl. IX, no. 2) map of Miami Fort. Miami Fort is about 275 feet above the elevation of the Great Miami River.

Above: USGS 7.5-minute series map showing location of Miami Fort. Map provided by MyTopo, annotation added.

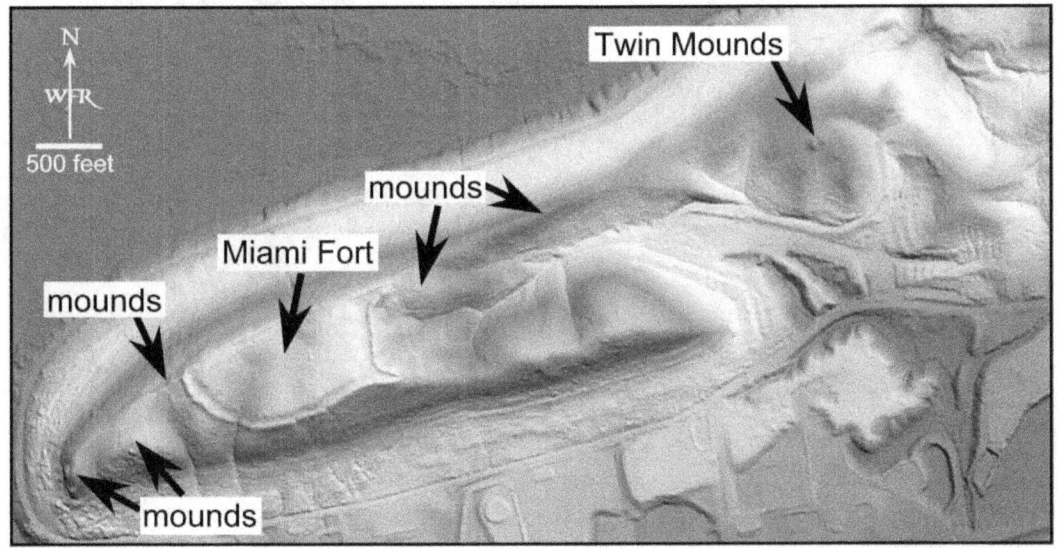

Above: LiDAR image showing Shawnee Lookout Park and several associated features. Mound locations based on Starr 1960, Fisher 1968.

Figure 9.14. Great Miami River Valley: Miami Fort Dimensions
The dimensions of Miami Fort appear based on a half-circle that incorporates the HMU.

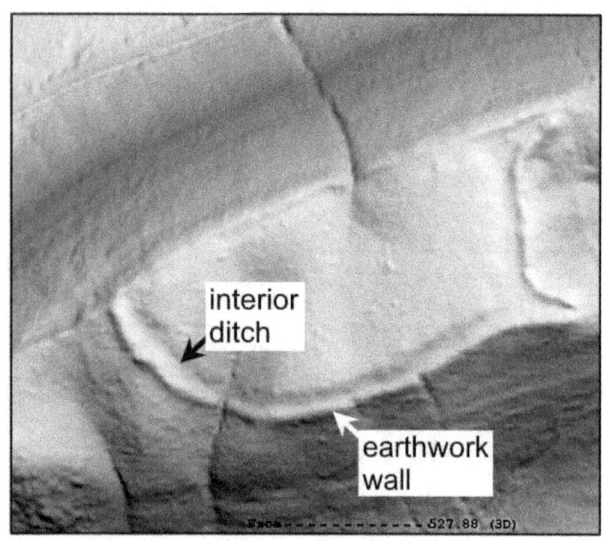
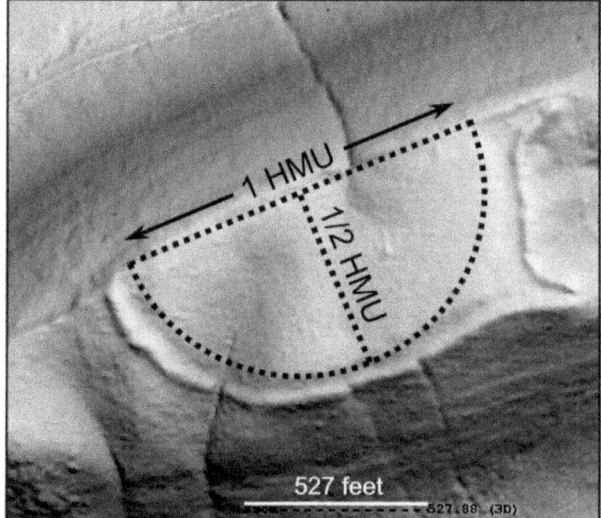

Above left: LiDAR image of Miami Fort showing perimeter walls and interior ditch. Above right: same image with a line equal to 1/2 HMU drawn. Using this line, a half circle is drawn. The diameter of the half circle is 1 HMU (or 1,054 feet); the radius of the half circle is 1/2 HMU (or 527 feet). Note how this geometric figure, based on the HMU, closely matches the size and shape of the interior area as defined by the inside edge of the ditch.

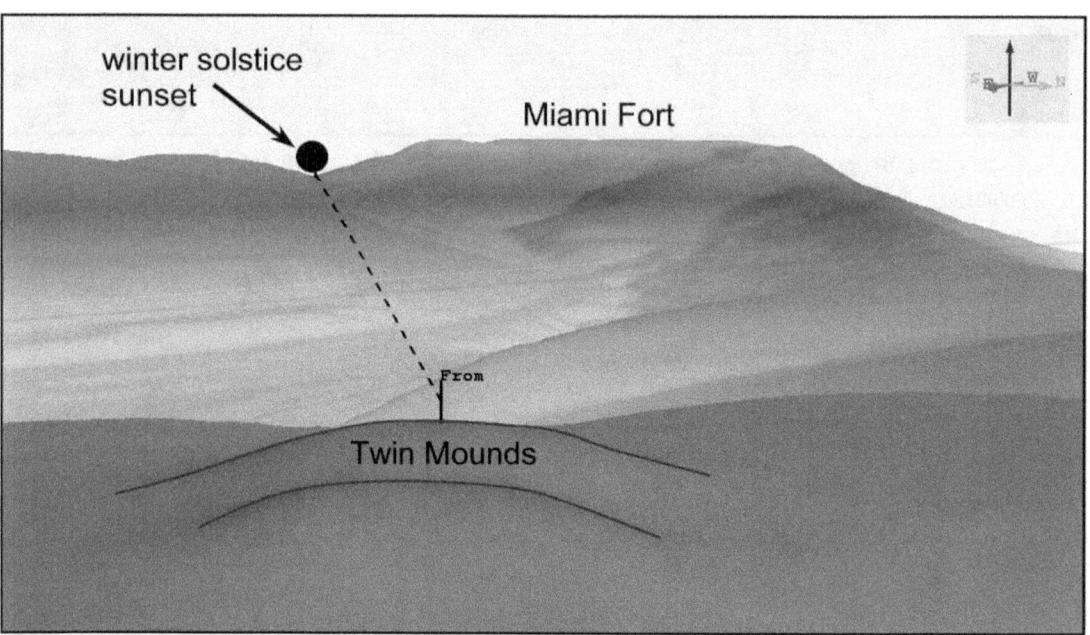

Above: LiDAR image showing the Twin Mounds - located about 1/2 mile east of Miami Fort. The Twin Mounds are situated so as to afford a view of the ravine that leads up to Miami Fort. Viewed from the Twin Mounds, the winter solstice sun will appear to set into the south side of Miami Fort (note 9).

Figure 9.15. Signal Mounds and the Great Miami River Valley
Of the mounds and enclosures located along the Great Miami River Valley, several could have been used for signaling purposes (note 10).

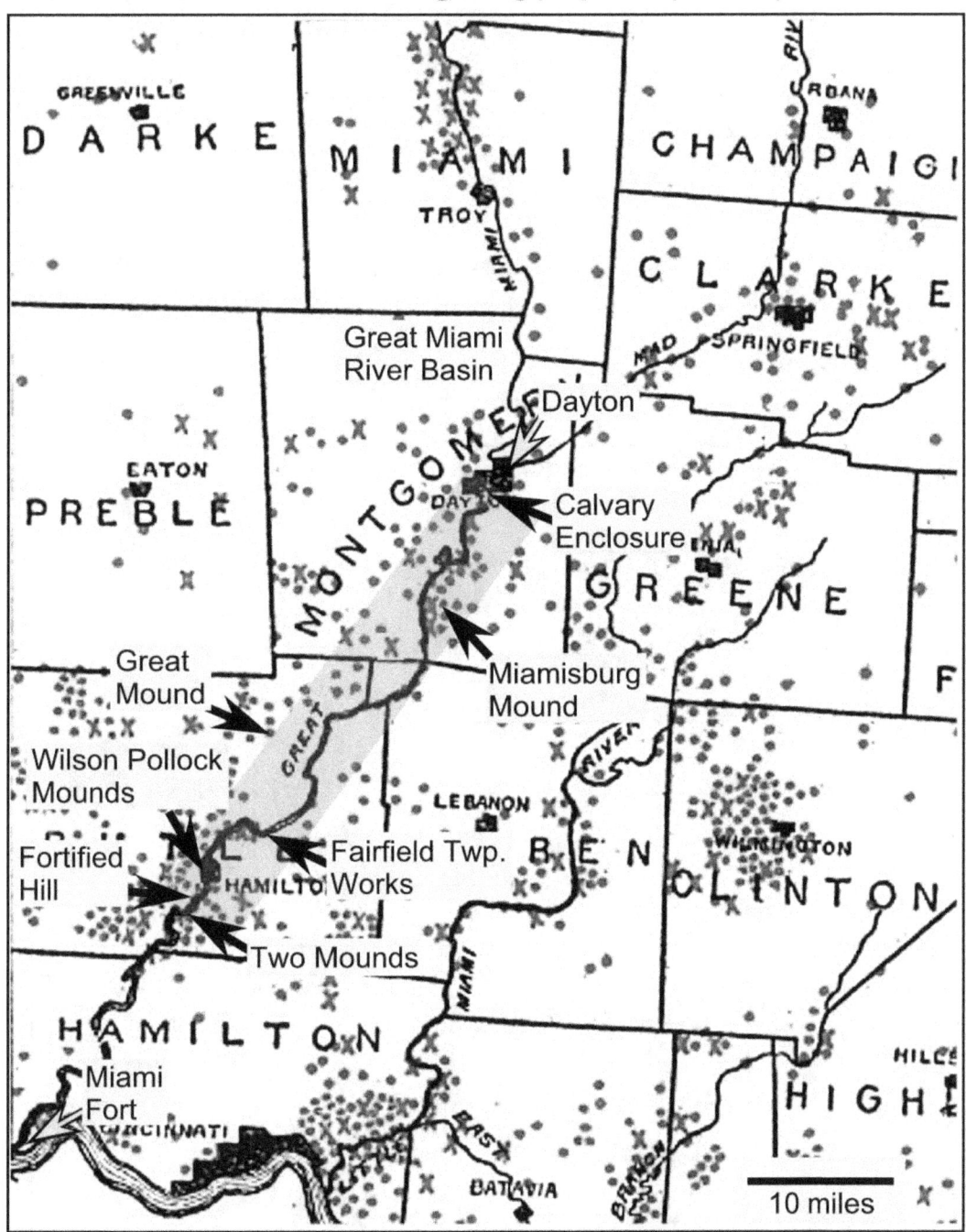

Above: shaded area shows length along the Great Miami River Valley where mounds are intervisible - from Two Mounds to Fortified Hill, to the Wilson Pollock Mounds, to the Great Mound, to the Hill-Kinder Mound, to the Miamisburg Mound, and to the Calvary Cemetery Enclosure - for a distance of about 40 miles. Composite map from Mills 1914, annotation added.

Figure 9.16. Great Miami River Valley: Site Intervisibility I
To assess the intervisibility between sites along the Great Miami River Valley, the line-of-sight feature of the mapping program Terrain Navigator Pro is used. For each instance below the program has generated a line of sight between selected mounds.

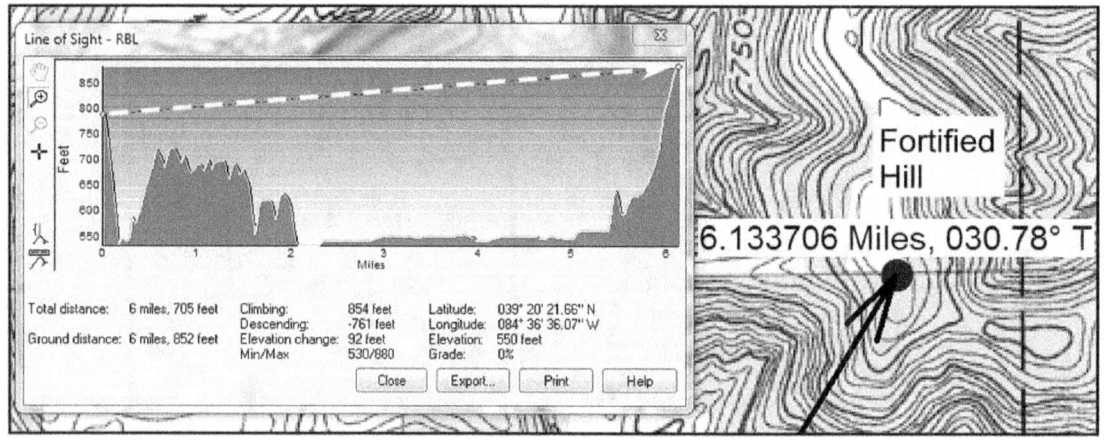

Above: line-of-sight between Two Mounds and Fortified Hill.

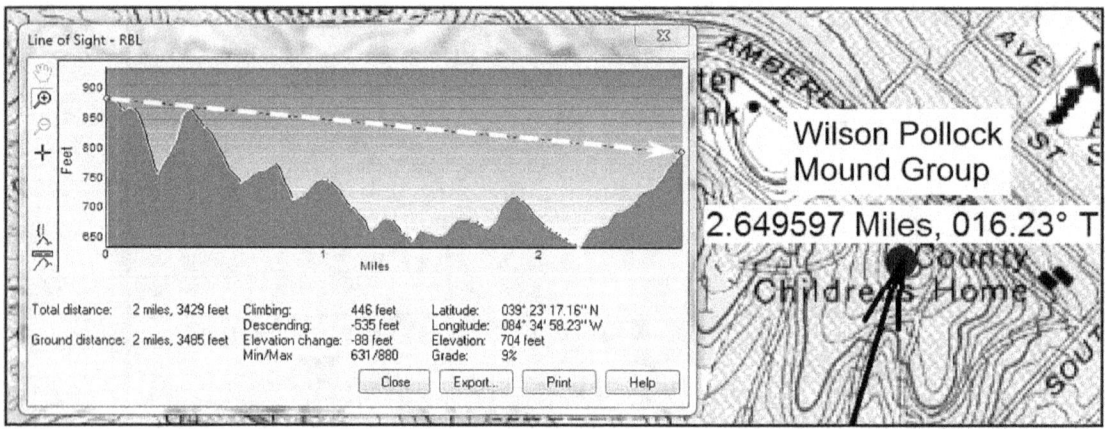

Above: line-of-sight between Fortified Hill and Wilson Pollock Mound Group.

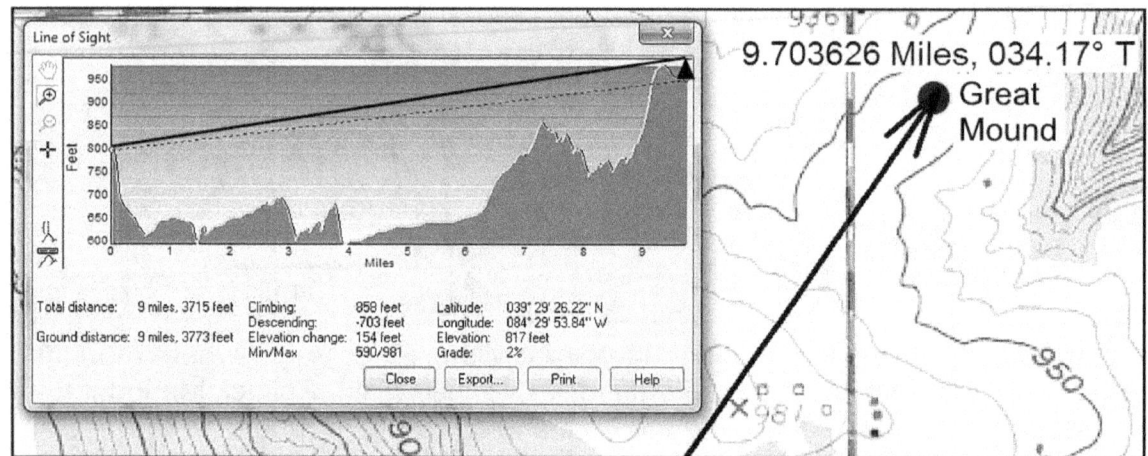

Above: line-of-sight between Pollock Mound Group and the Great Mound (note 11).

Figure 9.17. Great Miami River Valley: Site Intervisibility II

Continuing the line-of-sight assessement it is found that the total extent of inter-site visibility extends for 41 miles - from south of Hamilton to Dayton.

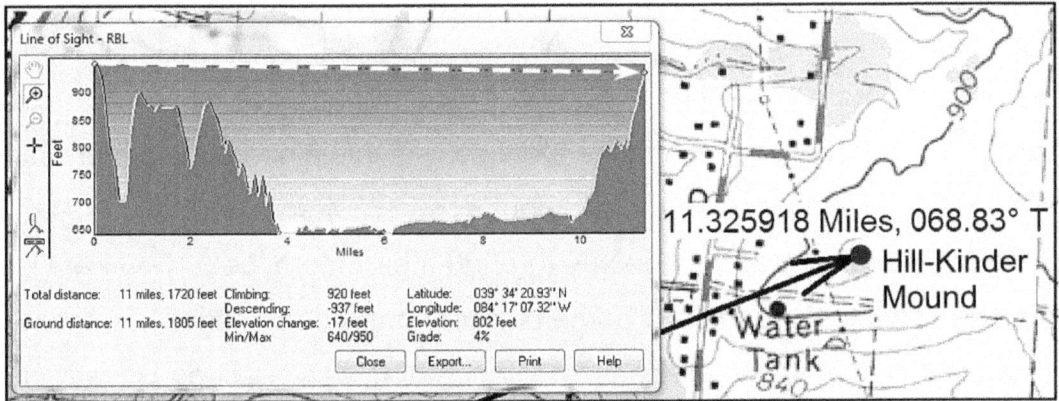

Above: line-of-sight between the Great Mound and Hill Kinder Mound.

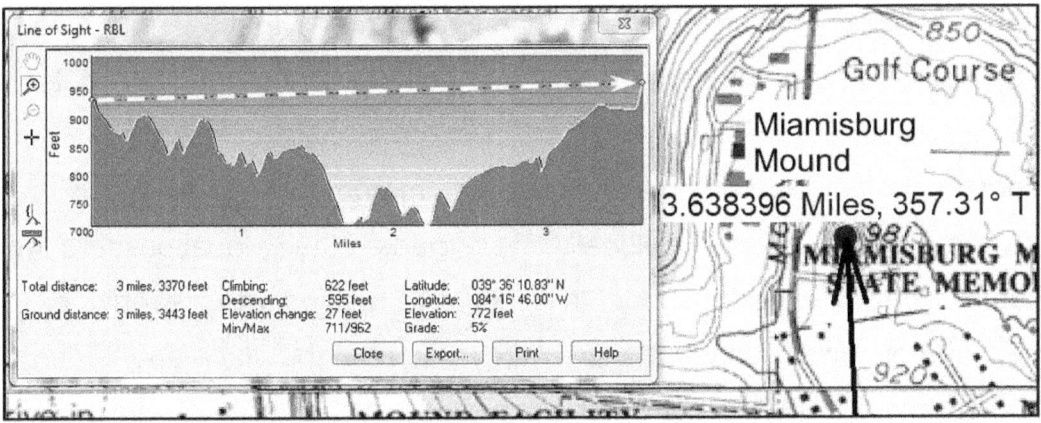

Above: line-of-sight between Hill-Kinder Mound and Miamisburg Mound.

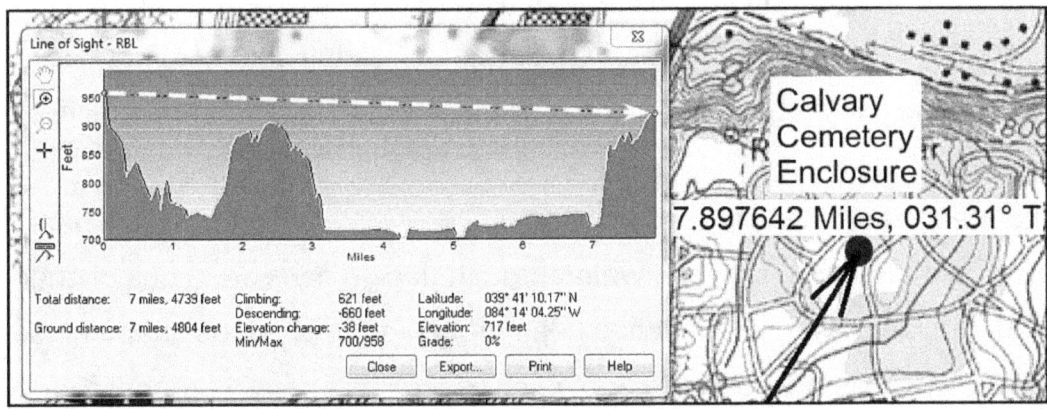

Above: line-of-sight between Miamisburg Mound and Calvary Cemetery Enclosure.

CHAPTER 10
DISCUSSION

Thousands of years ago, as Native Americans explored the Eastern Woodlands they occasionally came across special places. These were places that for a variety of reasons evoked emotional and spiritual responses. As explained by explorer Martin Gray (2007:1) "...these places had a mysterious power, a numinosity, and a spirit."

Some places resonated with people through their natural beauty. Others may have given rise to wariness or dread. Some places triggered memories of ancestors, or important past events. In some locations, the sun, moon, or stars lined-up with the earth in such a way as to inspire wonder and awe, or perhaps enable soul journeys to the Otherworld. In any case, what these places had in common is that they facilitated engagement with something other than self – i.e., something that manifested itself as numinous and sacred.

In these experiences, certain places became memorable. Grounded in experience and memory, Adena-Hopewell people marked these locations through the construction of earthworks and mounds. In a recursive fashion, the presence of earthworks and mounds enhanced the special character of the place and through ritual activity these places took-on even greater significance, with additional layers of memories and meanings. Thus an ever-expanding relational web was created that included people – living and dead, non-human agents, earthworks and mounds, celestial and terrestrial phenomena, myths, beliefs, memories, and ritual practices that together, provided meaning.

Where earthworks and mounds were built, they were typically built in harmonious relationships with the earth, sky, and water through design considerations of shape, size, and orientation. Based on what has been presented in previous chapters, several observations concerning these design variables as well as inferred purposes can be offered.

Place

Places are brought into being, or into conscious awareness, through the interaction between subjective human perceptions and the phenomenal world. The manner in which meanings for a place come about is explained by archaeologist Ruth Van Dyke (2007:238):

> "Places are made – meanings are inscribed onto the landscape – through ongoing meaning out of our own experiences, as well as the stories, traditions, and past experiences of others. Landscapes are the lived, sensual milieus in which the social and material worlds intersect and mutually recreate each other....Landscapes both shape and reflect worldviews and ideologies. Meanings are constructed over time, through layers of experiences, knowledge, and stories."

But there is more. In the relational world of the Moundbuilders, spirit, or the sacred, or numinous was present in special places in a way that could be tangibly experienced. At Newark, for example, the crisscrossing of solstice sightlines across the valley from hilltop to hilltop were not the result of human constructions. And yet that was the very point - i.e., the solstice sightlines between, for example, hilltops H1 and H4, or hilltops H2 and H3, were intrinsic to the landscape and were among the natural phenomena that made Newark special. They were visible expressions of the numinous. They were visible expressions of the sacred web that connected earth and sky.

In the Chillicothe area, multiple earthworks are connected by celestial azimuths to the most prominent mountain in the area - i.e., Sugarloaf Mountain. Indeed, Sugarloaf Mountain may have been a significant *axis mundi* for the Adena-Hopewell, linking important earthworks to cosmic realms.

Elsewhere, just south of Chillicothe, several earthworks are laid-out so they parallel the Scioto River while simultaneously extending in an orthogonal relationship to the moon's maximum north rise, thus connecting earth, sky, and water and providing an astro-geometric trajectory for an entire string of major earthworks.

At Serpent Mound one finds a gigantic serpent effigy nestled among faulted rocks and sinkhole entrances to the Lowerworld while at the same time connected through its body convolutions to the Milky Way constellation Scorpio.

These and other special places were marked by earthworks and mounds, or special places were themselves created through the building of earthworks and mounds. At such locations, through sight, sound, touch, and smell, people could enter into a conscious awareness of the sacred. Through such experiences, there emerged memories and stories, both real and mythic that forged connections to the past. Moreover, through the marking of such places by mounds and earthworks, the locations for future and continuing numinous experiences were identified.

For the Adena-Hopewell and probably for most Native Americans of traditional ways, to move through the landscape was to engage with a sacredness that could not be owned by a group of people, bounded by property lines, or specified in a deed. It was a sacredness that manifested itself to those who were sensitive to the world around them and moved to the rhythms of the earth, sky, and water. By physically entering into this sacredness Moundbuilder peoples re-affirmed their place in the cosmos and found stability, predictability, and meaning. For the Adena-Hopewell, the landscape was the very foundation for the field of relationships that was their life experience. For the Adena-Hopewell their way of being-in-the-world was inexorably tied to the land.

Of course one of the questions that comes to mind is: 'But why so many different earthworks?' If the object is to provide a location for an encounter with the sacred, wouldn't just one, or a maybe a few, centrally-located earthworks suffice?'

The answer, I believe, relates to how each earthwork mediates between human experience and the landscape. Each earthwork allows for a somewhat different experience. Each earthwork is, in effect, a unique consciousness-altering device that can be used to access and experience different nuances of the sacred, in much the same way as one would use different psychoactive substances, or different rituals depending on the desired interaction with the spirit world.

At the same time, earthworks are not and were not, static, unchanging objects. Each earthwork had its own life history. Although some were apparently built rather quickly, others were built, or added-to over generations. In either case, all were acted upon by various forces – e.g., fire, earthen deposition, rain, snow, wind erosion, sun baking, plant growth, animal burrowing, and so on. Thus humans, earthworks, and cosmic forces were intertwined in a matrix of dynamic relationships that was sensuous, ever-changing, and forever in the process of becoming. A single earthwork therefore could offer different experiences depending on the state of the matrix at any given time. Multiple earthworks offered many more varieties of experience.

Evidence presented in the preceding chapters suggests that each earthwork was fine-tuned to specific landscape features. Mounds, for example, might be located at entrances to valleys, or on high ridges overlooking a river valley. Hilltop enclosures take advantage of high cliffs, gullies, and surrounding streams to accentuate and close-off interior space creating the equivalent of earth islands. Geometric earthworks are typically situated at locations where earth, sky, and water trajectories intersect. Further, earthworks and mounds are often located at river confluences, near special outcroppings of rocks or minerals, at entry points into special areas, and in the case of hilltop enclosures, at locations that command panoramic views. In short, earthworks and mounds were situated at locations that memorialize a particular feature, or perspective of the world.

It is important to note, however, that the Moundbuilders were not simply marking special places in a passive manner. They were simultaneously and intentionally changing and modifying the world to suit their cultural needs. They were gathering and transporting special soils, stripping away topsoil, leveling areas, clearing forest, creating and enhancing water features; and on a massive scale, changing the visual landscape by adding earthworks and mounds. Architect John Hancock (2010:304) explains it this way: "It seems apparent that these projects were cosmological 'intensifications of the natural environment: completing water boundaries, leveling artificial horizons, foregrounding iconic hills and valleys, memorializing temples and the dead." In this, Moundbuilder activities reflect an on-going and dynamic interaction with the world, which is to say, earthworks and mounds do not simply reflect static cosmological ideas; they are not simply representative of beliefs, but rather, are dynamic places where on-going relationships with the landscape were engaged-in over generations.

With the understanding that there are always exceptions to the rule, the following observations can be made about Hopewell geometric earthworks relative to place:

1. Out of 19 major Hopewell geometric complexes analyzed (note 1), 18 are located in river valley settings (Romain 2004a:86). Other analytical categories where sites could have been located included: hill-ridge, plateau-plain, and floodplain topographic settings.

2. Hopewell geometric earthworks are associated with water in several ways. Physically, most are situated near rivers, streams, or river confluences. Statistical assessment of 19 major geometric earthwork complexes found that all are located less than 1 mile from a major water source (Romain 2004a:98). Several are immediately adjacent to, or include natural springs within their boundaries. Many have water features built into their design. Several earthworks, for example, have deep ditches around their perimeter walls. Many have large dug pits situated in close proximity that in addition to serving as borrow pits, may also have been pond features. Others have parallel-walled embankments that lead down slope to a nearby river. We cannot know exactly what Adena-Hopewell beliefs were about water. If we use ethnographic accounts as a guide, however, then bodies of water were likely believed inhabited by Lowerworld beings and considered as entrances to the Lowerworld.

3. Geometric earthworks are generally situated on level ground. This observation is based on a statistical analysis comparing the slope for 17 major complexes to the average slope values for the respective counties that each complex is located in (Romain 2004a:62-64, 92). Using a statistical measure known as the t-test for matched pairs, the results show a statistically significant occurrence of geometric complexes in areas that are more level than is the average for the county where each is located (obtained t ratio = 6.555, table t ratio = 2.120, df = 16, P = 0.05).

4. Geometric earthworks are located on well-drained soils. This finding is based on a statistical analysis comparing the drainage characteristics for 18 earthwork complexes

to the drainage values for the respective counties each complex is located (Romain 2004a:66-67, 94). Using the *t*-test for matched pairs, the results show a statistically significant occurrence of geometric complexes in areas that are better drained than is the average for each county where the complexes are located (obtained t ratio = 6.324, table t ratio = 2.110, df = 17, P = 0.05).

5. Geometric complexes are often located on soils that by today's standards are considered highly productive in terms of agricultural potential (Romain 2004a). Seventeen major geometric complexes were able to be assessed for wheat and hay yields based on soil types found within the enclosures (note 2). The yield data for wheat productivity for soils situated within the geometric complexes was generally higher than the average for the county where the complexes are located, but not high enough to be statistically significant. On the other hand, analysis showed that the soils on which the geometric enclosures are located are about 15% more productive for hay than is the average for the counties where they are situated (obtained t ratio = 4.543, table t ratio = 2.120, df = 16, P = 0.05). This is a statistically significant result.

Of interest is that hay is comprised of grasses, which in many ways are similar to cultigens that Hopewell people relied upon to include: goosefoot or chenopodium, marshelder, sunflower, knotweed, and maygrass. Indeed, one of the characteristics of the plants just mentioned is that they all thrive and favor the open, disturbed, river valley habitats characteristic of where the major geometric earthworks are located (Romain 2004a:171-173). Generally considered "weeds" by farmers today, these food plants would have been early successors in the interior areas cleared by earthwork construction.

Alignments

In many archaeoastronomic studies, statistical assessments are presented summarizing such things as the frequency of solar versus lunar alignments, ranges of sightline accuracy, tests for randomness of posited alignments, etc. Indeed, I have on occasion provided such assessments (e.g., Romain 2004a). Statistical assessments provide interesting and useful information. However, what statistical assessments fail to capture is what makes an earth- or sky-aligned site of interest in the first place – which is generally related to its uniqueness. Thus sites need to be considered relative to their uniqueness and the finding that a particular site alignment is unusual or one-of-a-kind is not necessarily a reason to consider that alignment as fortuitous.

That said, there are several statistical-based observations that provide useful insights into Adena-Hopewell astronomy. For example, more earthworks are solar-aligned than lunar-aligned. Summer and winter solstices apparently held special significance. Lunar-aligned earthworks tracked the 18.6-year cycle of the moon. Fewer earthworks are oriented or associated with star constellations, although there are important exceptions.

Lunar alignments found in Adena-Hopewell are of interest for a couple of reasons. First, alignments to the moon's maximum and minimum rise and set points demonstrate a commitment to long periods of detailed tracking across generations. Tracking the moon was not a casual or occasional exercise.

Second, from what I can tell, lunar alignments first appear in the Eastern Woodlands with the Adena-Hopewell. No doubt the monthly cycle of the moon was always important to hunting-gathering peoples, especially as related to nighttime hunting by the light of the moon. But interest in the 18.6-year lunar cycle as manifested in earthwork alignment seems to be something new at this time – at least in the Eastern Woodlands.

Notably, Adena-Hopewell lunar consciousness is apparent not only in earthwork alignments, but is also expressed in:

1. lunar counts found in Adena tablets (Romain 1991a);

2. crescent-shaped banner stones (e.g. Converse 2003);

3. cut mica crescents (e.g., Mills 1916:280);

4. copper crescents (e.g. copper crescent found in the Newark Eagle mound – see Chapter 3);

5. lunar alignments of spirit houses (often referred to as charnel houses) (Romain 2000); and

6. crescent-shaped earthworks (e.g. the crescent-shaped earthwork associated with the Newark Eagle mound, and the crescent-shaped earthwork found at the Yost earthworks).

In the natural world, the crescent moon is, by far, the most obvious and visible crescent shape, thus giving rise to the interpretation of the crescent shape as lunar inspired.

In any case, earthwork alignments are not limited to celestial events or bodies; but as demonstrated in previous chapters, earthworks are often simultaneously aligned along terrestrial and watercourse trajectories. The Newark Octagon, for example, is aligned through its major axis to the moon's north maximum rise. The Octagon earthwork is also and simultaneously aligned, however, parallel to the edge of the adjacent Raccoon River terrace.

The alignment of earthworks parallel to rivers and river terraces was noted for a number of Hopewell sites by Morgan (1980). This design feature is also found in later periods and cultures. Many Mississippian sites have earthen mounds situated so that their major or minor axes extend parallel to the lay of the land, generally established by the trajectory of nearby rivers (see e.g. Morgan 1980).

The fact is, there multiple alignment protocols evident in Adena-Hopewell earthwork constructions. In many cases earthworks are diagonally aligned to sky events. Other times, orthogonal relationships to earth, water, and/or sky phenomena are found. Some earthworks incorporate intersecting celestial alignments; while others include intersecting landscape and celestial alignments. Different earthworks incorporate combinations of alignments in unique ways. Perhaps each unique combination had a different intended ritual result as well as offering its own unique sensual experience. In all cases, however, through the alignment of earthworks to earth, sky, and water, Moundbuilders actively engaged with non-human entities and established for themselves, a participatory role in integrating cosmic realms.

One aspect of this engagement refers back to the recognition that meaning is mutually constructed. In other words, a celestial or terrestrial alignment only has meaning relative to human awareness of it; while in a reciprocal fashion, there is no alignment without the existence of the celestial object or terrestrial feature. So it is that two agents – one human and the other non-human, interact, thereby bringing into awareness an emergent and mutually constructed alignment phenomenon to which meaning is given.

The question that then arises is: 'What are the implications of earthwork alignments?' The answer I propose is that Adena-Hopewell alignments involve an understanding of the cosmos as revealed through pattern recognition. For example, we often seek to understand the essence of time as well as the relationship between time and death by first identifying spatial-temporal patterns. Cosmic temporal patterns are memorialized when we align earthworks to them.

On a more pragmatic level, astronomical alignments are useful for timing rituals and other events. In this regard, the comments of John Hancock (2010:269) are worth quoting:

> "The ceremonies are in part about subsistence activities such as planting or consecrating food and would necessarily evoke the rain and the soil and times of the seasons, among many other things. Astronomy then, for example, is no longer seen as an abstract branch of physics in which the Hopewell had expertise but as a linking process between the heavens and the earth, food production and the seasons."

This is not to say that astronomically-aligned Adena-Hopewell sites were astronomical observatories in the sense understood by Western astronomers. Nor were they calendar devices. Moundbuilders did not need astronomic alignments to know when to plant, harvest, gather, or hunt. There are other, better and more sensitive indicators that can be used. Rather, one important and likely reason for aligning earthworks to celestial events was to establish the proper or most auspicious time for ritual events and public ceremonies. People might, for example, agree to meet at a particular earthwork for a world renewal ritual on the night of the first full moon after the winter solstice. In setting these kinds of agreed-upon times, there is not much room for error. This kind of precision is useful in a world without modern time-keeping pieces, especially when traveling significant distances to arrive at a special event.

Of course the kind of time-reckoning just described does not require something as immense as the Newark Octagon. Nor does it explain the occurrence of simultaneous alignments of many earthworks to both earth and sky phenomena. Equally important as time-reckoning is that alignments integrated human activities into the fabric of the cosmos in a harmonious way. In other words, earthworks provided dedicated space where human beings could mediate cosmic forces that directly affected them.

Shape and Geometry

The orientation of an earthwork to a celestial event or terrestrial feature is often intended to direct conscious awareness to that phenomenon. Geometry functions in a different, but equally informative way. Geometry provides recognizable shapes that can be invested with particular meanings held in common by more than one person. Notably, geometric symbols can be of any size and still convey the same meaning. Thus if a crescent shape is used to symbolize the moon, the meaning of that shape holds whether it is made to the size of a huge earthwork, or a small copper piece.

Since geometric shapes can convey symbolic meanings, they are often used to influence thinking and emotional responses in a target audience. That is to say, the phenomenological experience of standing in the center of a circle earthwork is qualitatively different from that experienced within an octagon or square. Those experiences are further influenced by whatever meanings are culturally associated with those shapes. With reference again to the crescent symbol, if a crescent-shaped earthwork was intended as a moon symbol, then for people who know its meaning, when entering into proximity to a crescent-shaped earthwork, its shape would presumable guide thinking or awareness to the moon. In this way, the meaning-laden geometric symbol brings to mind, or references the moon just as assuredly as a direct alignment, albeit in a different way. The point is that geometrically shaped earthworks are not merely passive designs. They have the ability to influence thinking and behavior. Indeed, the ability of geometric symbols to influence human thinking and behavior is the essence of their seemingly "magical" power.

For Adena-Hopewell earthworks built in river valley areas, the preference was for symmetrical, geometrically shaped structures. Given that in these relatively flat areas any number of asymmetrical or symmetrical shapes could have been built, the preference for symmetry is of interest. As explained elsewhere (Romain 2009a, 2000), there are reasons to think that this preference relates to an understanding of the cosmos as dualistic, with complementary and opposite aspects. Geometric symmetry conveys the notion of balance between opposites. In this, Adena-Hopewell earthworks seem to recreate – albeit on a smaller scale, the symmetry found in the cosmos.

If pressed to interpret the dualistic symbolism of the earthworks, I would propose that circle earthworks replicate the circular shape of the flat earth with surrounding walls creating artificial horizons and perimeter ditches symbolizing the primal waters out of which the earth emerged (Romain 2000). No doubt there were exceptions to this. Some circle earthworks, especially those with crescent-shaped earthworks within, perhaps were moon symbols.

Hilltop enclosures seem concerned with the creation of a closed area delineating special space within. As is the case with circular earthworks, the intent may have been to create a miniature world.

Square and octagon shaped earthworks seem associated with sky world. Certainly it is a convenient fact that straight line walls as well as many of the diagonals across various square enclosures point to celestial events.

That said, it would not surprise me if circles and squares were interchangeable as either earth or sky symbols, depending upon the context and purpose of the rituals conducted within.

The chronology is not entirely clear, or straight-forward, but in general, the earliest geometric earthworks seem to have been small circles. More complicated forms resembling something in-between a circle and square occur in Late Adena. Large squares and octagons and other geometric shapes are typically Hopewellian.

As discussed earlier (Chapter 6; also see Romain 2000), Hopewell earthwork designers appear to have recognized four basic relationships between circles and squares: 1) square containing a circle - e.g. Hopeton; 2) square contained in a circle - e.g., Circleville, Milford, Works East; 3) square and circle with equal perimeters - e.g.,

Newark Wright Square and Newark Great Circle; and 4) square and circle with equal areas – e.g., Marietta Large Square and the Liberty Circle.

An interesting example of a structure that demonstrates Hopewell interest in the geometric relationships between circles and squares is provided by the recently-discovered Hopewell Mound Group Circle and Woodhenge (Burks 2013, 2014; Engberg 2014; Romain 2014b; Ruby 2014a). In this case, the 'square inside of a circle' concept finds expression. Shown in Figure 10.1, the Hopewell Circle and Woodhenge is located within the Hopewell Mound Group enclosure. Originally the Great Circle and Woodhenge consisted of a circle of wooden posts, surrounded by an circular earthen embankment with an exterior ditch. The embankment wall has been seriously degraded by plowing, the ditch is now filled-in, and the posts removed long ago. Still, evidence for the structure is revealed through geophysical survey. Figure 10.1 shows an annotated magnetic gradiometer image of the structure. In this image, the ditch and posthole footprints are visible.

Figure 10.1 further shows how the northwest gateway of the Circle is aligned to the summer solstice sunset as viewed from the center of the circle. Moreover, the size of the circle is based on its geometric relationship to an imaginary design square that uses a sub-multiple of the Hopewell Measurement Unit (1,054 feet) to establish its size. In particular, the circle ditch circumscribes an imaginary design square that is 1/4 HMU on each side. Thus the size of the circle is geometrically related to the square.

Noteworthy too is that the orthogonal of the solstice sightline (which forms one of the square's diagonals) is used to establish the location for a second gateway. The significance of this it that it clearly demonstrates the use of orthogonal relationships by the Hopewell, thus supporting the notion that the orthogonal relationships noted elsewhere (e.g., Newark and High Bank) were intentional.

Size and Mensuration

In order to build large-scale earthworks it makes sense that the Moundbuilders would have first built small models or design plans before attempting a full size structure. That way, relevant concepts to include astronomic alignments and geometric relationships between component parts could be visualized and communicated to those engaged in the actual construction. If that was the case, then, in order to translate small models into full-size earthworks, at least two things would have been needed: 1) a basic unit of length or agreed upon unit of measure; and 2) a method of counting.

Units of measure quantify 'how big' and provide an order to things. In the case of the Adena-Hopewell it has been shown how the 1,054-foot unit of length (also referred to as the Hopewell Measurement Unit, or HMU), or its submultiples, greater multiples, or other iterations are embedded in virtually every earthwork considered. Elsewhere (Romain 2000:93-96) I have proposed how the Hopewell Unit of Measurement may be based on the length of the human arm.

Especially interesting is that, regardless of what was used as a standard for the 1,054-foot unit of length, that unit was known to Adena-Hopewell people across a broad geographical area - further indicating that there was a detailed body of knowledge shared widely across time and space.

What Do Earthworks Do?

Having considered some of the design features incorporated in the earthworks, the next question is: 'To what ends were the earthworks and mounds created?' Undoubtedly, earthworks and mounds had multiple purposes, meanings, symbolism, and uses. In a summary chapter limited by space, it is not possible to discuss all the possibilities. What follows, however, is a list - in no special order, that identifies some of the more important things I believe earthworks 'do.' While the list is not all-inclusive, it does provide a beginning point for future conversations.

Identify Special Places

As previous chapters have documented, Adena-Hopewell earthworks and mounds were situated in special places that gathered celestial and terrestrial features in a way that facilitated particular sensory experiences and allowed people to enter into, or experience the sacred. In effect, earthworks and mounds stood at liminal locations, or places of intersection and convergence with other realms. In the case of geometrically-shaped earthworks, those designs brought to mind cosmological principles that helped interpret the phenomenological experience.

In many instances the landscape features that likely made a place special are obvious. The panoramic views, for example, from Spruce Hill, Fort Hill, Fort Ancient, or the Miamisburg mound are dramatic and awe-inspiring. The sensuous experiences offered by these locations makes them worthy of memorialization by earthwork construction.

So too the view from Mound City of the celestial events bracketing the Chillicothe mountains marks that location as special and again, worthy of commemoration. Similar visual heirophanies are witnessed at Fredericktown as the solstice sun rises in alignment with a small valley to the east. At Marietta, one cannot help but be moved by the winter solstice sunset as the sun sinks into the horizon in alignment with the Sacra Via. In these and other instances the phenomena that make the location special are obvious. Of equal interest, however, are those locations where the numinous is not quite so obvious, at least to the uninitiated.

Looking to the Newark Earthworks, for example, it would take years of *post hoc* observation to understand how the crisscrossing of solstice azimuths across the Newark valley contributes to making that area special. So too, recognition of how the 18.6-year cycle of the moon is incorporated in the design of the Octagon is not immediately apparent. It is also the case that many earthworks and earthwork complexes are so large that their geometric relationships are not apparent at ground level. When standing inside of the Newark Great Circle, for example, recognition that the earthwork is a circle is obvious enough. But at ground level it is virtually impossible to know by visual observation that the Great Circle and nearby Wright Square have circumferences of equal length. Similarly, it is impossible to know from visual observation at ground level, that the diameter of the Newark Great Circle is based on the length of the hypotenuse of a triangle that uses 1 HMU and ½ HMU for its sides.

What this suggests is that the details of what made certain earthworks special in terms of location and design variables may not have been known to everyone. First-time visitors or pilgrims probably needed to be introduced to the detailed nuances of some sites, just as is the case today. Consider, for example, that although known to archaeologists for more than 100 years, the lunar alignments of the Newark Octagon were only recently recognized. To me, this implies the possibility of shamanic initiatory levels. In other words, while the factors that made some earthworks special were probably known or accessible to everyone simply due to the obvious; information about the design factors that made more complicated sites special was perhaps limited to those who met some special criteria. Indeed, in some instances it may even be that only the original earthwork designer really knew what made an earthwork special and others had to rely upon that judgment, simply as a matter of faith.

In other words, while it is true that earthworks and mounds were focal points where cosmic realms, forces, and phenomena intersected with human consciousness, it was also the case that contextual information was needed to order to recognize and interpret the details of those intersections. The implication of this and why the point is important is that it speaks to a way of being-in-the-world that is not entirely centered on self, but rather, also relies on the accumulated wisdom and teachings of those who have gone before.

Symbols of Cosmological Truths

Reflected in the Adena-Hopewell earthworks are the most sophisticated understandings of the cosmos that the ancient Moundbuilders possessed, based on hundreds of years of observation and contemplation. Elsewhere (Romain 2000, 2009) I have discussed some of the cosmological ideas I believe represented in the earthworks. Among these concepts are a multi-layered cosmos connected vertically by an *axis mundi*; balanced dualism expressed in the horizontal plane by four quadrants or directions and in the vertical plane by an Upperworld and opposite Lowerworld, with This World balanced in-between; the idea of a center place at the intersection of the horizontal and vertical realms; and an understanding of time as cyclic.

In the Moundbuilder earthworks these cosmological ideas were physically manifested through design features and most importantly, could be phenomenologically experienced through peoples' movement among and through the earthworks, their visible expressions of geometry and celestial alignments, and invoked memories of events at those locations. Through earthwork design and layout, the Moundbuilders were able to memorialize the cosmological beliefs they found most relevant. The Newark Earthworks provide a useful example.

The Newark Observatory Circle and Octagon are comprised of two opposite geometric shapes connected by a parallel wall walkway. (The octagon shape is based on the more fundamental shape of a square.) In this configuration, there is a pairing of opposite shapes representing a dualistic use of space. Mediating between these two shapes are the parallel walls that connect geometric earthworks. Both spatially and experientially, this connecting wall segment creates a center place or point of balance between the two earthworks. Moreover, the major axis of the combined Observatory Circle and Octagon is aligned to the moon's maximum north rise thereby bringing the entire earthwork into harmony with a long-term lunar cycle.

If the alignment of this earthwork to the moon was intended at least in part, to coordinate the timing for periodic gatherings of people at the site, then this suggests that cosmological concepts of balanced dualism, center place, and cyclic time, were not only intellectually shared but further, these concepts were intended to be phenomenologically experienced as people moved between the Circle and Octagon, guided by the parallel-wall walkway along a celestial azimuth. In this, earthwork design was integrated into a web of relationships that included cosmological ideas, social interaction, and very likely, ritual activities intended to connect people with the sacred.

Integrate Cosmological Realms

Earthworks that are laid-out in simultaneous alignment to the course of rivers and celestial events integrate the forces of moving water with the movements of celestial entities. Examples of simultaneous alignments include the Newark Observatory Circle and Octagon, Hopeton, High Bank, Liberty, and Works East earthworks. For example, the major axis of the Newark Observatory Circle and Octagon earthwork is simultaneously aligned to the moon's maximum north rise as well as the course of Raccoon Creek. Through simultaneous alignments of this kind, terrestrial-bound earthworks mediate between cosmic realms of Upperworld and Lowerworld.

Another set of alignments are of similar interest. A significant number of earthworks include walled passageways that lead directly to water. Examples of this are found at the Newark, Portsmouth, High Bank, Hopeton, and Marietta earthworks. At the Hopeton earthworks, for example, the long parallel walls that lead from the earthwork toward the Scioto River are aligned to the winter solstice sunset. Similarly, the long parallel walls, or Sacra Via at Marietta leads to the Muskingum River and is aligned to the winter solstice sunset. The simultaneous alignment of passageways along celestial azimuths and leading to watercourses suggests an intentional effort to integrate or otherwise connect the aquatic and celestial realms through a humanly-created terrestrial vector. If, as seems reasonable to presume, these passageways were also intended to direct human movements or processions, then that suggests the ritual engagement of humans with the aquatic and celestial realms.

Burial mounds - to include those associated with geometric earthworks also integrate cosmic realms. In this case, the world of the living is integrated with the ancestor and spirit realm. Burial mounds extend the social network or web of relationships of this life into the Otherworld. Through the integration of the living with the ancestor and spirit realms, the Otherworld becomes more familiar and less foreboding.

Bring into Harmony

Congruent with traditional Native American religions, a fundamental principle that seems to have guided Adena-Hopewell life was a preference to live in harmony with the cosmic order. I make this claim based on the findings presented in previous chapters, as well as conclusions made by others (e.g., Carr and Case 2005a; Charles and Buikstra 2006; Hall 1979, 1997). This way of being-in-the-world originates in the recognition that there is a primal flow to the universe that pre-exists humans and it is the role of humans to harmoniously merge into this flow. As explained by Christopher Vecsey (1995:14) with reference to Native American religions in general:

> "This way of life can include ethical norms, modes of production, social organization, ritual activity, as well as sartorial, tonsorial, and other styles or behavior patterns. This way of life is based on an attitude of piety, reverence, acceptance, and affirmation concerning the cosmic order; the way of life is an attempt to live in harmony with cosmic principles or patterns."

The way of life described by Vecsey is not simply a romantic idealization. Looking to the pragmatic aspects of this life style it can be argued that, where peoples' lives depend on the migratory and seasonal movements of game, fish, and waterfowl, the birth cycles of animals, and the annual growth and availability of plant foods, it is vitally important that the activities of people are coordinated with their environment. To survive in a world where the availability and behaviors of one's food sources are contingent upon cycles of summer and winter, wet and dry seasons, night and day, solar and lunar cycles, and wind and weather patterns, it has always been necessary for humans to adapt to those rhythms. For the Adena-Hopewell, how their earthworks were situated with respect to the earth, sky, and water were reflections of this way of being-in-the-world.

Lend Stability

One gets the sense that regardless of whatever other purposes the earthworks may have served, for the Adena-Hopewell, their earthworks also provided a sense of stability in a world that may sometimes have seemed unpredictable, transitory, and beyond human control. Life in the Adena-Hopewell world was shorter and more precarious than it is today. However, by linking social order and ritual activities to earthen monuments, which in turn were linked through their alignments to the orderly movements of the cosmos, a significant element of predictability and stability was introduced into the human realm. Indeed, the creation of monuments that outlived generations of people no doubt lent a sense of constancy and regularity to life. The earthworks and mounds could be relied upon to always be there. Moreover, earthworks and mounds having alignments, spirit roads, and portals to the Otherworld provided cognitive mappings of unseen realms thereby contributing to a sense of security, order, and stability.

Materialize Myths

For most people, traditional stories and mythologies that explain how the universe came into existence, the role that humans have in the universe, and what happens to people when they die are their primary source of understanding. Such stories answer the most fundamental questions and help us interpret and make sense of life's most dramatic events to include birth and death. Furthermore, traditional stories help people understand the past and provide guidance for future actions.

Sometimes stories are written, such as found in the Bible or Mayan codices. Other times they are passed-down through the generations by oral teachings. Texts rely on writing systems. Oral traditions, however, are not without their own memory aids.

It may be, for example, that some Adena-Hopewell earthworks memorialized certain myths or stories. In particular I am referring to myths that tell of the world's creation and stories that tell of the soul's journey after death. For example, many Adena-Hopewell circle earthworks and especially those with perimeter ditches may have been intended to represent the landscape of creation, with the earth-island surrounded by and emergent from the primordial sea (also see Hall 1979; Romain 2000). In other words, these earthworks tell the story of the Earth-Diver myth and how This World came into existence. Perhaps this interpretation at least in part, accounts for several groups of earthwork complexes such as The Plains, Blackwater Group, and Junction Group that are mostly (but not entirely) comprised of small circle earthworks. Sometimes circle earthworks have central mounds, other time not; but in either case, the primal myth of This World having emerged from the primordial sea seems suggested by the shape and form of these particular earthworks.

On the center stage of circle earthworks - such as the Adena mound, Newark Great Circle, Marietta Conus mound, and others where burial mounds are found, rituals of life-renewal through death may have been re-enacted. Thus burials within sacred circles simultaneously occurred in two realms – within the physical earthwork itself, and within the context of a mythical story of creation, death, and renewal. If this interpretation is correct, then the tremendous number of circle earthworks found across the Adena-Hopewell landscape suggests that the creation story was re-enacted and told many times (note 3).

The effects of periodic re-enactments of the primordial creation event through earthwork construction would have been to impart a sense of stability to the community, explain through ritual participation the nature of the cosmos, as well as to re-vitalize and renew the resources the community relied upon. In this later purpose, it is interesting to observe how human beings presume to influence how the cosmos unfolds through ritual behaviors.

Portals to Other Realms

In addition to identifying or creating a place where the numinous or sacred can be experienced, I propose that some mounds and earthworks were intended as portals to the Otherworld (*sensu* Knight 1989). Burial mounds are liminal locations where remains of the Ancestors are physically situated and so it is at these locations that the living are presumably able to interact with the Ancestors through prayer, invocation, and various rituals. In other words, burial mounds are places of convergence where the realm of the dead intersects with the world of the living. In cosmological terms it makes sense to situate burial mounds at places in the landscape where celestial and/or terrestrial sightlines intersect because by such sightlines, direct and visible vectors to the Otherworld are identified.

In many cases, burial mounds are directly associated with earthwork enclosures. Indeed many earthwork enclosures have burial mounds within their perimeters. Where this becomes interesting is in the possibility that, if burial mounds themselves were considered as entranceways to the Otherworld, then it may be that some earthwork enclosures were intended to guard the portals between worlds. Alternatively, it may be that earthwork enclosures were in some cases intended to keep souls of the dead contained within, or to keep malevolent spirits out. In either case, some of the major enclosures that have burial mounds include: Circleville, Mound City, Hopewell, Liberty, Seip, Tremper, Turner, Frankfort, Shriver, Newark Ellipse, and Fort Ancient. Sometimes hundreds of burials are found in the mounds at these sites. To step into one of these enclosures was to step into a threshold or doorway between this world and the next.

If some earthworks were portals to the Otherworld, then surely the Serpent Mound is the most explicit of that idea. To begin with, the serpent effigy is situated adjacent to numerous sinkholes entrances to the Lowerworld. Burial mounds are located just a few hundred feet from the effigy. And in Eastern Woodlands mythology, the Great Horned Serpent was said to be the guardian of the entrance to the Land of the Dead (Lankford 2007b, 2007c).

Places for Ritual, Social, and Political Activities

Certain rituals enacted within designated earthwork places were likely intended to ensure the continued equilibrium of the cosmos. Other rituals were probably concerned with world renewal and the horticultural and hunting success of the community. Where burial mounds are found, mortuary rituals and processional movements opened gateways to the Otherworld. Through ritual activities at earthwork and mound locations, people came into emotional proximity with the Ancestors and engaged with various non-human entities. In some cases, special places identified by earthworks facilitated shamanic journeys to the Otherworld. In these instances and as described by Baires, Butler, Skousen, and Pauketat (2013:19): "Such experiences and the attendant associations and entanglements, *crossed boundaries* between the elements of earth and the sky, the past and present, and the dead and the living…" and "…coupled in powerful ways, if only momentarily, the seen and unseen dimensions of the cosmos."

At the same time, public performances and ceremonies such as marriages, funerary events, recognition of social and political accomplishments, trade activities, games, dances, and other community activities that took place within the enclosures certainly reaffirmed peoples' cultural identity and provided for the sanctioning of various political, social, and economic relationships. In this regard, it is interesting to speculate that perhaps the very large deposits of exotic materials sometimes found in the mounds may represent gifts brought from afar by people who wished to be part of the Hopewell phenomena or otherwise honor the deceased with whom these materials were often placed. Other times of course, large deposits of exotic materials may simply have been intended as offering or gifts to Otherworld entities.

Incorporate Memory

Earthworks transcend human life expectancy. A certain permanence is implied by their imposition on the landscape. Archaeologists Biloine Young and Melvin Fowler (2000:227) explain it this way:

> "Monuments anchor the present to a chain of historic events that are responsible for the current structure of reality. If the design has cosmic referents it can alter human perceptions and tie people to a timeless past, represented by the endless movement of heavenly bodies in a cycle that transcends time."

Accordingly, earthworks provide continuity with the past. Another way of saying this is that earthworks and mounds "...prevented the ritual and mythological significance of particular places being lost and forgotten" (Tilley 1994:204). This is an important point, especially in cultures where writing was not known.

Of course in addition to particular narratives that might be memorialized, mounds and in particular, burial mounds, are often built so that the dead people within might be remembered. Thomas (1996:80) explains, "in this way [mounds and earthworks]...can stand as evidence for past lives, identities and relationships." At one level, the Pyramids of Egypt, Imperial Tomb of China, and Moundbuilder mounds all express the same human impulse – i.e., the desire to remember loved ones and to be remembered. Thus mounds are memory devices as well as places where the remains of the ancestors reside. That said, it is a small step from a single burial mound intended for commemorative purposes to the addition of more mounds in close proximity, to embellishing a place like Mound City to where the enclosure perhaps comes to represent a mythic place – like the Land of the Dead. In any case, through the physical construction of a mound, a link is established between the living and the dead. The dead become part of the fabric of society by their inclusion in the landscape in mounds which forever after, trigger memories and emotions among the living. Mounds then, are very much about memory and bringing into conscious awareness, the trajectories of living and deceased persons and their stories.

Other Things Earthworks Do

While earthworks and mounds were certainly spaces for ritual activities that linked realms through sensual activities, many also likely served more mundane purposes. Surely some were territorial and route markers, lookouts, and signal mounds. By virtue of their location in river valley areas that were the preferred habitat for the plant species that the Adena-Hopewell are known to have consumed, some enclosures perhaps were – intentionally or not, special places where plant foods could be gathered. Although this posited use for the earthworks is speculative, there are several lines of evidence that are supportive:

1. Archaeological evidence indicates that the Ohio Hopewell cultivated a variety of plant foods including goosefoot, marshelder, cucurbits, sunflower, knotweed, and maygrass.

2. All of the plant foods grown by the Hopewell would have thrived in the river valley settings where the geometric complexes are located.

3. Hopewell geometric complexes are located on highly productive, well-drained, and level soils. These factors combine to make the areas where the geometric complexes are located, ideally suited for horticultural activities (but see note 2).

4. Archaeological evidence suggests that the Hopewell lived in small, scattered hamlets, in proximity to the geometric earthworks. A similar pattern of scattered hamlets was observed in early Historic times among the Indians of the Eastern Woodlands. These Historic Indian peoples engaged in communal horticultural efforts. In the same way, if the Hopewell complexes were used for horticultural purposes then, each enclosure could have served a nearby, surrounding population.

5. Very little in the way of cultural remains or artifacts are found inside a number of major geometric complexes. Although this finding can be indicative of ritual cleansing of special places, it is also consistent with use of the enclosures as horticultural areas. One would not expect to find much in the way of artifacts scattered about in horticultural areas.

6. High levels of Hopewell food plant pollens and several seed coats were recovered from Hopewell levels at Fort Ancient (McLauchlan 2000). And, both *Chenopodium* and *Polygonum* spp. seeds have been found in Hopewell contexts within the Marietta and Liberty geometric complexes (Wymer 1996:Table 3.3; Smart and Ford 1983:Table 5.3).

7. There is physical evidence from the Late Prehistoric Period indicating that on occasion, crops were grown by Native Americans within a geometrically shaped enclosure (e.g., Lee 1958:7).

This is not to say that entire earthworks like the Newark Octagon and Observatory Circle were solely dedicated to growing vast acres of chenopodium, maygrass, sumpweed, or other plants foods. But again, simply by virtue of their location, these weedy food plants would have thrived within the boundaries of the disturbed areas of the earthwork enclosures. To this point, Figure 3.13 shows an experimental garden of Hopewell cultigens planted at the Newark Great Circle. As evident, the plants are thriving and have reached substantial size.

The last point I wish to make in this regard is that, in discussions about the Adena-Hopewell, it is tempting to separate, divide, and classify activities associated with the earthworks into different functional categories, similar to what I have done in the preceding pages. This reductionist way of breaking-down a complex problem into manageable pieces is useful for analytical purposes. There is no reason to think, however, that the same way of thinking characterized Native American understandings, especially with regard to the geometric enclosures. More likely is that far less emphasis was placed on functional categories by the people who built the earthworks than those who would study them. I make this assertion based on the ethnohistoric observation that among many Native American peoples, mortuary rituals, world renewal ceremonies, and horticulture practices were often inter-woven in a complex web of beliefs and practices.

Creek, Cherokee, Iroquois, Huron, and Seneca stories, for example, tell of maize, beans, pumpkins, and tobacco having their origins in either the dead body, or grave of a mythical person known as "the old woman" (Lankford 1987:155-156). In these stories, the association between plant foods and the dead is clear. Life is regenerated from the dead. The implication is that life, death, and renewal comprise a cyclic process. As Robert Hall (1976:363) points out:

> "The world view of the Indian did not sharply distinguish between refleshing the bones of game animals respectfully treated, the reincarnation of the Indian in the Indian's turn, or germination and the appearance of life in buried seeds. Germination was associated with darkness, the underworld, among the Pawnee with the dark of

the new moon. Among the Mikmak there is even a tale "in which corn was brought back from the land of the dead by members of the tribe who had won it in gambling with Papkootparout, the master of souls" (Wallis and Wallis 1955:19-20).

I cannot prove that Hopewell beliefs were the same or similar to those just cited. The scenario of the Hopewell dead interred in association with horticultural plots is, however, consistent with this way of thinking. In this scenario, mound areas and enclosures bearing the Hopewell dead would come alive each spring with renewed plant foods – just as maize, beans, pumpkins, and tobacco were said to have sprung forth from the dead in the stories noted above. World renewal ceremonies would help ensure that the balance between life and death was maintained and that the cycle of life, reborn out of death continued. In short, I fully agree with Buikstra and Charles (1999:221) who propose that: "During Middle Woodland times, ancestral cults and earth renewal rites were inextricably linked...death and fertility were united." In this view, there would be no cognitive dissonance associated with the idea that the geometric earthworks could be places for individual revelation, group ceremonial activities, burial of the dead, and horticultural activities. In this worldview, plant fertility, the progression of seasons, annual cycle of life, death and rebirth, mortuary practices, and world renewal would have been brought together at one nexus, within the boundaries of the Hopewell enclosures, which were directly tied to celestial cycles.

Conclusion

For the Adena-Hopewell, theirs was a land of special places. Their special places included sacred mountains, river confluences, valleys where the sun rose or set in alignment, and in general, places where spirit was sensually present. Movement among and between these places was an experiential journey that connected people to earth, sky, and water realms, as well as to the past and future. Embedded in the Adena-Hopewell landscape were their cultural memories, stories of creation, connections to the past through ancestors buried, and continuing revelations and experiences of the sacred. For the Adena-Hopewell, to walk through the landscape was to engage with the sacred. By physically entering into the presence of this spirit, power, and memory, Moundbuilder peoples found existential meaning, order, and place in the cosmic scheme of things. For the Adena-Hopewell, the landscape was the very foundation of their life. In this world, Adena-Hopewell religion was not something 'believed-in'; but rather, was a lived experience. This entangled world was not static or monolithic, but was in a continuous state of change to include the movement of people between earthworks, movements of the dead from this world to the next, and the movement of the sun, moon, and stars relative to the earthworks. Human interactions and perceptions were also in a constant state of change; but what remained constant was that the earthworks identified special places where the sacred was sensuously present and where people could enter into the sacred.

From the experiences of people as they created and engaged the earthworks and mounds, a relational field was created that included people, ancestors, other-than-human beings, landscape features, memories, and other dimensions. For the Moundbuilders, theirs was life-world of entanglements and meaningful connections that was, in and of itself, sacred. As explained by religious scholar and Native American Vine Deloria Jr. 2003[1973]:65-66:

> "The structure of their [American Indian] religious traditions is taken directly from the world around them, from their relationships with other forms of life. Context is therefore all-important for both practice and the understanding of reality. The places where revelations were experienced were remembered and set aside as *locations where, through rituals and ceremonials, the people could once again communicate with the spirits*....It was not what people believed to be true that was important but what they experienced as true. Hence revelation was seen as a continuous process of adjustment to the natural surroundings and not as a specific message valid for all times and places" (emphasis mine).

In conclusion, I believe that, for the Moundbuilders, their way of being-in-the-world was fundamentally and inexorably tied to the landscape. The Adena-Hopewell were part of the land; and the land was integral to the identity of the people. Everything was connected.

Figure 10.1. Hopewell Mound Group Woodhenge

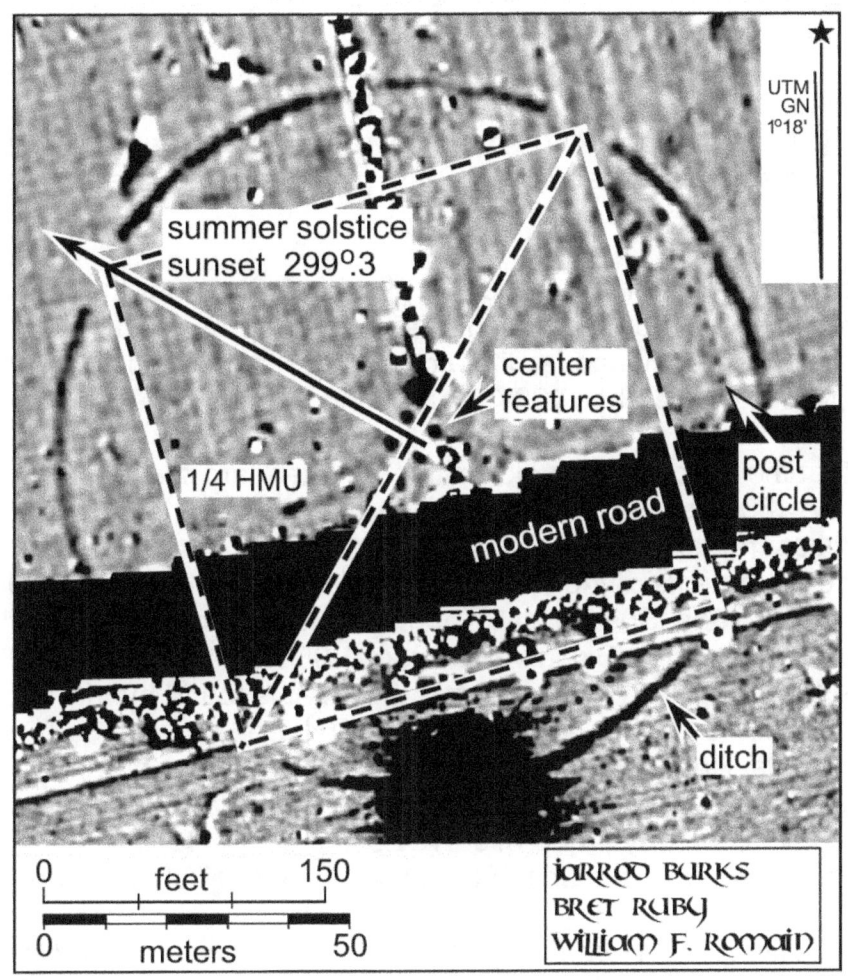

Mag data collected and processed by Jarrod Burks, provided courtesy of Bret Ruby, National Park Service. Summer solstice alignment first suggested by Ruby (2014a; pers. comm. 6-26-2014); confirmed by Romain 6-29-2014. Solstice calculations, geometric, and mensuration relationships by Romain. Mag image corrected to true north. Summer solstice sightline calculated for A.D. 100, 2.14 degree corrected for horizon elevation and lower limb tangency.

NOTES

Chapter 1: Introduction

1. The observation that certain of the Adena-Hopewell earthworks were modeled after geometric shapes is not new. This fact was one of the primary reasons why the earthworks rose to prominence in the first place (e.g., Atwater 1820; Squier and Davis 1848; Thomas 1894).

2. In this book units of length are generally given in feet, rather than meters, although in some cases, both are given. The data are presented in this way to better maintain consistency and facilitate cross-referencing to older reports that typically use feet (e.g., Marshall 1987; Squier and Davis 1848; Thomas 1894).

3. The idea that the Ohio Moundbuilders used a "standard of measurement" was proposed in the 1800s by Squier and Davis (1848:48-49). The idea that this standard of measurement is equal to 1,054 feet (321.26 m) originates in pre-restoration measurements of the Newark Observatory Circle published in the late 1800s by Cyrus Thomas. Specifically, Thomas (1894:16) reported that the diameter of the Observatory Circle varied from 1,050 feet [320 m] to 1,059 feet [332.2 m]. Further, Thomas (1894:16) stated that, "it is found by trial that the nearest approximate circle has a diameter of 1,054 feet."

In the 1980s, Ray Hively and Robert Horn (1982) corroborated Thomas's findings at the Newark Octagon and Observatory Circle relative to the 1,054-foot (321.3 m) mean diameter by physical survey using a steel tape and transit. Hively and Horn referred to the unit of length as the OCD (Observatory Circle Diameter).

In 1984, Hively and Horn published a second article. This article concerned the High Bank earthwork. In this article they concluded that the High Bank Circle closely approximates a true circle having a diameter of 320.6 meters [1,051.84 feet]. They referred to this unit as the High Bank Diameter, or HBD. Hively and Horn did not derive the HBD from a ground survey of High Bank. Rather, they used Thomas's (1894) data. Following upon this, Hively and Horn (1984:S99) averaged the Newark OCD and High Bank HBD and proposed that the mean of those two values – i.e., 1,053 feet (321 m) was the "fundamental" Hopewell unit of length.

In earlier publications I have used both the 1,054-foot (321.3 m) and 1,053-foot (320.9 m) lengths for earthwork analyses. The difference derives from how the number is calculated. For the present discussion I favor the 1,054-foot (321.3 m) length as the "standard" if only because it is based on measurements of the Newark Observatory Circle – which in the late 1800s, had higher walls and better definition than the High Bank Circle. My own survey experience is that, the higher the earthwork wall, the easier it is to visually determine its crest. According to Thomas (1894:462-463), when surveyed, the Newark Observatory Circle embankment was averaged 4 – 5 feet in height and was no lower than 3 feet.

The embankment averaged 39 feet wide. The High Bank Circle on the other hand, was 2 feet high and averaged 34 feet in width (Thomas 1894:476-477).

Having said that, and as a practical matter, the immense scale of the earthworks renders the 1-foot difference between the OCD and HBD numbers inconsequential. Moreover, it seems that for the Hopewell, the fundamental unit of length was somewhat fluid. That is to say, the 1,054-foot (321.3 m) (or alternatively, the 1,053-foot [320.9 m]) unit seems to represent an ideal length. On-the-ground measurements of earthworks other than Newark and High Bank often present dimensions that vary from the 1,054-foot or 1,053-foot lengths by as much as 2 percent.

Another unit of length that has been proposed for the Hopewell earthworks is 187 feet (Marshall 1978, 1987). Marshall (1978:31) claimed this unit of length was introduced to the Ohio Moundbuilders by a group of Indians who traveled from Mexico. Marshall suggests that, if a grid is made comprised of squares that are each 187 feet on a side, then, depending on how that grid is rotated, it will in some cases, correspond the dimensions of several earthworks.

Consideration of Marshall's proposal, however, suggests that the 187-foot length is likely an iteration of the HMU length based on a right triangle. Specifically, if a 45°- 45°- 90° triangle is constructed using the 1/4 HMU length as its hypotenuse, the result is that the other two sides will each be 186.3 feet in length – which is close to Marshall's 187-foot length and certainly within the limits of accuracy of the survey maps that Marshall used for his calculations.

Worth noting is that other units of length for Moundbuilder earthworks have been proposed. John Clark (2004) posited an 86.63-meter length (52 X 1.666 m) for several Archaic sites in the Southeast and a 1.544 meter length for several Mesoamerican sites. Like Marshall, Clark (2004:205) proposed that Woodland peoples "…used a measurement system and geometry derived, at least in part, from, Formative Mesoamerica…." Interestingly, if the 1.544-meter length is multiplied by 52 (a number that figures prominently in Clark's calculations), the result is close to the ¼ HMU length.

Alternatively, Sherrod and Rolingson (1987) proposed a 47.5-meter unit of length for dozens of Southeastern sites ranging across several time periods. Notably the 47.5-meter length finds expression at the Mississippian site of Cahokia (Romain 2012, 2013; Sherrod and Rolingson 1987).

4. Archaeoastronomic hypotheses for Ohio earthworks have been long been proposed. As early as 1820, Caleb Atwater (1820:237) opined that the Moundbuilders had "knowledge of astronomy." More than one hundred years later, archaeoastronomic hypotheses were posited by: Eddy (1978); Essenpreis and Duszynski (1989); Hardman and Hardman (1987); Hively and Horn (1982, 1984); Romain (1987, 19991, 1995, 1998); and Turner (1983). In retrospect, the early work needs to be approached with caution. Some is quite good for its day. Other efforts, however, including my own, were sometimes flawed by their reliance on old or poor quality maps. Most of these early studies provide little cultural context for posited findings. Others plot every conceivable alignment through gateways, along axes, between internal features, and between sites -

resulting in dozens of posited alignments with no apparent thought given as to what might be fortuitous. Some are simply wrong in their calculations.

Chapter 2: Background and Methods

1. Soul concepts among Native Americans of traditional belief can be complicated. There are free-souls, breath-souls, body-souls, ego-souls, intellect-souls, wish-souls, life-souls, guardian-souls, dream-souls, and shadow-souls. There can also be souls associated with a particular body part or organ. And, there are shadows and ghosts of various sorts. Sometimes, souls can have their own ghosts. For a detailed discussion see Hultkrantz 1997.

2. For purposes here, *astronomic north* is considered synonymous with *true north*. It should be noted, however, that true north can also refer to geodetic north. The difference between astronomic and geodetic north relates to the fact that astronomic and geodetic azimuths use different vertical references. Astronomic azimuths use gravity at the instrument location for reference. Geodetic azimuths use the center of a theoretical earth spheroid for reference. For archaeoastronomic analyses, the quantitative difference between astronomic north and geodetic north is so minimal that it can be ignored.

3. Most refraction tables are based on standard atmospheric conditions. Variations from standard conditions, however, will result in a corresponding change in azimuth for the body being observed (Schaefer and Liller 1990). For this reason, it is generally not practical to quote alignment values to a precision greater than $0°.1$ (or six minutes of arc) (Ruggles 1999:227).

4. Lunar parallax varies depending upon the linear distance of the moon from earth. It ranges from about $0°.90$ to $1°.01$. The value of $0°.95$ represents a mean value.

5. For the sun, maximum and minimum declinations are calculated using a value of $23°.43$ for the mean obliquity of the ecliptic for year A.D. 2000 which is then corrected for A.D. 100. This yields a value of $23°.677$ for year A.D. 100.
    For the moon, maximum and minimum declinations are obtained by reference to standard formulae (e.g. Ruggles 1999; Wood 1978):

maximum north limit = $+ (\varepsilon + i)$;
minimum north limit = $+ (\varepsilon - i)$;
minimum south limit = $- (\varepsilon - i)$;
maximum south limit = $- (\varepsilon + i)$;
where,

    ε is the obliquity of the ecliptic for year A.D. 100 (equal to $23°.677$) as obtained above; and

    i is equal to $5°.15$ which is the average inclination of the orbit of the moon.

6. Although some of the geometric enclosures appear to have been located in prairie openings (Lepper 1998; Wymer 1997) that does not mean the surrounding areas for miles around forming the local horizon for the earthworks were devoid of trees. Certainly, prehistoric peoples relied heavily on trees for fuel and building materials. However, the total Adena-Hopewell population at any given time in south-central Ohio seems to have been fairly small, perhaps numbering in the low thousands. Accordingly, there is little reason to believe that the Moundbuilders would have totally denuded the surrounding forests that formed the distant horizons. Even today, with millions of people living in south-central Ohio, there are millions of acres of forests. And today, when looking out from the nearly all of the earthworks, one sees a horizon that has trees. Given that, I believe it is reasonably and more representative of Adena-Hopewell times, to calculate distant horizon elevations taking tree height into account.

Based on analysis of vegetation maps for Ohio at time of contact, it appears that the tableland and hilltop areas that comprised the horizon elevations for the lowland geometric enclosures were mostly covered by Oak-Hickory forests. The 80-foot tree height added to the foresight locations for calculation of horizon elevations is based on the maximum height given by Petrides (1998) for a variety of "indicator" trees representative of Oak-Hickory forests (Kricher 1998:81-85) including: White Oak (80 ft.), Southern Red Oak (80 ft.), Black Oak (80 ft.), Mockernut Hickory (80 ft.), and Pignut Hickory (90 ft.).

Chapter 3: Muskingum River Watershed

1. As mentioned in Chapter 2, the mensuration feature of the LiDAR program locks-on to the closest ground coordinate points that it finds data for. The program then measures the distance between those two points. So, for example, in Figure 3.3, I wish to draw a line that is 263.5 feet in length. The LiDAR computer program locks-on to the ground coordinate points closest to that length, which is this case results in a length of 263.13 feet. The difference between 263.5 feet and 263.13 feet is 4 inches and at map scale is *diminimus*.

2. For Fredericktown, the theta correction is 0.0218 degrees. Summer solstice sunrise of 60.6 degrees for Stackhouse is calculated for A.D. 100, map-measured horizon elevation corrected for refraction, lower limb tangency and foresight tree height of 80 feet = corrected h value of 2.49 degrees. Winter solstice sunset azimuth was also calculated. Based on a corrected map-measured horizon elevation of - 0.33 degrees, the winter sunset azimuth is 238.46 degrees. Of the two azimuths - i.e., sunrise and sunset, the diagonal of the earthwork more closely aligns to the summer solstice sunrise.

3. For Kandel the summer solstice sunrise is calculated for A.D. 100, map-measured elevation corrected for refraction, lower limb tangency and foresight tree height of 80 feet = corrected elevation of 8.38 degrees. This high horizon elevation results from the proximity of the site to a nearby hill in the direction of the solstice sunrise.

4. Calculated summer solstice sunrise from the Rowley Mound based on corrected map-measured horizon elevation = 1.1 degrees.

5. Urban expansion has destroyed much of the Newark Earthworks Complex. Restored earthworks that are visible include the Octagon and Observatory Circle, Great Circle, and part of the Wright Square. A small extant section of the Great Hopewell Road is still visible in a woodlot just north of the Newark-Heath Airport. Discussions relative to the accuracy of restoration efforts can be found elsewhere (e.g., Romain 2004). Based on pre-restoration data, and except for the length of one Octagon wall, restoration work does not appear to have materially affected the azimuths or the lateral dimensions of the Octagon and Observatory Circle, or Great Circle earthworks.

6. The concept of cross-valley alignments is from Hively and Horn (2013). Solar azimuths calculated by Romain; lunar azimuths per Hively and Horn (2013). Hively and Horn (2013) posit several solar and lunar alignments in addition to those shown here. The present discussion is limited to those I believe most plausible.

7. The major axis alignment of the Newark Octagon and Observatory Circle to the moon's maximum north rise was discovered by astronomer John Eddy (1978:149). Eddy's findings started a decades-long search for alignments by Hopewell researchers.

   Lunar azimuths calculated by Romain for A.D. 100, 0-degree horizon elevation, corrected for parallax, refraction, and lower limb tangency. Wall azimuths per Romain ground survey.

   Hively and Horn (2013, 2010) propose that for Newark, lunar azimuths were determined by observations from surrounding hilltops. As viewed across the Newark Valley from any of the posited Hively and Horn hilltop observation points (H1 - H4), the distant horizon elevation is - as they suggest, close to 0 degrees. Hively and Horn refer to this as the H-zero hypothesis. The alternative is to establish horizon elevations for various sightlines from earthwork level. Hively and Horn refer to this as the H-local horizon.

   I have calculated the Octagon's lunar azimuths using measured horizon elevations from the Octagon (H-local) and a zero-degree elevation (H-zero). I find that while the difference between the two sets of calculations is not great, it is significant enough to favor the H-zero hypothesis, at least for the layout of the Octagon's walls. This does not mean, however, that all alignments at Newark or other sites were necessarily established using a zero-degree horizon elevation. Use of either a zero-degree, or actual local horizon elevation seems to have varied, depending on the earthwork. The Wright Square, for example, seems to have used the local horizon for alignment purposes. The Great Circle used an artificial horizon created by its entrance walls.

8. Hopewell earthworks vary in their geometric precision. The Great Circle is a good example of that. According to Thomas's (1894:464) data, the diameter of the Great

Circle varies between 1,163 - 1,189 feet - for a difference of 26 feet, depending on the direction measured. The mean of the two lengths just given, however, is 1,176 feet - which is to within 2 feet of the ideal length of 1,178 feet.

9. Newark Great Circle moon minimum north rise azimuth calculated for A.D. 100, with horizon elevation calculated using an entrance wall height = 15 feet, horizontal distance = 589 feet, 0.35 refraction correction, 0.25 lower limb correction, 0.95 parallax correction

It is interesting that the Great Circle uses its own perimeter walls to establish an artificial horizon. Thus at Newark, the earthworks appear to have used three different horizon elevations - i.e., local, zero-degree, and artificial. It may be that different horizon elevations were selected-for based on whether or not actual observations were to be made, or alternatively for symbolic purposes and based on a pre-existing template.

10. Several mound-like features are found on the summit of Geller Hill. The Salisburys (1862) posited that these features were burial mounds. In September, 2013, the author in company with G. William Monaghan of the Glenn Black Laboratory of Archaeology and Edward Herrmann, Indiana State University, conducted Geoprobe coring into five of the mound-like features. Multiple cores were placed into each feature. Examination of the retrieved cores (to a depth of eight feet) failed to reveal any evidence that these features are man-made mounds.

11. There are two knolls on the summit of Geller Hill that could have served as useful reference points. Vantage point 1 (VP1) appears to have been used for geometric design purposes; VP2 may have been used for lunar observations.

12. The Newark Great Circle - Observatory Mound summer solstice sunset azimuth calculated for A.D. 100, map-measured distant horizon elevation corrected for refraction, lower limb tangency and estimated foresight tree level of 80-foot foresight tree level = corrected horizon elevation of 0.98 degrees.

The Observatory Mound was dug into during the 1800s. No human remains or cultural materials were found. In 2005 the present writer conducted coring operations into two of the Octagon's vertex mounds with similar results. A total of four cores were made into Octagon vertex mounds D and E. These cores were made from the tops of the mounds and penetrated through their bases. No evidence was found for burials, special soils, or cultural materials (Romain 2005b).

13. For the immediate Newark Earthworks area, the theta correction to convert LiDAR grid north (State Plane NAD 83) to true (astronomic) north is negligible - i.e., equal to + 0.03 degrees. For all practical purposes therefore, at Newark, grid north and true north can be considered equivalent.

14. Due to the concave nature of the Great Hopewell Road and its path across flat, low land, it was most likely wet or muddy much of the time. Walking the Road would have been easy enough on its flanking walls. However, walking down the centerline of the Road could have been a muddy mess. Cosmologically, however, the wet and concave

associations for the Road might be considered appropriate for a Lowerworld structure, or counterpart to the celestial Milky Way Path, visible in the opposite celestial realm.

15. The exact location for the beginning of the long, straight segment of the Great Hopewell Road is not known with certainty, as that part of the Road was obliterated before aerial photographs were taken. The point of origin for the Road shown here is an estimated position based on the Salisbury (1862) map, as well as known location and trajectory of the Road through the woodlot just north of the Newark-Heath airport.

In the original, full-size LiDAR image, small marker pins were placed so that significant points could be identified when the image was reduced to fit the page. These pins are visible in Figure 3.21 at the Octagon vertex, inside entrance to the Great Circle, in the woodlot at the known location of the Road, and at the center of the Upham Mound. The Upham Mound is on private property. There is no record of its having been excavated.

Celestial azimuths are calculated for a zero-degree horizon elevation. For the summer solstice sunset azimuth the reader will note that the zero-degree horizon elevation value used in Figure 3.22 differs by about 1 degree azimuth using the measured horizon elevation for the Great Circle-Observatory Circle solstice alignment shown in Figure 3.19. As mentioned, it may be that the mound builders used different horizon elevations for different purposes. For ideal or symbolic design purposes they may have used zero-degree horizon elevations. For cases where actual observations were anticipated, they used either actual, or artificial horizon elevations. Since the origin point for the Road was a design point, a zero-degree horizon elevation was appropriate.

16. The observation that the trajectory of the Great Hopewell Road intersects Sugarloaf Mountain was noted in the late 1990s by Joseph M. Knapp. It appeared in an Internet posting by Mr. Knapp (1998). To my knowledge Mr. Knapp's finding has not been reported in the professional literature. I happened upon the Knapp article after I concluded the same thing independently based on LiDAR analysis. Mr. Knapp and I differ slightly in our azimuth values. Mr. Knapp used USGS maps and aerial photos to determine the azimuth for the Road. As indicated, my findings rely on LiDAR data to establish the trajectory for the Road.

17. In A.D. 100 the summer solstice occurred on June 23rd. Nightfall occurred at about 10:05 pm local time. Nightfall occurs after sunset and after civil, nautical, and astronomical twilight. Nightfall occurs when the sun reaches an elevation of 18 degrees below the horizon. It is at this time, that all the stars that can be seen by the naked eye become visible (sixth magnitude).

18. The likelihood that the 5 HMU distance is intentional is indicated by similar findings for the distance between the Octagon and Wright Square and Observatory Circle and Great Circle. In those instances too, the distances separating the earthworks are multiples of the HMU.

19. I have intentionally flipped the daVinci illustration to better show the resemblance of the lunar crescent and earthshine phenomena to the Newark small circles. DaVinci

shows the crescent facing left, or to the west. The concept remains the same, however, whether the crescent faces east or west.

Earthshine occurs during, before, or after the new moon, during both its waxing and waning phases. It results from light reflected from the earth onto the moon. It is brightest during the spring (Goode et al. 2001). As shown, the crescent moon is visible. Additionally though, the dark part of the moon appears to faintly glow – giving rise to the illusion of a crescent inside of a circle.

20. It is often assumed that small circle earthworks are Adena in origin. This assumption may not always be true. Small circle and crescent earthworks appear to have been built by Adena and Hopewell. A case in point may be provided by the small circle-crescents near the Newark Octagon. The manner in which these small earthworks flank the walls of the parallel walkways suggests they were built contemporaneous with, or after the parallel walls linking the large Hopewell earthworks (also see Lepper 2010:115).

Several lines of evidence suggest that these crescent-circles were associated with the moon. First is their shape. To begin with, their overall shape resembles the moon's crescent shape; and in many instances the outside wall of the earthworks are accentuated by a second embankment inside the earthwork that is clearly crescent-shaped.

Although the circle-crescent earthworks appear diminutive in size when presented at map scale and are now almost all destroyed, they were visible in the early 1800s. The Salisbury manuscript describes several of these earthworks in good detail. With reference to one of the circles, the Salisbury's (1862:18) provide the following account:

".... 1g and 1h...each 72 feet in diameter and containing small depressions or pits near their centers. In making excavations in these pits, 1h disclosed nothing of importance, but in 1g we found at the depth of 3 and 4 feet...decayed animal or vegetable matter, amongst which were irregular masses of tough potters clay, plates of mica much decayed, burnt stone, some arrow points and chips of flint, and many large fragments of pottery...."

The small depressions or "pits" found in the centers of some of the small circle earthworks are shown on the Salisbury map as small dots. What is especially interesting is that the Salisburys' excavations of the two depressions they noted uncovered clay nodules, flint chips, and "plates of mica." If, by virtue of their silvery color the mica plates were associated with the moon this would further suggest a lunar association for the circle-crescent earthworks.

Further it is noted that the majority of the small circle-crescents are oriented toward the northeast or east. That is to say, the gateways in these earthworks open toward the northeast or east. A sole surviving small circle (i.e., the one at the southeast corner of the Octagon) likewise has its opening oriented to the northeast. In fact, this circle is oriented in precisely the same direction as the Newark Great Circle – which is to the moon's minimum north rise. Thus we have two extant circle earthworks oriented to the moon.

Next it is noted that the Wyrick map of Newark shows the small circles more or less clustered in two groups. One group of 18 - 21 circles (depending on which earthworks are included in the count) extends from the Observatory Circle eastward along one of the parallel walkways. The second group which includes the Great Circle extends from along both sides of that earthwork, northeast to the Ellipse and is comprised of 19 circle earthworks. Both groups thus approximate the number of years in the moon's 18.6-year standstill cycle. Admittedly the correspondence is perilous, given that we do not know how accurate the Wyrick map is, and the count of the first group is not an exact match. Nevertheless, the possibility exists that each circle might have been associated with each of the years in the 18.6-year lunar cycle.

The last reason for thinking the Newark small circles are lunar associated relates again to their shape. Based on the observation that the openings into the circles are to the east or northeast, I believe these earthworks could represent either a waxing, or waning crescent moon. The waxing crescent moon is best seen in the west just after sunset. In many cultures this moon phase is used to mark the beginning of the month. The waning crescent moon, however, appears in the pre-dawn hours in the northeast region of the sky. Since the individual and group orientations of the Newark crescents are oriented to the northeast, I am inclined to believe these are waning crescent moons. As discussed elsewhere, the northeast region of the sky is significant because that is where the Milky Way Path appears to originated from, on the night on the summer solstice. I suspect that the waning crescent moon was one of the elements involved in the mythology surrounding the journey of the soul to the Land of the Dead.

21. The perimeter around the main Newark earthworks was not an absolutely solid contiguous barrier. There were three main entrances and several additional gaps in the perimeter. What seems to have been intended was more of a symbolic demarcation of special space.

22. Theta correction for Hazlett Mound area = +0.14 degrees. Summer solstice sunset azimuth for Hazlett is for A.D. 100, map-measured horizon elevation corrected for refraction, lower limb tangency, and foresight tree height of 80 feet = corrected h value of -0.159 degrees.

23. The significance of this area and its relationship to Flint Ridge is revealed by the following quote from Moorehead (1897:176): "Never have we witnessed so many chips and discs of Flint Ridge material (except at the Ridge itself) as occur upon the Perry County sites. Nearly every knoll was a workshop. Boys and farmers find thousands of arrows, spears, knives, and scrapers of this material and yet the supply does not seem to be exhausted. Little other material occurs."

24. The size of the Yost Circle can also be established another way. A 30-60-90 degree triangle that has a hypotenuse of 263.5 feet (1/4 HMU) and opposite side of 131.75 feet (1/8 HMU) will have a base leg equal to 228.2 feet. A 228.2-foot diameter circle will closely match the outside edge of the Yost Circle.

An unpublished manuscript by the Salisburys shows the Yost earthworks to be more extensive than the figure provided by Martzolff (1902). Also see Moats (2012) for a detailed assessment of the larger site and its astronomic alignments.

25. I wish to thank Rich Moats, President of the Flint Ridge Chapter of the Archaeological Society of Ohio for pointing-out to me, the east-west alignment between Yost and the Roberts Mound.

26. There are a number of other significant mounds in Licking and Perry counties - e.g., the Salisbury Hill forts, Reservoir Mound, Fairmount Mound, and Large Stone Mound west of Hazlett. Indeed, nearly every hilltop in the Newark area has some sort of prehistoric structure. Looking at these features on a map, it is tempting to drawn lines between them and read into those lines intentional relationships. We need to keep in mind, however, that not all celestial alignments, angles, and distances are intentional or were recognized by the Moundbuilders.

27. Theta correction for Marietta is 0.66. Summer solstice sunrise azimuth for Conus Mound - Hill # 2 is for A.D. 100, map-measured horizon elevation corrected for refraction, lower limb tangency, and foresight tree height of 80 feet = corrected h value of 4.40 degrees.

28. Of interest is that this alignment serves as a counterpart to the Conus Mound-hill # 2 solstice sunrise alignment and helps establish the lateral location of the Conus Mound on the river terrace.

Summer solstice sunset azimuth for Conus Mound is for A.D. 100, map-measured horizon elevation corrected for refraction, lower limb tangency, and foresight tree height of 80 feet = corrected h value of 1.12 degrees.

29. Winter solstice sunrise azimuth from Quadranaou Mound to Conus Mound is for A.D. 100, horizon elevation corrected for refraction and lower limb tangency = corrected h value of 0.76 degrees.

30. Winter solstice sunset azimuth calculated for A.D. 100, horizon elevation corrected for refraction, lower limb tangency, and horizon tree height of 80 feet = corrected h value of 5.3 degrees. This relatively high horizon elevation is the result of a high ridge located across from the Quadranaou Mound, extending to the Muskingum River.

31. Excavated materials from the Capitolium Mound identify it as Hopewell (Pickard 1996). In about 1916, the Capitolium Mound was modified by construction of the Washington County library building on top of the mound. Landscaping residue from that construction is described by (Pickard 1996:279) as ranging from 8 inches to 3 feet in depth.

One of the interesting things about this mound is its unique shape. On the southeast side of the mound, there is a ramp that originally led from ground level to the top of the mound. Unlike most other earthen ramps, which generally extend outward from the mound structure they are connected to, the southeast Capitolium Mound ramp was cut into the mound. The result was a ramp with walls on either side formed by the mound. The concept of a ramp cutting into the earth and thereby forming earthen walls is repeated elsewhere at Marietta - e.g., in the construction of the Sacra Via.

32. Winter solstice sunset azimuth for Capitolium Mound is for A.D. 100, map-measured horizon elevation corrected for refraction, lower limb tangency, and foresight tree height of 80 feet = corrected h value of 4.43 degrees.

33. Winter solstice sunset azimuth for Sacra Via measured from Sacra Via and Third Street is for A.D. 100, map-measured horizon elevation corrected for refraction, lower limb tangency, and foresight tree height of 80 feet = corrected h value of 6.85 degrees.

Chapter 4: Hocking River Watershed

1. For the Rock Mill Works, summer solstice sunset azimuth is for A.D. 100, map-measured horizon elevation corrected for refraction, lower limb tangency, and foresight tree height of 80 feet = corrected h value of -0.53 degrees; winter solstice sunset azimuth calculated using corrected h value of 0.38 degrees.

2. When I suggest that The Plains circles are based on the 186.3-foot unit of length, I do not mean to say that the circles are of equal size. Sometimes the 186-foot length is found in the crest to crest dimensions across an earthwork, other times the length is incorporated as the outside edge-to-outside edge diameter. Further, the 186-foot length is expressed on the ground to an accuracy of only about 10 percent. In the case of The Plains circles, the unit of length was not exact. Confounding the matter is that the circle embankment walls are often less than 2 feet high, but 20 - 30 feet in width - making determination of exact distances - either crest to crest, or edge to edge subject to variability in interpretation.

Chapter 5: Scioto River Watershed

1. The chimney rock formation itself would not have been visible from the Stitt Mound. It is too distant and too small. Chimney Rock Mountain, however, upon which the rock formation of the same name is located, forms part of the visual backdrop for the valley. Chimney Rock Mountain is one of the highest mountains in Ohio.

2. Stitt Mound winter solstice sunrise calculated for A.D. 100, map-measured horizon elevation corrected for refraction, lower limb tangency, and foresight tree height of 80 feet = corrected h value of 0.50 degrees.

3. The 1938 aerial photo of Dunlaps Earthwork is the best surviving documentation we have for this structure. Agricultural plowing has destroyed it.

4. The Austin Brown Mound is a 23-foot high, conical mound presumed to be Adena in origin. The mound is situated on a river terrace, between streams that feed into Deer

Creek. The mound was partially excavated in 1897. Remnants of a log tomb were found within (Moorehead 1899:137)

5. During World War I, Mound City was the location for Camp Sherman, a U.S Army training base. A considerable number of buildings and other structures were built for the base, with the result that most of the mounds and other archaeological features were damaged or destroyed. The mounds within the enclosure as they appear today are reconstructions based on early records and subsurface features found during later restoration work. The Mound City perimeter wall is also mostly restored. However, detailed review shows that the restored walls faithfully follow the course of original wall segments observed both above ground and below ground (see Romain 2000; Lynott, Mandel, and Brown 2010 for further details).

6. The Hopewell used several different alignment schemes in their earthwork designs to include: orthogonal alignments, along major or minor earthwork axes, and diagonally from corner to opposite corner. At Mound City, Hopeton, and similar earthworks, a question often asked is: 'How is it possible to determine the azimuth for a diagonal sightline across the earthwork when the structure has rounded corners?'
     The answer is that there several methods can be used. Lines that follow the straight-line trajectory of the walls can be extended to where they intersect - thus providing a reference point where they cross; or a geometrically perfect square can be drawn inside the earthwork with opposite corners of the drawn square used as reference points. Another way is to measure the arc degrees of the opposite corners of interest and then bisect those arcs.

7. Summer solstice sunset azimuth for Mound City as measured from the southeast outside corner of the structure is for A.D. 100, map-measured horizon elevation corrected for refraction, lower limb tangency, and foresight tree height of 80 feet = corrected h value of 2.0 degrees. Milky Way azimuth per calculations made earlier for the Great Hopewell Road.

8. Calculations for nautical twilight at about 9:25 pm local time.

9. There is an undeniable resemblance between the Cygnus and the outline of a bird in flight. Indeed, according to the ancient Greeks, Cynus was a swan. That said, I know of no Native American traditions that refer to Cygnus as a swan. George Lankford (2011, 2007c, 2004), however, suggests that the constellation could represent an eagle, or other raptor; and in that guise may be associated with the Milky Way Path of Souls, particularly as represented in Mississippian symbolism. The notion that the star Deneb might be a "celestial referent" for the Milky Way antagonist or judge of souls (in the guise of an eagle, dog, old man, or old woman) was likewise suggested by George Lankford (2004:212).
     Alternatively this constellation is sometimes called the Northern Cross. In this regard it is interesting to note the resemblance between the cross pattern in the constellation and the crossed solstice and Milky Way azimuths in the square shape of Mound City.

10. Square-shaped earthworks having rounded corners are fairly common. In addition to Mound City, the Dunlaps Square, Anderson Square, and Hopeton Square have rounded corners. In many cases too, perimeter walls deviate from being a straight lines. Indeed, the concept of squares with rounded corners and ballooned-out walls is found in quite a few small earthworks to include Stackhouse, Kandel, the Davis Works, Johnson Works, and Jackson Square. Burks (2006b:19) refers to this geometric shape as a 'squircle' - i.e., something in-between a circle and a square. In any case, the concept is also expressed in the design of smaller, 'spirit house' structures found under many mounds to include those at Mound City and Seip (e.g., Brown 2004). In these instances, the outlines of spirit houses (referred to in older literature as charnel houses) are documented by post holes. Thus where they occur, deviations from sharp-edged corners and straight-line walls appear intentional rather than the result of poor design execution.

11. The solstice and lunar relationships relative to Sugarloaf and Mount Logan were first published by Hively and Horn (2010:145). Azimuths provided here were calculated by the present author for A.D. 100, map-measured horizons corrected as follows. Summer solstice rise corrected for refraction, lower limb tangency, and foresight tree height of 80 feet = 59.2 degrees. Winter solstice rise corrected for refraction, lower limb tangency, and foresight tree height of 80 feet = 121.8 degrees. Moon minimum north rise corrected for refraction, center of moon, parallax, and foresight tree height of 80 feet = 67.8 degrees. Moon minimum south rise corrected for refraction, center of moon, parallax, and foresight tree height of 80 feet = corrected h value of 116.8 degrees.

12. To the best of my knowledge, the equinox heirophany was discovered by Richard Conway. I wish to thank Mr. Conway for pointing-out this phenomenon to me.

Today, the east horizon at Mound City is partially blocked by trees. To photograph the equinox the vantage point for this photo was about 3,500 feet west of Mound City.

The equinox hierophany is not an alignment, *per se*. Neither the major, minor, nor diagonal axes of Mound City are oriented to Bunker Hill. I would characterize the Mound City event as an equinox-related phenomenon - the point being the visually observable, balanced relationship between earthwork, sun, and mountains (also see Romain 1993:figure 11).

13. Squier and Davis (1848:55) describe the Shriver Circle perimeter wall as 5 feet in height and 25 feet wide at its base; with the center mound as 5 feet high and 40 feet in diameter. By the time Squier and Davis observed the Shriver earthwork, however, it had been seriously degraded. A map made sometime between 1800 and 1809 and found in the collections of the Wisconsin Historical Society gives the center mound height as 10 feet (see Burks 2010:Figure 8.5a).

Squier and Davis (1848:156-157) excavated the Shriver Mound. Within it they found a large 'altar' paved with small round stones. On the altar they found a deposit that included copper bracelets, mica, and burned human bones.

14. Shriver-Sugarloaf summer solstice sunrise calculated for A.D. 100, map-measured horizon elevation corrected for refraction, upper limb tangency, and foresight tree height of 80 feet = corrected h value of 0.96 degrees. In this case upper limb tangency is used since the angle of the mountain along its south side allows the sun to be visible before it reaches its lower limb tangency elevation with the top of Sugarloaf Mountain.

15. Generally, it is not a good idea to use Squier and Davis's maps for archaeoastronomical analyses. The maps are simply not reliable enough in terms of orientation to rely upon. In the present case, however, since we have no alternatives, it is instructive consider the implications of the Squier and Davis-shown gateways - once their map has been properly oriented relative to existing features. In Figure 5.13 the Squier and Davis map has been oriented by referencing it to the north-south road shown in the Burks-Cook image.

16. Hopeton summer solstice sunrise calculated for A.D. 100, map-measured horizon elevation corrected for refraction, lower limb tangency, and foresight tree height of 80 feet = corrected h value of 2.59 degrees.

17. Hopeton summer solstice sunset calculated for A.D. 100, map-measured horizon elevation corrected for refraction, lower limb tangency, and foresight tree height of 80 feet = corrected h value of 0.72 degrees. Also see Turner 1983 for earlier independent discovery of this alignment. Based on radiocarbon dating results from two samples obtained from within the east embankment wall of the Hopeton Square, Lynott and Mandel (2009:172) propose that "...it seems likely that at least part of this wall segment was built or significantly modified about 800 years after the southern and western walls of the enclosure were built." Of course, late use, modification, or even rebuilding of part of the east wall during Late Prehistoric times does not negate the summer solstice sunset sightline findings since the wall section that Lynott and Mandel are referring to is not part of the alignment scenario. In the same publication, Lynott and Mandel (2009:171) provide a series of radiocarbon dates indicating that the walls of the Hopeton earthwork were built during the Middle Woodland, between A.D. 100 and A.D. 300. Based on soil coring results, Dempsey (2010:3) reports: "...this study found that the confluence of the circular and rectangular enclosures was built as a single unit." Further, "The data from Core Set 3...indicate that the two large enclosures were constructed at approximately the same time" (Dempsey 2010:4). Whatever the events that resulted in the anomalous late dates noted by Lynott and Mandel, it seems certain that the location, size, shape, and orientation of the Hopeton Square were established during Hopewell times.

18. Hopeton parallel wall winter solstice sunset calculated for A.D. 100, map-measured horizon elevation corrected for refraction, lower limb tangency, and foresight tree height of 80 feet = corrected h value of 1.15 degrees.

19. Hopeton South Circle summer solstice sunrise calculated for A.D. 100, map-measured horizon elevation corrected for refraction, lower limb tangency, and foresight tree height of 80 feet = corrected h value of 2.59 degrees.

20. It is often presumed that less than perfect geometric regularities among Hopewell earthworks are the result of either poor ground survey or haphazard construction by the Hopewell. Such presumptions are probably not always warranted. For example, the peculiar offset of the Hopeton Large Circle relative to the major axis of the Hopeton Square may have been intentional - so that the Large Circle would not impinge on the northwest corner of the Square - thereby maintaining the integrity of the Square's solstice sightline.

Similarly, the curvature of the east wall of the Square may have been intentional - with the embankment wall in this case following the curve of the terrace edge - which about midway along the length of the Square, begins to arc toward the southwest.

In these and similar cases, it appears that design modifications to ideal geometric shapes could be made if these changes were needed to facilitate either important terrestrial or celestial alignments, or to accommodate certain terrain features.

21. Adena Mound moon maximum north rise azimuth calculated for A.D. 100, map-measured horizon elevation corrected for refraction, lower limb tangency, parallax, and foresight tree height of 80 feet = corrected h value of 0.8 degrees.

Based on several distinguishing attributes, Mills (1902:6) concluded that the Adena Mound "had been built at two different periods." Trait analysis by Webb and Snow (1945:219) concluded that the mound dates to Late Adena times, when Adena and Hopewell overlapped chronologically and geographically. Recent radiocarbon dating of curated materials recovered from the central burial yielded median date of ca. 50 B.C. (Leone et. al. 2013).

22. In addition to the unfortunate fact that no visible remnants of the Works East Earthwork remain, identifying the exact location for the earthwork is made difficult by the fact that Squier and Davis's (1848:Pl. II) map showing the earthwork is incorrect in at least one major respect. Squier and Davis represent the streets of Chillicothe as extending north-south. In fact, however, they are skewed from that direction by about 10 degrees. Further, the course of the Scioto River at Chillicothe has changed somewhat since the early 1800s, due to natural causes and man-made changes.

23. Works East - Rattlesnake winter solstice rise calculated for A.D. 100, map-measured horizon elevation corrected for refraction, lower limb tangency, and foresight tree height of 80 feet = corrected h value of 0.8 degrees.

24. The finding that Liberty, High Bank, and the Works East extend in a straight line and that this line is parallel to a remnant of the Teays River - or more simply put, to the current trajectory of the Scioto River Valley, was first noted by Hively and Horn (2010). The findings that this straight line trajectory is orthogonal to the moon's maximum north rise and that the 33RO257 sightline to Works East is also on a lunar sightline are my contributions to the discussion. Also my contributions are the notions that the Mills Square may be part of the overall alignment scheme; and that the location for the Works East at the head of the Paint Creek Valley is at intersecting lunar trajectories. The

foregoing presumes that locations for the Works East and Mills Square have been correctly identified.

25. The Mills Square is no longer visible on the ground or in LiDAR imagery. The location shown here is based solely on Mills's (1914) representation.

26. The circle earthwork at site 33RO257 is no longer visible to the naked eye or by LiDAR. The enclosure was documented, however, by Squier and Davis (1848:Pl. II) and Mills (1914). Of interest is that Squier and Davis and Mills indicate that in addition to the circle earthwork, the site included five mounds.

27. It is interesting to compare Squier and Davis's representation of the terrain surrounding High Bank with the LiDAR imagery showing same. Notably, Squier and Davis show the river terrace further west than indicated by LiDAR.

28. Celestial azimuths for High Bank shown in this figure as posited by Hively and Horn (1984) using Aveni's (1972) tables for A.D. 250, lower limb tangency, map-measured horizon elevations, no correction for trees at foresight locations.

29. High Bank moon maximum north rise azimuth calculated for A.D. 100, map-measured horizon elevation corrected for refraction, lower limb tangency, parallax, and foresight tree height of 80 feet = corrected h value of 2.6 degrees.

    For High Bank, the azimuth I provide for the moon's maximum north rise differs from Hively and Horn's (1984) by about one-half of one degree. I attribute this to likely differences in calculated horizon elevations. As my horizon elevations are based on map data that have benefited from 30 years of increased accuracy since Hively and Horn's work, as well as site-specific calculations rather than data interpolated from Aveni's tables, my preference is for the newer azimuth data I have provided.

30. I have noted elsewhere (Romain 2004a, 2008a, 2008b, 2009b, 2009c) that Hopewell geometric earthworks often incorporate alignments to the lay of the land, or 'strike' as that term is used in geology. Early on, I also noted that many of the inter-site lunar alignments found for the Chillicothe-area sites extend parallel to major watercourses to include Paint Creek, the North Fork of Paint Creek, and the Scioto River (Romain 1992b:4-5).

    In the case of High Bank, Hively and Horn (2010:139-141) recently proposed that the major axis of that earthwork extends along the trajectory of the preglacial Teays River Valley. The Teays Valley was formed 2 - 5 million years ago by the now extinct, Teays River. After retreat of the last glaciers from Ohio, new river systems were formed. Some of these rivers, such as the Scioto River, commandeered sections of the old Teays Valley as they made their own channels. Near High Bank and Liberty, the Scioto River cuts along the base of hills that define what remains of part of the Teays Valley.

31. The correspondence in size between the High Bank Circle, Newark Observatory Circle, and Hopeton Circle was noted by Squier and Davis (1848:71).

32. The longitudinal axes of the Newark Octagon and Observatory Circle and High Bank earthworks are orthogonal to each other to within about one degree.
   Moon minimum south rise azimuth of 115.6 degrees is calculated for A.D. 100, based on a nominal or assumed horizon elevation of 1.0, center of moon, corrected for refraction and parallax = corrected h value of 1.55 degrees.

33. The remains of nearly 200 people and associated grave goods were found within the Harness Mound. Unfortunately, the mound has been obliterated by farming. For reports on excavations of the Harness Mound see Moorehead (1897), Mills (1907) and Greber (1983).

34. Liberty maximum north moon rise calculated for A.D. 100, horizon elevation corrected for lower limb tangency, refraction, and parallax = corrected h value of 2.2 degrees.

35. Liberty maximum south moon set calculated for A.D. 100, horizon elevation corrected for lower limb tangency, refraction, and parallax = corrected h value of 2.6 degrees. (Also see Romain 1991b:5-6.)

36. Liberty minimum south moon rise calculated for A.D. 100, horizon elevation corrected for lower limb tangency, refraction, and parallax = corrected h value of 2.4 degrees.

37. Johnson Works summer solstice sunrise calculated for A.D. 100, map-measured horizon elevation corrected for refraction, lower limb tangency, and foresight tree height of 80 feet = corrected h value of 0.23 degrees.

38. This earthwork is also sometimes called the Barnes Works. My preference in the present case is for the name Seal Township Works - if only to maintain consistency with Squier and Davis (1848).

39. There is disagreement concerning the configuration of Small Work A at the Seal Township Works. Differing from Squier and Davis, Fowke (1902:179) states: "A is a circular ditch embankment around a square level area; but the ditch, so far from being narrow and touching the circle only at the corners...reaches the outer embankment all around. In other words, the outside line of the ditch is a circle, while the inside is a square." Without physical evidence it is not possible to determine which description is more accurate. What both seem to agree upon, however, is that some combination of nested geometric figures was represented.

40. As shown, there is no error in the alignment of the north-south walls of the Seal Township Works Square to true north. Further, the east-west wall of the Barnes Square is aligned east-west to within 0.6 degrees.
   For the analysis shown in this figure, to establish true, a series of solar observations using a total station were made on the ground, near the Square. The aerial

photo was imported into TurboCAD and rotated according to the ground-truthed, astronomically-determined azimuth data (for further details see Romain 2004).

41. The Tremper Mound was originally more than eight feet in height. Excavation of the Tremper Mound by Mills (1916) found more than 600 postmolds indicating a large multi-chambered building roughly 200 feet long by 100 feet wide. Four communal graves were situated on the floor of the structure. The communal graves are estimated to have held the cremated remains of 375 individuals. The Tremper Mound was restored after excavation.

42. Mills (1916:270) gives the maximum height of the Tremper Mound as 8 1/2 feet, with an average height of about 5 feet. Mills (1916:270) notes that after excavation, the mound "was restored to its original height and dimensions."

43. The Portsmouth Group A Square looks like a typical Hopewell square with rounded corners - similar to Mound City, for example. Excavations at the Square, however, revealed Adena pottery and boatstones (Henderson et. al. 1988). As suggested by Railey (1996:108): "The Adena-like artifacts from the Old Fort Earthworks and Biggs suggest early Middle Woodland construction dates and underscore the significant overlap between Adena and Hopewell."

44. The mouth of the Scioto River was not always where it is currently. In the 1800s the flow of the river at Portsmouth was redirected in order to better accommodate the Ohio and Erie Canal. The present confluence is the result of this change. The former river channel and old mouth, however, can still be seen in detailed topographic maps. In these maps the old channel and mouth are shown as Slab Run.

45. Winter solstice sunrise calculated for A.D. 100, map-measured horizon elevation corrected for refraction, lower limb tangency, and foresight tree height of 80 feet = corrected h value of 7.4 degrees. This high horizon elevation results from the mountains to the immediate southeast of the Portsmouth Group A Square.
    Interesting to note is that the northeast and southwest walls of the Square are not parallel to each other. The LiDAR azimuth for the southwest wall is 126.7 degrees, the northeast wall is 130.68 degrees.

46. Portsmouth Group B lunar max north rise azimuth calculated for A.D. 100, map-measured horizon elevation corrected for refraction, lower limb tangency, parallax, and foresight tree height of 80 feet = corrected h value of 7.5 degrees.
    Group B lunar max north set azimuth calculated for A.D. 100, map-measured horizon elevation corrected for refraction, lower limb tangency, parallax, and foresight tree height of 80 feet = corrected h value of 2.6 degrees.

47. It is interesting to note that the section of the parallel walls on the Kentucky side that lead to the Portsmouth Group C Earthwork do not extend in a straight line; but rather, curve in several places. The theta-corrected azimuth for the general trajectory of the walls is about 303.8 degrees.

48. Summer solstice sunrise calculated for A.D. 100, map-measured horizon elevation corrected for refraction, lower limb tangency, and foresight tree height of 80 feet = corrected h value of 1.6 degrees. No evidence for this posited section of parallel walls is likely to be found, since the area has been covered-over by a major railroad yard.

Chapter 6: Paint Creek Watershed

1. Limited excavation of one of the Junction Group mounds by Squier and Davis (1848:62) revealed three human skeletons, charcoal, and pieces of an 'altar' within the mound. No mention of the Steel Group or 33RO257 is made by Squier and Davis. Several of the enclosures at Junction and Steel are visible in LiDAR data. However, no trace of the structures at 33RO257 can be seen. The designation 33RO257 is the site number provided by the Ohio Historic Preservation Office.

2. Comparison of Squier and Davis's map for Junction and the results of a magnetic gradiometer survey completed by Jarrod Burks shows several differences. One difference reflects Burks's discovery in the magnetic data of a small, quatrefoil-shaped earthen embankment and ditch - where Squier and Davis show a squaroid earthwork and ditch. The difference is significant because as Burks (2006:10) points out, although other four-sided structures are found in Ohio - such as the Tarlton Cross, the Junction Group four-lobed enclosure, or quatrefoil, is the only one known for Ohio.

3. Chillicothe Country Club grounds superintendent Brian Martin advises that two long-time members of the Club recall that the mound was dug into in the 1920s. No record of that event is known, however. The mound is documented in the Ohio Archaeological Inventory as site # 33RO1176.

4. Anderson summer solstice sunset azimuth sunrise calculated for A.D. 100, map-measured horizon elevation corrected for refraction, lower limb tangency, and foresight tree height of 80 feet = corrected h value of 1.0 degrees.

5. From the east, the easiest way into Paint Valley would have been via the Anderson-Hopewell sites and then, from either of those sites, southwest into the valley. That route would have avoided the Paint Creek - Alum Cliffs Gorge. The Alum Cliffs Gorge is located in one of the most rugged areas of Ross County. The gorge is narrow and travel through it can be treacherous, with steep cliffs, narrow paths, falling rocks, and danger of flash flooding. Today the Alum Cliffs Gorge is part of the Ross County Park District's Buzzard's Roost Preserve.

6. The Hopewell Mound Group derives it name from Mordecai Hopewell, the former property owner of the land on which the Hopewell Mound Group is located.
   In 1980 Mark Michel (1980:9) reported that, "...a line bisecting the Hopewell mounds also bisects Mound City, 5 miles away. We may yet discover that Mound City was a kind of staging area for the major ceremonial events practiced at Hopewell."

7. Hopewell Mound Group winter solstice sunrise azimuth calculated for A.D. 100, map-measured horizon elevation corrected for refraction, lower limb tangency, and foresight tree height of 80 feet = corrected h value of 1.26 degrees.

    Looking in the other direction, from the southeast corner through the northwest corner, due to the high horizon elevation (corrected h = 2.27 degrees), the summer solstice sunset azimuth deviates from the azimuth of the diagonal sightline by 3.96 degrees - hence the winter solstice sightline is more accurate.

8. Squier and Davis (1848:Pl. XXI, no. 4) give the lengths of the walls that comprise the Frankfort Square as 1,080 feet. I suspect that measurement is across the Square from outside edge to outside edge. If the walls were each 20 feet at base, then the crest to crest length across the earthwork would be 1,080 - 20 = 1,060. The difference of 4 feet between 1,060 feet and 1,054 feet (or 1 HMU) is negligible at earthwork scale.

    The orientation of Frankfort Works as provided here is estimated based on the relationship between the earthwork and nearby topographic features illustrated by Squier and Davis. In this regard it should be noted that Squier and Davis's representation of the city street grid is incorrect. Squier and Davis show the street grid extending approximately north-south and east-west. In actuality, however, the grid extends northeast-southwest. Thus the city streets as shown by Squier and Davis cannot be reliably used to establish the orientation of the earthwork. For that, we must rely on topographic relationships.

9. The grave goods found by Moorehead (1892:113-143) in Porter mounds 15 and 17 at Frankfort include typical Hopewell items such as: platform pipes, copper plates, earspools, drilled bear and wolf teeth, cut mica, shell drinking cups, and hundreds of pearl beads.

10. Frankfort winter solstice sunrise azimuth calculated for A.D. 100, map-measured horizon elevation corrected for refraction, lower limb tangency, and foresight tree height of 80 feet = corrected h value of 1.26 degrees.

11. From Frankfort, the Hopewell Mound Group would not have been visible due to the intervening hills. Nevertheless, the alignment is an objective fact. (Also see DeBoer (2010:182) - who, although he notes the intersecting trajectory, does not note the solstice alignment). The fact that the alignment cuts to the center of the Hopewell Mound Group Great Enclosure increases the likelihood of its intentionality. Further, the Frankfort - Hopewell Great Enclosure alignment is reminiscent of the straight line alignment between the Works East, High Bank, Liberty, and Mills Square earthworks. Thus this would not be the only example of this kind of inter-site alignment.

12. Baum winter solstice sunset azimuth calculated for A.D. 100, map-measured elevation corrected for refraction, lower limb tangency and foresight tree height of 80 feet = corrected horizon elevation = 1.9 degrees. (Also see Hively and Horn 1982 who earlier found that Baum is oriented to the winter solstice sunset.)

13. A number of excavations have been carried-out at Seip. In fact, the Pricer Mound that visually dominates the site today is largely a reconstruction. Restoration efforts, however, were based on the dimensions of the original structure and comparison of before and after photos shows that the size and shape as seen today is an accurate representation of the original.

14. Thomas (1894:488-489) gives the average length of the Seip Square walls as about 1,126 feet. This differs by 72 feet from the 1 HMU length and by 46 feet from the 1,080-foot lengths shown by Squier and Davis in their illustration of the site. Some question exists, however, as to the accuracy of the Middleton-Thomas survey data, because as explained by Thomas (1894:488-489): "The southeastern and southwestern corners [of the Square], which are now obliterated, were placed back to correspond with the ratio of the lines in Messrs. Squier and Davis's figure....This resurvey, therefore, is to be accepted as reliable only as far as it relates to the northern wall, and the eastern and western walls so far as the later extend."

Given the above, I am inclined to accept Squier and Davis's figures - which predate, by almost 50 years, the Middleton-Thomas survey. As may be the case with other Squier and Davis figures, the 1,080 length could represent outside wall-to-outside wall measurements; thus bringing the center-to-center or inside measurements closer to the posited 1 HMU length.

15. Seip winter solstice sunrise azimuth calculated for A.D. 100, map-measured horizon elevation corrected for refraction, lower limb tangency, and foresight tree height of 80 feet = corrected h value of 2.27 degrees.

16. The Pricer Mound (also known as Seip Mound 1) was enormous. It is one of the largest Hopewell mounds known for Ohio - at 240 feet in length, 160 feet wide, and 30 feet high. The Pricer Mound was excavated in the late 1920s (Shetrone and Greenman 1931). No less than 122 interments were found in the mound.

The other very large mound within the Large Circle is known as the Seip Mound, or Seip Mound 2. Seip Mound was about half the size of the Pricer Mound and was actually a set of three connected or conjoined mounds. A series of three Spirit houses were found within this mound, along with numerous cremated and non-cremated burials (Mills 1909).

17. It should be noted that although the winter solstice sunrise appears over Spruce Hill when viewed from the Bourneville Circle, that solstice azimuth is not coincident with the major axis of the Bourneville Circle. The major axis of the Bourneville Circle is skewed from the winter solstice azimuth by several degrees. Nor is the solstice sightline to any particular foresight location on Spruce Hill. These observations bring into question the intentionality of the Bourneville solstice alignment. However, the alignment is worth noting because of the complementary opposite summer solstice sunrise view from nearby Seip and the

visual resemblances as already noted between Spruce Hill as seen from the Bourneville Circle and the Pricer Mound at Seip.

Also worth noting are the comments of Squier and Davis (1848:12) concerning the presence of "strong traces of fire...visible at many places on the line of the wall, particularly at F...." at Spruce Hill. Point F on Squier and Davis's map of Spruce Hill is located along the western edge of Spruce Hill, only 600 feet from where the Bourneville winter solstice sightline intersects Spruce Hill. Given this, it seems possible that fires on Spruce Hill at point F could have been related to winter solstice observations made from the Bourneville Circle. That said, however, it is also the case that the fires could date to anytime.

Bourneville Circle winter solstice sunrise azimuth calculated for A.D. 100, map-measured horizon elevation corrected for refraction, lower limb tangency, and foresight tree height of 80 feet = corrected h value of 3.57 degrees.

18. Pricer Mound - Spruce Hill summer solstice sunrise azimuth calculated for A.D. 100, map-measured horizon elevation corrected for refraction, lower limb tangency, and foresight tree height of 80 feet = corrected h value of 0.7 degrees.

19. Inter-site alignments for the Chillicothe area earthworks have been suggested by several researchers (e.g., Romain 1992a,1992b, De Boer 2010, Hively and Horn 2010). From the many possibilities possible or proposed, I have limited my selection of plotted alignments to those I believe were either intentional, or recognized by the Adena-Hopewell. I tend to not include in this group sightlines that posit locations where there are no documented sites - such as Mount Prospect (Hively and Horn 2010), or Jester Hill (Hively and Horn 2013b).

Seip - Baum - Works East alignment after Romain 1992b. Works East - Anderson - Frankfort alignment based on Hively and Horn 2010. Liberty - High Bank - Works East alignment after Hively and Horn 2010, with Mills Square added by Romain, this volume. High Bank - Newark orthogonal by Romain, this volume.

All lunar azimuths re-calculated by Romain for A.D. 100, using a 1 degree horizon elevation. The inter-site alignments presented in this figure were, in most cases, probably not intended for observation purposes, as the distances are too great and intervening terrain obstructs most views. More likely is that they were important for their symbolic meanings and associations.

Chapter 7: Brush Creek Watershed

1. The Fort Hill perimeter wall "was built slightly below the more or less level top of the hill, with the top of the wall in some places being about even with the summit of the hill and in other places somewhat below it....The wall, which is approximately one and five-eights miles in total length is broken by thirty-three openings or gateways which are from fifteen to twenty feet in width and which are irregularly spaced....The height of the wall varies from about six to fifteen feet, and its basal width is around forty feet" (Potter and Thomas 1970:33).

There are three depressions or "ponds" within the enclosure. The largest apparently held water until it was drained in the early 1800s (Squier and Davis 1848:15). Squier and Davis (1848:15) suggest that "when full, the water must have covered very nearly an acre."

In 1964, Olaf Prufer (1997) conducted an excavation into a section of wall near the south end of the enclosure. Prufer found that the wall had been constructed in two stages and was comprised of earth and slabs of limestone, the largest being about 2.8 feet square (Prufer 1997:318). Small test excavations within the enclosure failed to reveal any cultural materials.

2. Fort Hill summer solstice sunrise azimuth calculated for A.D. 100, map-measured horizon elevation corrected for refraction, lower limb tangency, and foresight tree height of 80 feet = corrected h value of - 0.3 degrees.

3. Investigations by Baby (1954) just south of Fort Hill revealed two earthen circle enclosures, and the postmold pattern of a large wooden rectangular structure. The rectangular structure measured roughly 120 feet long and 60 feet wide. Underneath one of the circle enclosures Baby found postmold patterns suggestive of a circular wooden building that had been disassembled and buried beneath the circle earthwork. The immediate area also revealed fire pits, flint debitage, part of an obsidian biface, pottery, bladelets, and mica fragments.

4. The 186.3-foot unit of length is derived by constructing a 45-45-90 degree triangle having two sides equal to 131.75 feet. The resulting hypotenuse will be 186.3 feet. Recall that 131.75 feet is equal to 1/8 HMU (where 1 HMU = 1,054 feet).

5. LiDAR image and model by William F. Romain. Data download from Ohio Statewide Imagery Program by William F. Romain and Jeffrey Wilson. Vertical exaggeration 2.0. Crater dimensions based on Milam (2010), Hansen (1994), and Reidel, Koucky, and Stryker (1982). Of interest is that the Serpent Mound is located within the impact crater zone. This does not mean that the builders of Serpent Mound knew that a meteorite impact occurred there. The meteorite impact happened millions of years before occupation of area by humans. The Serpent Mound builders most likely did, however, recognize that the topography of the area is unusual - as the meteorite impact cause huge blocks of bedrock to be displaced both upward and downward from their normal positioning.

6. Radiocarbon dates recently obtained (Herrmann, et al. 2014; Romain, et al. 2013) indicate that the Serpent Mound was built during the Early Woodland Period. Five radiocarbon samples extracted by GeoProbe coring from five different locations on the Serpent's body yielded an OxCal median date of 321 B.C., with a 2 sigma-range of 381 - 44 B.C. (Herrmann, et al. 2014:121). Two additional samples yielded dates about 200 years earlier. Based on these data, it appears that Serpent Mound was built by people of the Adena culture and later repaired or renewed by Fort Ancient peoples. Physical

evidence shows that the Serpent Mound ridge was occupied by Archaic, Adena, and Fort Ancient peoples.

The effigy was partially restored in the late 1800s by Frederic W. Putnam (1890). What Putnam did was scrape up earth from the immediate area alongside the serpent and heap this earth back onto the body of the effigy - thus building it up to what he believed its original height. He did not materially change the outline of the effigy from what was first represented in print by Squier and Davis (1848:Pl. XXXV).

7. Charles Willoughby (1919) was the first to suggest that Serpent Mound represents the Great Serpent of Native American mythology. Following upon this, Romain (1988; 2000:252-253) documented a number of physical resemblances between the Serpent Mound and ethnographic descriptions of the Great Serpent. More recently, George Lankford (2007a) likewise concluded that Serpent Mound represents the Great Serpent of Native American mythology. Perhaps Lankford's (2007a, 2011) most valuable contribution to the discussion, however, has been to connect the Great Serpent of Eastern Woodlands mythology and the Serpent Mound effigy to the star constellation Scorpius. As Lankford (2007a:132) explains: "...the identification of Scorpio and Antares as the Serpent and his red eye points to a significant ethnoastronomical belief complex which cut across tribal and linguistic lines." Further Lankford (2007b) posits that in this context, at the end of the Milky Way Path, the Great Serpent was the guardian of the "Realm of the Dead."

8. Solar and lunar alignments calculated for 300 B.C., map-measured horizon elevations corrected for refraction, lower limb tangency, and foresight tree height of 80 feet. Lunar alignments also corrected for parallax.

The Serpent Mound map shown in this figure was made from a radial traverse survey oriented to true north by a series of solar and Polaris observations (Romain 2000:233-253, 1987). Additional LiDAR analyses and ground survey in 2011 by the author confirm the accuracy of the original map.

9. Photo by the author, June 21, 1990.

10. The serpent pictured was a commercially purchased rattlesnake specimen on store display. I purchased it with regret, as I do not support the killing of animals for display purposes. The sun-disk was a circle cut from cardboard. When rotated as shown it presents the appearance of an oval.

11. For illustration purposes, the daylight was removed from this illustration otherwise it would not be possible to show the stars and sun at the same time. Once the sun sets along the projected trajectory shown, then of course Scorpius becomes visible. Since only a very few minutes occurs between sunset and the visibility of stars, it would have been fairly straightforward to deduce that Scorpius was chasing the sun.

12. Serpent Mound is situated on promontory that overlooks a wide expansive valley to the south and east - ideal for viewing Scorpius.

Scorpius, also known as Scorpio is a summer constellation, visible in the northern latitudes during summer months in the southern sky. Although the constellation bears a Western name, it does resemble a serpent in its long curving form and the tail in particular is reminiscent of the coiled tail of Serpent Mound.

Scorpius is not a well-documented constellation in Native American ethnographic literature. Among the tribes who recognized the constellation Scorpius as a serpent, however, were the Skidi Pawnee. The Pawnee called the constellation "Real Snake" (Chamberlain 1982:132-134).

That said, many Native American stories tell of different animals and monsters biting or otherwise attacking the sun (Romain 1988). Several, such as the Iroquois, Kiowa, and Yurok describe the attacking monster as a fire-dragon, serpent, or rattlesnake.

It should be noted that, although the declination values for stars have changed over the years due to precession, their positions relative to each other have not. The same star configuration and rise-set pattern for Scorpius would have been seen throughout the Early Woodland through Mississippian times and is still the case today.

In this figure, the rotation of Scorpius is shown at one-month intervals at the same local time each month. The same rising, rotation, and setting phenomena, however, also occurs each night over the course of several hours, during the summer months. Thus the rotation phenomenon occurs at two different time scales.

With reference to Figure 7.11, the reader will notice that in order to show the correspondence between the body convolutions and monthly rotation of Scorpius, the map image of Serpent Mound has been flipped so that the Serpent's head is pointing to the upper right, equivalent to northeast – consistent with the star constellation in real life. On the ground, however, the Serpent's head points in the exact opposite direction – i.e., to the northwest. I believe there is a plausible explanation for this flipped imagery.

As is well-known, many Native American tribes across the Eastern Woodlands considered the Otherworld to be the reverse of This World - so, for example, when it is daytime in This World it is nighttime in the Otherworld, or what is right-side up in This World, is upside-down in the Otherworld. By laying-out the Serpent Mound using a flipped azimuth it may be that this concept of reversed cosmological realms is implied. In This World the Serpent points in one direction, while in the Sky-Otherworld the grid points to its mirror image direction. Neither direction is the 'correct' direction, as both are appropriate to their respective worlds and dependent upon the viewer's perspective.

13. Azimuths for Scorpius and Antares plotted for midnight local time. Midnight is the conceptual opposite of noon - hence a symbolically meaningful time. Antares

is the brightest star in Scorpius. It is bright red and is the only red star in the summer sky.

The location of Serpent Mound appears selected-for based on the azimuthal relationship between Sugarloaf Mountain, the Milky Way, Scorpius, and the effigy mound. The azimuthal relationships pointed out are too mutually supportive to have been due to chance.

Chapter 8: Little Miami River Watershed

1. Squier and Davis (1848:34) reported that the Pollock isthmus wall was 10 feet high and 30 feet wide at its base.

Pollock has been the focus of long-term investigations by archaeologist Robert Riordan. Riordan (1995) determined that the enclosure walls were built in stages between A.D. 50 to A.D. 225.

2. Williamson Mound winter solstice sunrise azimuth calculated for A.D. 100, map-measured horizon elevation corrected for refraction, lower limb tangency, and foresight tree height of 80 feet = corrected h value of 0.6 degrees. Calculated winter solstice az = 122.07 + 0.89 theta corr = LiDAR az = 122.9.

3. Research by archaeologist Robert Connolly shows that the Fort Ancient walls were built over a period of years from 100 B.C. - A.D. 300 (Connolly 2004:220). The walls range from 4 - 23 feet in height. More than 60 breaks occur in the walls. Several mounds are located inside and outside of the enclosure. Other features include a set of parallel embankments, ponds, stone circles, ditches, cobble-paved areas, ramps, and terraces outside of the perimeter walls.

Excavations conducted by Moorehead (1890:33) resulted in dozens of burials found within the enclosure, as well as on outside terraces. Given that a Fort Ancient village was located within the 'South Fort' area, many of these burials may date to that occupation.

4. Fort Ancient summer solstice sunset azimuth from center stone circle calculated for A.D. 100, map-measured horizon elevation corrected for refraction, lower limb tangency, and foresight tree height of 80 feet = corrected h value of 1.05 degrees.

Location for center stone circle is plotted based on total station ground survey using the North Fort stone mounds as reference points (Romain 2004b:fig. 6.1).

5. Documentary research (Romain 2004b:66-67) suggests that although the mounds that comprise the North Fort Square were surfaced with limestone slabs in historic times, their original location is unchanged.

Located nearly in the center of the major axis for the site, the center stone ring is about 35 feet in diameter.

6. North Fort Square summer solstice sunset azimuth calculated for A.D. 100, on-site measured near horizon elevation corrected for refraction, lower limb tangency, to include observed tree height = corrected h value of 3.01 degrees (see Romain 2004b:69-70 for further discussion).

7. It is interesting to note that Atwater's map shows the azimuth for the parallel walls quite accurately as determined by comparison with the 1930s aerial photograph.

8. Fort Ancient parallel walls summer solstice sunrise azimuth calculated for A.D. 100, map-measured horizon elevation corrected for refraction, lower limb tangency, and foresight tree height of 80 feet = corrected h of 0.07 degrees.

I am not the first to suggest that the parallel walls were oriented to the summer solstice sunrise. Essenpreis and Duszynski (1989) also made that suggestion some years ago. The present analysis corroborates their discovery.

My assessment that the parallel walls are aligned to the summer solstice sunrise differs from an earlier assessment I made suggesting a winter solstice sunset alignment for this feature (Romain 2004b:70-71). The present analysis benefits from use of a higher resolution aerial photograph than originally used.

9. Thanks to a report by Charles Willoughby (1922), the location and dimensions of the Turner Earthworks are known. The dimensions for the Milford and West Milford earthworks are documented by Squier and Davis (1848:Pl. XXXIV, no. 1) and Roberdeau's annotated map (McCullough 2010 [1991]), respectively. The location for Milford is fairly certain based on analysis of Squier and Davis's map. The location for the West Milford Work, however, is not as certain. Squier himself surveyed Milford. However, for West Milford he appears to have relied on maps by others. Unfortunately, these early sources do not agree with each other in their particulars. The most authoritative source for the location of Milford and West Milford appears to have been a map made in 1823 by Isaac Roberdeau (McCullough 1996:fig. 6). However, the Roberdeau map has inconsistencies in its orientation. By far, the most exhaustive analysis of the location of these earthworks was made by J. Huston McCullogh (1996, 2010 [1991]). I agree with McCullough's (1996:figure 3) estimate of where Milford West was located. I depart slightly from McCullough in my placement of the Milford Square, as I am inclined to locate it a bit further south than he does - so that the Chillicothe-Milford Pike cuts through the earthwork in the manner represented by Squier and Davis - based upon Davis's ground-based survey.

10. Summer solstice sunset azimuth for Turner Ellipse calculated for A.D. 100, map-measured horizon elevation corrected for refraction, lower limb tangency, and foresight tree height of 80 feet = corrected h value of 2.8 degrees. Calculated summer solstice sunset az = 298.6 degrees.

11. The illustration of the East Fork Works provided by Squier and Davis (1848:Pl. XXXIV, no. 2B) is not based on a survey they made; rather, it is a

second or possibly third generation map derived from a survey made in 1823 by Major Isaac Roberdeau (see McCullough 1996, 2010 [1991] for a detailed discussion). Although Squier and Davis did not survey the site, it is possible that they visited the East Fork Works. This is suggested by an interesting addition to their illustration showing the bluff upon which the earthwork is purportedly located. This feature is not shown by Roberdeau, or other maps of the time (e.g., Williamson, Warden) but it does correspond fairly well to an actual bluff at the location where I believe the site is located.

12. The water flow arrow shown by Squier and Davis perhaps indicates the directional flow of the small rivulet that leads from the earthwork to the East Fork Little Miami River.

Hugh McCullough (personal communication 2011) has advised me that the location I propose for the East Fork Works was also posited in 1999 in unpublished work by Mr. John Browning of Loveland, Ohio. I have not seen Mr. Browning's unpublished work and to date, neither McCullough nor I have been able to locate Mr. Browning.

Chapter 9: Great Miami River Watershed

1. Foerste (1915:142) provides a survey map of the Alexandersville Works. Foerste's measurements for the Large Circle and Small Circle are the same as those provided by McBride (in Squier and Davis 1848:Pl.XXIX, no. 1). Further, the Foerste map distinguishes between wall segments that can still be traced on the ground (in 1915) and those shown on the McBride map, but no longer traceable. The interesting thing is that, if the dimensions for the earthwork as given by McBride and Foerste are correct, then the dimensions of the earthwork do not appear based on any obvious iteration of the HMU. Further, if the diameter of the Large Circle was in fact 1,950 feet, that would make the Alexandersville Circle the largest of all Hopewell circles. Worth noting too is that this is the only tripartite earthwork known for the Great Miami River Valley.

2. Summer solstice sunrise for Alexandersville Works calculated for A.D. 100, map-measured horizon elevation corrected for refraction, lower limb tangency, and foresight tree height of 80 feet = corrected h value of 0.9 degrees. Calculated summer solstice sunrise az = 59.4 degrees.

3. In the late 1800s, employees of Springboro College dug into the Hill-Kinder Mound. They reportedly found copper beads and what may have been a burial chamber within.

4. The Great Mound is the largest mound in Butler County. Excavation into the mound in 1879 substantially reduced its height.

5. The circular embankments at Carlisle Fort are not circles, but rather, arcs that for much of their course follow the trajectory of the drawn circles. The long

circular embankment in particular deviates from an ideal circle along its south side.

6. Summer solstice sunrise for Carlisle Fort for A.D. 100, map-measured horizon elevation corrected for refraction, lower limb tangency, 0 feet tree height = corrected h value of -0.09 degrees. Calculated summer solstice sunrise az = 58.5 degrees.

7. MacLean (1879:182) describes the entrance to the Fairfield Township Works thusly: "The ends of the wall [i.e., the outside perimeter wall] curve inwardly, forming a true circle ninety-eight feet in diameter. Within the circle thus formed is another circle, with no opening, fifty-eight feet in diameter. The external gateway *e* is seventy-five feet in width, covering which is a mound five feet high and forty feet in diameter at the base. The internal gateway (*d*) is twenty feet wide. The passage way between the mound and the embankment, and between the walls of the circles is about six feet wide."

A human skeleton was reported by McBride in one of the mounds within the enclosure (quoted by Murphy 1978:11). Grave goods accompanying this burial included diagnostic Adena artifacts to include three drilled quadriconcave gorgets made from dark slate, as well as an Adena stemmed blade made from Harrison County, Indiana hornstone.

8. The Pollock Wilson Mounds I-III are designated as site # 33BU98 in the Ohio Archaeological Inventory. The location for the mound group shown here is based on the annotated USGS 7.5-minute series map that accompanies the site OAI site form for the site. The mounds are described by MacLean (1879:215) thusly: "A long narrow ridge of land, running nearly east and west, in section 31, and belonging to Pollock Wilson, is the most prominent point close to the city of Hamilton. On this ridge are three small mounds. The one located in the rear of the dwelling is now about eighteen inches high, and used for a flower bed. The other two are in a grove, one forty feet and the other forty-eight feet in diameter. Both are three and a half feet high, four feet apart."

9. This is a difficult calculation to make with precision due to the nearness of the horizon and not knowing if there where trees or not at the end of the sightline. Because the sightline is so short, the presence of trees affects the azimuth of the sunset significantly - ranging from 237.8 degrees (no trees, lower limb tangency) to 235.9 (with 80-foot trees, lower limb tangency).

Excavation into one of the Twin Mounds (33HA105) revealed a cremated burial with copper breastplate and celt, drilled bear teeth, and marine shell and pearl beads and other items (Lee and Vickery 1972:6).

10. The notion that mounds were used as signal stations for communication by fire, smoke, or mirrors - whether up and down the Scioto River Valley, or Great Miami River Valley has been around for a considerable time (e.g., Squier and Davis 1848:141; MacLean 1879:59). As remarked by early observers, many

mounds located on high vantage points exhibit traces of fire. Of course we need to keep in mind that just because a mound or series of mounds 'could have been' used for signaling, does not necessarily mean that they were used that way. Still, given the occurrence of ethnographic accounts that tell of Native Americans using signaling methods across great distances, the matter of inter-visibility of sites especially along the Scioto and Great Miami River valleys is something that needs to be seriously considered.

In 1990 a group of 20 persons coordinated by The Archaeological Conservancy, tested the signal hypothesis for a section of the Great Miami River Valley (Archaeological Conservancy 1990:4). Volunteers positioned themselves on the Great Mound, Hill-Kinder Mound, and Miamisburg Mound. Using mirrors, flashes of light sent from the Great Mound were seen by the group on the Hill-Kinder Mound. Likewise, the mirror signals send from the Hill-Kinder Mound were seen by those at the Miamisburg Mound.

11. At first it appears that the summit of the hill that the Great Mound is situated on blocks the line of sight from the Pollock Wilson Mound Group to the Great Mound. Notice, however, that the topographic map shows the elevation not for the top of the Great Mound; but rather, for the level flat ground at the base of the mound. If we add to that 950-foot ground level, the height of 43 feet to account for the original height of the Great Mound, plus a minimum of 5 feet to account for observer height, then the elevation is 998 feet - which is greater than 981 feet shown for the summit of the hill. (We can also add 5 feet to the elevation at the Pollock end of the sightline which results in even more favorable results.) Thus a person standing on top of the Great Mound would in fact be able to see the Pollock Wilson Mound Group (assuming cleared trees at the Wilson Mound Group), and the reverse would also be true. To further illustrate this, the small black triangle and the black line in Figure 9.16 represent the height of the Great Mound and resulting line of sight between the two mounds. Notice that the line of sight clears the summit of the hill upon which the Great Mound is situated.

Chapter 10: Discussion

1. The criteria used to select the earthworks included in locational analyses are discussed in detail elsewhere (see Romain 2004a). Selected Hopewell geometric complexes included: Newark Great Circle and Wright Square; Newark Octagon and Observatory Circle; Circleville; Hopeton; Works East; High Bank; Liberty; Seal; Hopewell; Frankfort; Baum; Seip; Marietta; Turner; Milford; Stubbs; Alexandersville; Portsmouth Group B; and Portsmouth Group A.

2. The evidence on this point is controversial, but based on excavations at Hopeton, archaeologists Mark Lynott and Rolfe Mandel (2009) are of the opinion that the topsoil within some geometric earthworks was completely stripped, or removed, perhaps to help build the enclosure walls. If that was the case then, the soils found within the earthworks today developed subsequent to Hopewell times and yield potentials may have been different.

3. In the case of the Junction Group, there are several small earthworks that look like squares with rounded corners - affectionately known as "squircles" (Burks 2006b:19). I suspect these were simply iterations on the theme expressed by earthen circles – i.e., world creation and renewal. In their directionality, however, squircles - through their diagonal axes, provide a way to reference the horizontal division of space into four quadrants. Where large circles and squares occur together, it may be that those geometric forms were multivalent and interchangeable. Where found together perhaps they represented the dual aspect of the cosmos – i.e., earth and sky.

REFERENCES

Anderson, Jerrrel C.
2011 The Circleville Earthwork and Hopewell. *Ohio Archaeologist* 61(1):18-29.

Andrews, E. B.
1877 Report of Explorations of Mounds in Southeastern Ohio. *Tenth Annual Report of the Peabody Museum of American Archaeology and Ethnology* 2(1):48-74.

Applegate, Darlene, and Robert C. Mainfort, Jr. (editors)
2005 *Woodland Period Systemics in the Middle Ohio Valley*. University of Alabama Press, Tuscaloosa.

Archaeological Conservancy
1990 Ohio Mound Signal Test Successful. In *The Archaeological Conservancy Newsletter*, Summer, 1990, p. 4.

Atwater, Caleb
1820 Description of the Antiquities Discovered in the State of Ohio and Other Western States. *Archaeologia Americana* 1:105-267.

Baby, Raymond S.
1954 Archaeological Explorations at Fort Hill. *Museum Echoes* 27:86-87.

Baby, Raymond S., and Susan M. Langlois
1979 Seip Mound State Memorial: Non-Mortuary Aspects of Hopewell. In *Hopewell Archaeology: The Chillicothe Conference*, edited by David S. Brose and N'omi Greber, pp. 16-18. Kent State University Press, Kent.

Bacon, Willard S.
1993 Factors in Siting a Middle Woodland Enclosure in Middle Tennessee. *Midcontinental Journal of Archaeology* 18(2):245-281.

Baires, Sarah E.
2014 Cahokia's Origins: Religion, Complexity and Ridge-top Mortuaries in the Mississippi River Valley. Ph.D. dissertation, University of Illinois at Urbana-Champaign.

Baires, Sarah E., Amanda J. Butler, B. Jacob Skousen, and Timothy R. Pauketat
2013 Fields of Movement in the Ancient Woodlands of North America. In *Archaeology After Interpretation: Returning Materials to Archaeological Theory*, edited by Benjamin Alberti, Andrew M. Jones, and Joshua Pollard, pp. 197-218. Left Coast Press, Walnut Creek, California.

Basso, Keith H.
1996 *Wisdom Sits in Places: Landscape and Language Among the Western Apache*. University of New Mexico Press, Albuquerque.

Beauchamp, William M.
1922 *Iroquois Folklore, Gathered from the Six Nations of New York*. Kennikat, New York.

Blazier, Jeremy, AnnCorinne Freter, and Eliot M. Abrams
2005 Woodland Ceremonialism in the Hocking Valley. In *The Emergence of the Moundbuilders: The Archaeology of Tribal Societies in Southeastern Ohio*, edited by Eliot M. Abrams and AnnCorinne Freter, pp. 98-114. Ohio University Press, Athens.

Boivin, Nicole
2004 Mind Over Matter? Collapsing the Mind-Matter Dichotomy in Material Culture Studies. In *Rethinking Materiality: The Engagement of Mind with the Material World*, edited by Elizabeth DeMarrais, Chris Gosden, and Colin Renfrew, pp. 63-71. McDonald Institute for Archaeological Research, Cambridge University, Cambridge.

Bowser, Brenda J., and Maria Nieves Zedeño
2009 *The Archaeology of Meaningful Places*. University of Utah Press, Salt Lake City.

Brine, Lindesay
1894 *Travels Amongst American Indians: Their Ancient Earthworks and Temples*. Sampson, Low and Marston, London.

Brinkley, S. H.
1889 Carlisle Fort. *American Antiquarian* 2:174-184.

Brown, James A.
2013 Place, Practice, and Process in Hopewell Culture. Keynote presentation. Midwest Archaeological Conference, October 26, 2013. Columbus, Ohio.

2012 *Mound City: The Archaeology of a Renown Ohio Hopewell Mound Center*. Midwest Archeological Center Special Report No. 6. National Park Service, Midwest Archeological Center, Lincoln, Nebraska.

2004 Mound City and Issues in the Developmental History of Hopewell Culture in Ross County Area of Southern Ohio. In *Aboriginal Ritual and Economy in the Eastern Woodlands: Essays in Memory of Howard Dalton Winters*, edited by Anne-Marie Cantwell, Lawrence A. Conrad, and Jonathan E. Reyman, pp. 147-168. Illinois State Museum Scientific Papers, Vol. 30, Springfield.

1992 Closing Commentary. In *Cultural Variability in Context: Woodland Settlements of the Mid-Ohio Valley*, edited by Mark F. Seeman, pp. 80-82. Midcontinental Journal of Archaeology Special Paper 7, Kent State University Press, Kent.

1997 The Archaeology of Ancient Religion in the Eastern Woodlands. *Annual Reviews in Anthropology* 26:465-485.

Brown, Joseph Epes

1953 *The Sacred Pipe: Black Elk's Account of the Seven Rites of the Ogala Sioux*. University of Oklahoma Press, Norman.

Buikstra, Jane E. and Douglas K. Charles
1999 Centering the Ancestors: Cemeteries, Mounds and Sacred Landscapes of the North American Midcontinent. In *Archaeologies of Landscape: Contemporary Perspectives*, edited by Wendy Ashmore and Barbara Knapp, pp. 201-228. Blackwell, Oxford, UK.

Buikstra, Jane E., Douglas K. Charles, and Gordon F.M. Rakita
1998 *Staging Ritual: Hopewell Ceremonialism at the Mound House Site, Greene County, Illinois*. Kampsville Studies in Archeology and History, No. 1. Center for American Archeology, Kampsville, Illinois.

Burks, Jarrod
2014 Recent Large-Area Magnetic Gradient Surveys at Ohio Hopewell Earthwork Sites. International Society for Archaeological Prospection *ISAP News* 39:11-13.

2013 Large Area Magnetic Survey at the Hopewell Mound Group Unit, Hopewell Culture National Historical Park, Ross County, Ohio. OVAI Contract Report #2012-52-1. Contract P12PX15855. Prepared for Midwest Archaeological Center, National Park Service, Lincoln, NE. Prepared by Ohio Valley Archaeology, Inc., Columbus, Ohio.

2010 Rediscovering Prehistoric Earthworks in Ohio, USA: It All Starts in the Archives. In *Landscapes Through the Lens: Aerial Photographs and the Historic Environment*, edited by D. C. Cowley, R. A. Standring, and M. J. Abicht, pp. 77-87. Oxbow Books, Oxford.

2006 Geophysical Survey of the Junction Group (33RO28) Earthworks in Ross County, Ohio 2005: A Progress Report. Ohio Valley Archaeological Consultants Contract Report # 2005-15.

Burks, Jarrod, and Robert Cook
2011 Beyond Squier and Davis: Rediscovering Ohio's Earthworks Using Geophysical Remote Sensing. *American Antiquity* 76(4):667-689.

Byers, A. Martin
2004 *The Ohio Hopewell Episode: Paradigm Lost and Paradigm Gained.* University of Akron Press, Akron.

1996 Social Structure and the Pragmatic Meaning of Material Culture: Ohio Hopewell As An Ecclesiastic-Communal Cult. In *A View From the Core: A Synthesis of Ohio Hopewell Archaeology*, edited by Paul J. Pacheco, pp. 174-193. Ohio Archaeological Council, Columbus.

Carr, Christopher, and D. Troy Case (editors)
2005a *Gathering Hopewell: Society, Ritual, Ritual Interaction*. Kluwer Academic/Plenum Publishers, New York.

Carr, Christopher and D. Troy Case
2005b The Gathering of Hopewell. In *Gathering Hopewell: Society, Ritual, and Ritual Interaction*, edited by Christopher Carr and D. Troy Case, pp. 19-50. Kluwer Academic/Plenum Publishers, New York.

Carskadden, Jeff, and Donna Fuller
1967 The Hazlett Mound Group. *Ohio Archaeologist* 17(4):139-143.

Chamberlain, Von Del
1982 *When Stars Came Down to Earth: Cosmology of the Skidi Pawnee Indians of North America*. Ballena Press, Menlo Park, California and Center for Archaeoastronomy, College Park, Maryland.

Charles, Douglas K., and Jane E. Buikstra (editors)
2006 *Recreating Hopewell*. University Press of Florida, Gainesville.

Clark, John E.
2004 Surrounding the Sacred: Geometry and Design of Early Mound Groups as Meaning and Function. In *Signs of Power: The Rise of Cultural Complexity in the Southeast*, edited by Kenneth E. Sassaman and Michael J. Heckenberger, pp. 162-213. University of Alabama Press, Tuscaloosa.

Clay, R. Berle
2002 Deconstructing the Woodland Sequence from the Heartland: A Review of Recent Research Directions in the Upper Ohio Valley. In *The Woodland Southeast*, edited by David G. Anderson and Robert C. Mainfort, Jr., pp. 162-184. University of Alabama Press, Tuscaloosa.

1986 Adena Ritual Spaces. In *Early Woodland Archaeology*, edited by Kenneth B. Farnsworth and Thomas E. Emerson, pp. 581-595. Center for American Archeology, Kampsville, IL.

Connelley, William E.
Notes on the Folk-lore of the Wyandots. *Journal of American Folklore* 12:116-125.

Connolly, Robert P.
2004 Time, Space, and Function at Fort Ancient. In *The Fort Ancient Earthworks: Prehistoric Lifeways of the Hopewell Culture in Southwestern Ohio*, edited by Robert P. Connolly and Bradley T. Lepper, pp. 217-222. Ohio Historical Society, Columbus.

1996 Prehistoric Land Modification at the Fort Ancient Hilltop Enclosure: A Model of Formal and Accretive Development. In *A View from the Core: A Synthesis of Ohio Hopewell Archaeology*, edited by Paul J. Pacheco, pp. 260-273. Ohio Archaeological Council, Columbus.

Connolly, Robert P., and Ted S. Sunderhaus
2004 Rules for "Reading" Fort Ancient Architecture. In *The Fort Ancient Earthworks: Prehistoric Lifeways of the Hopewell Culture in Southwestern Ohio*, edited by Robert P. Connolly and Bradley T. Lepper, pp. 51-65. Ohio Historical Society, Columbus.

Converse, Robert N.
2003 *The Archaeology of Ohio*. Archaeological Society of Ohio, Columbus.

Count, Earl W.
1952 The Earth-Diver and the Rival Twins: A Clue to Time Correlation in North-Eurasiatic and North American Mythology. In *Indian Tribes of Aboriginal America*, edited by Sol Tax. University of Chicago Press, Chicago.

Dancey, William S.
1984 The 1914 Archaeological Atlas of Ohio" Its History and Significance. Paper presented at the 49th Annual Meeting of the Society for American Archaeology, Portland, Oregon.

Dancey, William S., and Paul J. Pacheco
1997 A Community Model of Ohio Hopewell Settlement. In *Ohio Hopewell Community Organization*, edited by William Dancey and Paul Pacheco, pp. 3-40. Kent State University Press, Kent.

Deloria, Vine, Jr.
2003 [1973] *God is Red: A Native View of Religion*. Fulcrum Publishing, Golden, Colorado.

1991 *Sacred Lands and Religious Freedom*. Association on American Indian Affairs, New York. Also see electronic document, https://eee.uci.edu/clients/tcthorne/Socec15/SLReader.pdf, assessed September 26, 2013.

Dempsey, Erin C.
2010 Small Scale Geoarchaeological Investigations of Earthen Wall Construction at the Hopeton Earthworks (33RO26). *Hopewell Archaeology: The Newsletter of Hopewell Archaeology in the Ohio River Valley* 7(2):1-5.

Dracup, Joseph F.
1974 Fundamentals of the State Plane Coordinate Systems. U.S. Department of Commerce, National Oceanic and Atmospheric Administration, National Ocean Survey. Electronic document, http://www.ngs.noaa.gov/PUBS_LIB/FundSPCSys.pdf, accessed September 9, 2013.

Dutcher, James
1988 C-14 Dating Results From the Glenford Stone Mound Site #33-PE-3. *Ohio Archaeologist* 38(3):24-26.

Eddy, John
1978 Archaeoastronomy of North America: Cliffs, Mounds, and Medicine Wheels. In *In Search of Ancient Astronomies*, edited by E. C. Krupp, pp.133-163. McGraw-Hill, New York.

Eliade, Mircea
1987 *The Sacred and the Profane: The Nature of Religion*. Harcourt Brace and Company, San Diego.

Engberg, Tom
2014 This Week in Hopewell Culture Archaeology Week 6. Blog posting at http://www.nps.gov/hocu/photosmultimedia/this-week-in-hopewell-culture-archeology.htm, accessed January 29, 2015.

Emerson, Thomas E.
1989 Water, Serpents, and the Underworld: An Explanation into Cahokia Symbolism. In *The Southeastern Ceremonial Complex: Artifacts and Analysis*, edited by P. Galloway, pp. 45-92. University of Nebraska Press, Lincoln.

Emerson, Thomas E., Susan M. Alt, and Timothy R. Pauketat
2008 Locating American Indian Religion at Cahokia and Beyond. In *Religion, Archaeology, and the Material World*, edited by Lars Fogelin, pp. 216-236. Center for Archaeological Investigations, Occasional Paper No. 36, Southern Illinois University, Carbondale.

Essenpreis, Patricia S., and David J. Duszynski
1989 Possible Astronomical Alignments at the Fort Ancient Monument. Paper presented at the 54th Annual Meeting of the Society for American Archaeology, Atlanta.

Fischer, Fred W.
1968 A Survey of the Archaeological Remains of Shawnee Lookout Park. Unpublished manuscript in the possession of the author.

Foerste, Aug. F.
1915 *An Introduction to the Geology of Dayton and Vicinity With Special Reference to the Great Ridge Area South of the City, Including Hills and Dales and Moraine Park*. Hollenbeck Press, Indianapolis.

Fowke, Gerard
1902 *Archaeological History of Ohio: The Moundbuilders and Later Indians*. Ohio State Archaeological and Historical Society, Columbus.

Goode, P. R., J. Qui, V. Yurchyshyn, J. Hickey, M. C. Chu, E. Kolbe, C. T. Brown, and S. E. Koonin
2001 Earthshine Observations of the Earth's Reflectance. *Geophysical Research Letters* 28(9):1671-1674.

Gosden, Chris
2004 Aesthetics, Intelligence and Emotions: Implications for Archaeology. In *Rethinking Materiality: The Engagement of Mind with the Material World*, edited by Elizabeth DeMarrais, Chris Gosden, and Colin Renfrew, pp. 33-40. McDonald Institute for Archaeological Research, Cambridge.

Gray, Martin
2007 *Sacred Earth: Places of Peace and Power*. Sterling, New York.

Greber, N'omi B.
2005 Adena and Hopewell in the Middle Ohio Valley. To Be or Not to Be? In *Woodland Systematics in the Middle Ohio Valley*, edited by Darlene Applegate and Robert C. Mainfort, pp. 19-39. University of Alabama Press, Tuscaloosa.

1983 Recent Excavations at the Edwin Harness Mound, Liberty Works, Ross County, Ohio. *Midcontinental Journal of Archaeology*, Special Paper No. 5, Kent State University Press, Kent.

Greenman, Emerson F.
1928 Field Notes on the Excavation of the Eagle Mound. Manuscript on file, Department of Archaeology, Ohio Historical Society, Columbus.

Griffin, James B.
1996 The Hopewell Housing Shortage in Ohio, A.D. 1-350. In *A View From the Core: A Synthesis of Ohio Hopewell Archaeology*, edited by Paul Pacheco, pp. 4-15. Ohio Archaeological Council, Columbus.

Hall, Robert L.
1997 *An Archaeology of the Soul: North American Indian Belief and Ritual*. University of Illinois Press, Chicago.

1976 Ghosts, Water Barriers, Corn, and Sacred Enclosures in the Eastern Woodlands. *American Antiquity* 41(3):360-364.

1979 In Search of the Ideology of the Adena-Hopewell Climax. In *Hopewell Archaeology: The Chillicothe Conference*, edited by David S. Brose and N'omi B Greber, pp. 258-265. Kent State University Press, Kent.

Hallowell, A. Irving
1975 [1960] Ojibwa Ontology, Behavior, and Worldview. In *Teachings from the American Earth*, edited by Dennis Tedlock and Barbara Tedlock, pp. 141-178. W. W. Norton, New York.

Hamilakis, Yannis
2011 Archaeologies of the Senses. In *Oxford Handbook of the Archaeology of Ritual and Religion*, edited by Timothy Insoll, pp. 208-225. Oxford University Press, Oxford.

Hancock, John E.
2010 The Earthworks Hermeneutically Considered. In *Hopewell Settlement Patterns, Subsistence, and Symbolic Landscapes*, edited by A. Martin Byers and DeeAnne Wymer, pp. 263-275. University of Florida Press, Gainesville.

Hansen, Michael C.
1994 Return to Sunken Mountain: The Serpent Mound Cryptoexplosion Structure. *Ohio Geology* 1:3-7.

Hardman, Clark, Jr., and Marjorie Hardman
1987 The Great Serpent and the Sun. *Ohio Archaeologist* 37(3):34-40.

Harvey, Graham
2006 *Animism: Respecting the Living World*. Columbia University Press, New York.

Heidegger, Martin
1993 [1954] Building Dwelling Thinking. In *Martin Heidegger. Basic Writings: From Being and Time (1927) to the Task of Thinking (1964)*, edited by D. F. Krell, pp. 347-363. Harper Collins Publishers, New York.

Henderson, A. Gwynn, David Pollack, and Dwight R. Cropper
1988 The Old Fort Works, Greenup County, Kentucky. In *New Deal Archaeology and Current Research in Kentucky*, edited by David Pollack and Mary Lucas Powell, pp. 64-81. Kentucky Heritage Council, Frankfort.

Hewitt, John Napolean Brinton
1918 Seneca Fiction, Legends, and Myths. In *Thirty-second Annual Report of the Bureau of American Ethnology*. Smithsonian Institution, Washington, D.C.

Herrmann, Edward W., G. William Monaghan, William F. Romain, Timothy M. Schilling, Jarrod Burks, Karen L. Leone, Matthew P. Purtill, Alan C. Tonetti
2014 A New Multistage Construction Chronology for the Great Serpent Mound, USA. *Journal of Archaeological Science* 50:117-125.

Hildreth, Samuel P.
1842 Conus Mound, Marietta Report. *American Pioneer* 1:340.

Hively, Ray, and Robert Horn
2013a A New and Extended Case for Lunar (and Solar) Astronomy at the Newark Earthworks. *Midcontinental Journal of Archaeology* 38(1):83-118.

2013b Hopewell Geometry and Astronomy in the Scioto and Paint Creek Valley. Paper presented at the Midwest Archaeological Conference, October 25, 2013, Columbus, Ohio.

2010 Hopewell Cosmography at Newark and Chillicothe, Ohio. In *Hopewell Settlements and Symbolic Landscapes*, edited by Martin Byers and Dee Anne Wymer, pp. 128-164. University Press of Florida, Gainesville.

1984 Hopewellian Geometry and Astronomy at High Bank. *Archaeoastronomy* (Supplement to Vol. 15, *Journal for the History of Astronomy*) 7: S85-S100.

1982 Geometry and Astronomy in Prehistoric Ohio. *Archaeoastronomy* (Supplement to Vol. 13, *Journal for the History of Astronomy*) 4: S1-S20.

Hodder, Ian
2011 *Entangled: An Archaeology of the Relationships between Humans and Things*. Wiley-Blackwell, West Sussex, England.

2006 *The Leopard's Tale: Revealing the Mysteries of Catahoyuk*. Thames and Hudson, London.

Holmes, William H.
1886 A Sketch of the Great Serpent Mound. *Science* 8 (204):624-628.

Hudson, Charles
1976 *The Southeastern Indians*. University of Tennessee Press, Knoxville.

Hultkrantz, Åke
1953 *Conceptions of the Soul Among North American Indians*. Monograph Series No. 1. Museum of Sweden, Stockholm.

Ingold, Tim
2011 *Being Alive: Essays on Movement, Knowledge and Description*. Routledge, London.

2007 *Lines: A Brief History*. Routledge, London.

2006 Rethinking the Animate, Re-Animating Thought. *Ethnos* 71(1):9-20.

Jones, Owain, and Paul Cloke
2008 Non-Human Agencies: Trees in Place and Time. In *Material Agency: Towards a Non-Anthropocentric Approach*, edited by Carl Knappett and Lambros Malafouris, pp. 79-96. Springer, New York.

Kearney, Michael
1984 *World View*. Chandler & Sharp Publishers, Novato, California.

Knapp, Joseph M.
1998 On the Great Hopewell Road. Electronic document, http://coolohio.com/octagon/onroad.htm, accessed January 15, 2014.

Knight, Vernon J., Jr.
1989 Symbolism of Mississippian Mounds. In *Powhatan's Mantle: Indians in the Colonial Southeast*, edited by P. H. Wood, G. A. Waselkov, and M. T. Hatley, pp. 279-291. University of Nebraska Press, Lincoln.

Kricher, John
1998 *A Field Guide to Eastern Forests, North America*. Peterson Field Guide Series. Houghton Mifflin, New York.

Lankford, George E.
2011 The Raptor on the Path. In *Visualizing the Sacred: Cosmic Visions, Regionalism, and the Art of the Mississippian World*, edited by George E. Lankford, F. Kent Reilly III, and James F. Garber, pp. 240-250. University of Texas Press, Austin.

2007a *Reachable Stars: Patterns in the Ethnoastronomy of Eastern North America*. University of Alabama Press, Tuscaloosa.

2007b The Great Serpent in Eastern North America. In *Ancient Objects and Sacred Realms: Interpretations of Mississippian Iconography*, edited by F. Kent Reilly III and James F. Garber, pp. 107-135. University of Texas Press, Austin.

2007c The "Path of Souls": Some Death Imagery in the Southeastern Ceremonial Complex. In *Ancient Objects and Sacred Realms: Interpretations of Mississippian Iconography*, edited by F. Kent Reilly III and James F. Garber, pp. 174-212. University of Texas Press, Austin.

2007d Some Cosmological Motifs in the Southeastern Ceremonial Complex. In *Ancient Objects and Sacred Realms: Interpretations of Mississippian Iconography*, edited by F. Kent Reilly III and James F. Garber, pp. 8-38. University of Texas Press, Austin.

2004 World on a String: Some Cosmological Components of the Southeastern Ceremonial Complex. In *Hero, Hawk, and Open Hand: American Indian Art of the Ancient Midwest and South*, edited by Richard F. Townsend and Robert V. Sharp, pp. 206-217. The Art Institute of Chicago and Yale University Press, Chicago and New Haven.

Lankford, George E. (editor)
1987 Native American Legends of the Southeast: Tales from the Natchez, Caddo, Biloxi, Chickasaw, and Other Nations. August House, Little Rock.

Latour, Bruno
2005 *Reassembling the Social: An Introduction to Network-Actor-Theory*. University of Oxford Press, Oxford.

Lee, T. E.
1958. The Parker Earthwork, Corunna, Ontario. *Pennsylvania Archaeologist* 28:5-32.

Lee, Alfred M., and Kent D. Vickery
1972 Salvage Excavations at the Headquarters Site, A Middle Woodland Village Burial Area in Hamilton. *Ohio Archaeologist* 22(1):3-11.

Leone, Karen L., Bradley T. Lepper, Kathryn A. Jakes, Linda L. Pansing, and William H. Pickard
2013 Radiocarbon Dates from the Central Grave of the Adena Mound. Paper presented at the Midwest Archaeological Conference, October 25, 2013. Columbus, Ohio.

Lepper, Bradley T.
2010 The Ceremonial Landscape of the Newark Earthworks and the Raccoon Creek Valley. In *Hopewell Settlement Patterns, Subsistence, and Symbolic Landscapes*, edited by A. Martin Byers and DeeAnne Wymer, pp. 97-127. University of Florida Press, Gainesville.

1998 The Archaeology of the Newark Earthworks. In *Ancient Earthen Enclosures of the Eastern Woodlands*, edited by Robert C. Mainfort, Jr. and Lynn P. Sullivan, pp. 114-134. University Press of Florida, Gainesville.

Locke, John
1838 Geological Report on Southwest District. *Ohio Division of Geological Survey, Second Annual Report* 2:201-286.

Lynott, Mark J., and Rolfe D. Mandel
2009 Archaeological and Geoarchaeological Study of the Rectangular Enclosure at Hopeton Works. In *Footprints: In the Footprints of Squier and Davis: Archaeological Fieldwork in Ross County, Ohio*, edited by Mark J. Lynott, pp. 159-177. U.S. Department of the Interior, National Park Service, Midwest Archaeological Center, Special Report No. 5. Lincoln.

Lynott, Mark J., Rolfe Mandel, and James A. Brown
2010 Earthen Monument Construction at Mound City, Ohio: 2009-2010 Investigations. Paper presented at the 56th Annual Meeting of the Midwest Archaeological Conference, Bloomington, Indiana.

MacLean, John P.
1879 *The Mound Builders: Being an Account of a Remarkable People that Once Inhabited the Valleys of the Ohio and Mississippi, together with an Investigation into the Archaeology of Butler County, Ohio*. Robert Clarke, Cincinnati.

Marshall, James A.
1987 An Atlas of American Indian Geometry. *Ohio Archaeologist* 37(2):3-49.

1978 American Indian Geometry. *Ohio Archaeologist* 28(1):29-33.

Martzloff, Clement L.
1902 *History of Perry County*. Ward & Weiland, New Lexington, Ohio.

Maslowski, Robert F., and Mark F. Seeman
1992 Woodland Archaeology in the Mid-Ohio Valley: Setting Parameters for Ohio Main Stem/Tributary Comparisons. In *Cultural Variability in Context: Woodland Settlements of the Mid-Ohio Valley*, edited by Mark F. Seeman, pp. 10-14. Mid-Continental Journal of Archaeology Special Paper No. 7. Kent State University Press, Kent.

McCormac, Jack C.
1983 *Surveying Fundamentals*. Prentice Hall, Englewood Cliffs, New Jersey.

McCulloch, J. Huston
2010 (1991) Isaac Roberdeau on the East Fork, Milford, and West Milford Works. Electronic document, http://www.econ.ohio-state.edu/jhm/arch/efw.html, accessed September 21, 2013.

1996 Ohio's Hanukkah Mound. *Ancient American* 14:28-37.

Meskell, Lynn
2003 Memory's Materiality: Ancestral Presence, Commemorative Practice and Disjunctive Locales. In *Archaeologies of Memory*, edited by R. M. Van Dyke and S. E. Alcock, pp. 34-55. Blackwell, Oxford.

Michel, Mark
1980 Archaeological Conservancy Acquires Hopewell Mounds. *Ohio Archaeologist* 30(3):9-10.

Milam, Keith A.
2010 A Revised Diameter for the Serpent Mound Impact Crater in Southern Ohio. *Ohio Journal of Science* 111(3):34-43.

Miller, Dorcas S.
1997 *Stars of the First People: Native American Star Myths and Constellations*. Pruett, Boulder, Colorado.

Mills, William C.
1922 *Certain Mounds and Village Sites in Ohio. Volume 3, Part 3: Flint Ridge*. F. J. Heer, Columbus, Ohio.

1921 Flint Ridge. In *Certain Mounds and Village Sites in Ohio*. Ohio State Archaeological and Historical Society Publications 30:91-161.

1916 Exploration of the Tremper Mound. *Ohio Archaeological and Historical Quarterly* 25:262-398.

1914 *Archaeological Atlas of Ohio, Showing the Distribution of Various Classes of Prehistoric Remains in the State with a Map of the Principal Indian Trails and Towns*. Ohio State Archaeological and Historical Society, Columbus.

1907 The Exploration of the Edwin Harness Mound. *Ohio Archaeological and Historical Quarterly* 16:113-193.

1902 Excavations of the Adena Mound. *Ohio Archaeological and Historical Quarterly* 10:452-479.

Moats, Richard D.
2012 The Reconstruction and Archaeoastronomy of a Hopewell Hilltop Earthwork in Ohio. *Ohio Archaeologist* 62 (4):4-12.

2011 A Summary of Fort Glenford Hilltop Enclosure and Mound. *Ohio Archaeologist* 61(1):30-35.

Moorehead, Warren K.
1899 Report of Field Work in Various Portions of Ohio. *Ohio State Archaeological and Historical Quarterly* VII:110-203.

1897 Field Work During the Spring and Summer of 1896. *Ohio Archaeological and Historical Quarterly* V:169-274.

1890 *Fort Ancient: The Great Prehistoric Earthwork of Warren County, Ohio*. Robert Clarke, Cincinnati.

Morgan, William N.
1980 *Prehistoric Architecture in the Eastern United States*. MIT Press, Cambridge, Massachusetts.

Moser, Claudia, and Cecelia Feldman
2014 Introduction. In *Locating the Sacred: Theoretical Approaches to the Emplacement of Religion*, edited by Claudia Moser and Cecelia Feldman, pp. 1-12. Oxbow Books, Oxford.

Murphy, James L.
1989 *An Archaeological History of the Hocking Valley*. Revised ed. Ohio University Press, Athens.

1978 An Adena Component Associated With the Fairfield Township Earthworks, Butler County, Ohio. *Ohio Archaeologist* 28(2):11-13.

Norberg-Schulz, Christian
1980 *Genius Loci: Towards a Phenomenology of Architecture*. Rizzoli, New York.

Ortmann, Anthony L., and Tristram R. Kidder
2013 Building Mound A at Poverty Point, Louisiana: Monumental Public Architecture, Ritual Practice, and Implications for Hunter-Gatherer Complexity. *Geoarchaeology: An International Journal* 28:66-86.

Otto, Martha P., and Brian G. Redmond
2008 *Transitions: Archaic and Early Woodland Research in the Ohio Country*. Ohio University Press, Athens.

Parker Pearson, Mike
2001 Death, Being, and Time: The Historical Context of the World's Religions, In *Archaeology and World Religion*, edited by Timothy Insoll, pp. 203-219. Routledge, London.

1999 *The Archaeology of Death and Burial*. Sutton, London.

Pauketat, Timothy R.
2013a *An Archaeology of the Cosmos: Rethinking Agency and Religion in Ancient America*. Routledge, London.

2013b Bundles of/in/as Time. In *Big Histories, Human Lives*, edited by John Robb and Timothy R. Pauketat, pp. 35-56. School for Advanced Research Press, Santa Fe, New Mexico.

2009 Of Leaders and Legacies in Native North America. In *The Evolution of Leadership: Transitions in Decision Making from Small-Scale to Middle-Range Societies*, edited by Kevin J. Vaughn, Jelmer W. Eerkens, and John Kanter, pp. 169-192. School for Advanced Research Press, Santa Fe.

Penny, David W.
1985 Continuities of Imagery and Symbolism in the Art of the Woodlands. In *Ancient Art of the American Woodland Indians*, edited by David S. Brose and David W. Penny, pp. 147-198. Harry N. Abrams, New York.

Petrides, George A.
1998 *A Field Guide to Eastern Trees: Eastern United States and Canada, Including the Midwest*. Peterson Field Guide Series. Houghton Mifflin, New York.

Petro, James H., William Shumate, and Marion F. Tabb
1967 Soil Survey of Ross County, Ohio. U.S. Department of Agriculture, Soil Conservation Service. Government Printing Office, Washington D.C.

Pickard, William H.
2009 Falling Through a Crack in the Core: The Surprise and Demise of Anderson Earthwork. In *Footprints: In the Footprints of Squier and Davis: Archaeological Fieldwork in Ross County, Ohio*, edited by Mark J. Lynott, pp. 67-75. United States Department of Interior, National Park Service, Midwest Archaeological Center, Lincoln.

1996 1990 Excavations at Capitolium Mound (33WN13), Marietta, Washington County, Ohio: A Working Evaluation. In *A View From the Core: A Synthesis of Ohio Hopewell Archaeology*, edited by Paul Pacheco, pp. 274-285. Ohio Archaeological Council, Columbus.

Potter, Martha A., and Edward S. Thomas
1970 Fort Hill. [Booklet.] The Ohio Historical Society, Columbus.

Pratt, Christina
2007 *An Encyclopedia of Shamanism*. Rosen, New York.

Prufer, Olaf H.
1997 Fort Hill 1964. New Data and Reflections on Hilltop Enclosures in Southern Ohio. In *Ohio Hopewell Community Organization*, edited by William S. Dancey and Paul J. Pacheco, pp. 311-327. Kent State University Press, Kent.

1965 *The McGraw Site: A Study in Hopewellian Dynamics*.
Cleveland Museum of Natural History Scientific Publications, 4(1), Cleveland.

Putnam, Frederic W.
1890 The Serpent Mound of Ohio. *Century Illustrated Magazine* 39:871-888.

Railey, Jimmy A.
1996 Woodland Cultivators. In *Kentucky Archaeology*, edited by R. Barry Lewis, pp. 79-125. University Press of Kentucky, Lexington.

Reidel, Stephen P., Frank L. Koucky, and J. Roger Stryker
1982 The Serpent Mound Disturbance, Southwestern Ohio. *American Journal of Science* 282:1343-1377.

Reilly, F. Kent, III
2011 The Great Serpent in the Lower Mississippi Valley. In *Visualizing the Sacred: Cosmic Visions, Regionalism, and the Art of the Mississippian World*, edited by George E. Lankford, F. Kent Reilly III, and James F. Garber, pp. 118-134. University of Texas Press, Austin.

2004 People of Earth, People of Sky: Visualizing the Sacred in Native American Art of the Mississippian Period. In *Hero, Hawk, and Open Hand: American Indian Art of the Ancient Midwest and South*, edited by R. F. Townsend and R. V. Sharp, pp. 125-137. Art Institute of Chicago and Yale University Press, New Haven.

Riley, Thomas J., Richard Edging, and Jack Rossen
1990 Cultigens in Prehistoric Eastern North America: Changing Paradigms. *Current Anthropology* 31(5): 525-541.

Riordan, Robert V.
2010 Enclosed by Stone. In *Hopewell Settlement Patterns, Subsistence, and Symbolic Landscapes*, edited by A. Martin Byers and DeeAnne Wymer, pp. 215-229. University Press of Florida, Gainesville.

1995 A Construction Sequence for a Middle Woodland Hilltop Enclosure. *Midcontinental Journal of Archaeology* 20:62-104.

Rodaway, Paul
1994 *Sensuous Geographies: Body, Sense, and Place.* Routledge, London.

Romain, William F.
2014a Moonwatchers of Cahokia. In *Medieval Mississippians*, edited by Tim Pauketat and Susan Alt, pp. 32-41. School for Advanced Research Press, Santa Fe, New Mexico.

2014b Newly-Discovered Hopewell Mound Group Woodhenge Found to be Solstice Aligned. Electronic document, http://www.ancientearthworksproject.org, accessed January 29, 2015.

2012 Moonrise Over Cahokia: Sacrificed Women, Mound 72, and Lunar Alignments. Paper presented at the Southeastern Archaeological Conference, Baton Rouge, Louisiana.

2011 LiDAR Views of the Serpent Mound Impact Crater. Ohio Archaeological Council, Current Research in Ohio Archaeology 2011. Electronic document http://www.ohioarchaeology.org/joomla/index.php, accessed September 19, 2013.

2009a *Shamans of the Lost World: A Cognitive Approach to the Prehistoric Religion of the Ohio Hopewell.* AltaMira, Lanham, Maryland.

2009b Hopewell Archaeoastronomy and Geomancy: New Discoveries Using LiDAR. Invited presentation for the Sunwatch Village 2009 Lecture Series: Archaeoastronomy in the Americas, April 4, 2009, Dayton, Ohio.

2009c LiDAR Assessment of the Ohio Hopewell Earthworks. Invited presentation for the National Center for Great Lakes Native American Culture: Academic Conference on Native American Astronomy, April 18, 2009, Portland, Indiana.

2008a LiDAR Assessments of Hopewell Earthworks. Paper presented at the Ohio Archaeological Council Fall Membership Meeting, November 1, 2008, Newark, Ohio.

2008b LiDAR Imaging of Ohio Hopewell Earthworks: New Images of Ancient Sites. Invited presentation for the Ohio State University, American Indian Studies Department, October 23, 2008, Columbus, Ohio.

2005a Summary Report on the Orientations and Alignments of the Ohio Hopewell Geometric Enclosures. In *Gathering Hopewell: Society, Ritual, and Ritual Interaction,* edited by Christopher Carr and Troy Case, Appendix 3. Kluwer Academic Publishers, New York.

2005b Results of Coring at Newark Octagon Mounds D and E. Report on file, Archaeology Department, Ohio Historical Society, Columbus.

2005c Newark Earthworks Cosmology: This Island Earth. National Park Service, *Newsletter of Hopewell Archaeology in the Ohio River Valley* 6(2):article 1. Electronic document, http://www.nps.gov/MWAC/hopewell/v6n2/one.htm, accessed February 6, 2014.

2005d Hopewell Astronomy, Geometry, and Cosmology. Invited presentation for 'Ohio's Ancient Earthworks: A Public Symposium' sponsored by the Ohio Archaeological Council and Newark Earthworks Center, Ohio State University, November 19, 2005, Ohio State University, Newark, Ohio.

2004a Hopewell Geometric Enclosures: Gatherings of the Fourfold. Unpublished Ph.D. dissertation, School of Archaeology and Ancient History, University of Leicester, Leicester, England.

2004b Journey to the Center of the World: Astronomy, Geometry, and Cosmology of the Fort Ancient Enclosure. In *The Fort Ancient Earthwork: Prehistoric Lifeways of the Hopewell in Southwestern Ohio*, edited by Robert P. Connolly and Bradley T. Lepper, pp. 66-83. Ohio Historical Society, Columbus.

2000 *Mysteries of the Hopewell: Astronomers, Geometers, and Magicians of the Eastern Woodlands*. University of Akron Press, Akron.

1998 Winter Solstice Alignments at Marietta. *Ohio Archaeologist* 48(1):16-17.

1996 Hopewellian Geometry: Forms at the Interface of Time and Eternity. In *A View from the Core: A Synthesis of Ohio Hopewell Archaeology*, edited by Paul Pacheco, pp. 194-209. Ohio Archaeological Council, Columbus.

1995 In Search of Hopewell Archaeoastronomy. *Ohio Archaeologist* 45(1):35-41.

1993a Further Notes on Hopewellian Astronomy and Geometry. *Ohio Archaeologist* 43(3):48-52.

1993b Hopewell Ceremonial Centers and Geomantic Influences. *Ohio Archaeologist* 43(1):35-44.

1992a Hopewellian Concepts in Geometry. *Ohio Archaeologist* 42(2):35-50.

1992b Hopewell Inter-site Relationships and Astronomical Alignments. *Ohio Archaeologist* 42(1):4-5.

1992c Azimuths to the Otherworld: Astronomical Alignments of Hopewell Charnel Houses. *Ohio Archaeologist* 42(4):42-48.

1991a Calendric Information Evident in the Adena Tablets. *Ohio Archaeologist* 41(4):41-48.

1991b Possible Astronomical Alignments at Hopewell Sites in Ohio. *Ohio Archaeologist* 41(3):4-16

1988 The Serpent Mound Solar Eclipse Hypothesis: Ethnohistoric Considerations. *Ohio Archaeologist* 38(3):32-37.

1987a The Serpent Mound Map. Ohio Archaeologist 37(4):38-42.

1987b Serpent Mound Revisited. *Ohio Archaeologist* 37(4):5-10.

Romain, William F., and Jarrod Burks
2008a LiDAR Imaging of the Great Hopewell Road. Ohio Archaeological Council, Current Research in Ohio Archaeology 2008. Electronic document, http://www.ohioarchaeology.org/joomla/index.php, accessed August 24, 2013.

2008b LiDAR Assessment of the Newark Earthworks. Ohio Archaeological Council, Current Research in Ohio Archaeology 2008. Electronic document, http://www.ohioarchaeology.org/joomla/index.php?option=com, accessed August 24, 2013.

2008c LiDAR Analyses of Prehistoric Earthworks in Ross County, Ohio. Ohio Archaeological Council, Current Research in Ohio Archaeology 2008. Electronic document, http://www.ohioarchaeology.org/joomla/index.php, accessed August 24, 2013.

Romain, William F., G. William Monaghan, Edward Herrmann, Timothy Schilling, Jarrod Burks, Karen Leone, Matthew Purtill, Al Tonetti
2013 Serpent Mound Revealed: New Discoveries, New Dates.
Paper presented at the Archaeological Society of Ohio Spring Meeting, March 24, 2013, Columbus.

Ruby, Bret J.
2014a Current Research in the Park: The Great Circle Project at Hopewell Mound Group. Presentation for the National Park Service Harness Lecture Series, June 26, 2014, Hopewell Culture National Park, Chillicothe, Ohio.

2009 Spruce Hill Earthworks: The 1995-1996 National Park Service Investigations. In *Footprints: In the Footprints of Squier and Davis: Archaeological Fieldwork in Ross County, Ohio*, edited by Mark J. Lynott, pp. 49-66. United Stated Department of the Interior, National Park Service, Midwest Archaeological Center, Special Report Number 5. Lincoln.

Ruggles, Clive
2005 *Ancient Astronomy: An Encyclopedia of Cosmologies and Myth*. ABC-CLIO, Santa Barbara.

1999 *Astronomy in Prehistoric Britain and Ireland*. Yale University Press, New Haven.

Salisbury, James, and Charles Salisbury
1862 Accurate Surveys and Descriptions of the Ancient Earthworks at Newark, Ohio. Transcribed from the original by B. T. Lepper and B. T. Simmons. Manuscript on file, American Antiquarian Society, Worcester, Massachusetts.

Schaefer, Bradley, and William Liller
1990 Refraction Near the Horizon. *Publications of the Astronomical Society of the Pacific* 102:796-805.

Schilling, Timothy
2013 The Chronology of Monks Mound. *Southeastern Archaeology* 32(1):14-28.

Seeman, Mark F.
2004 Hopewell Art in Hopewell Places. In *Hero, Hawk, and Open Hand: American Indian Art of the Ancient Midwest and South*, edited by Richard F. Townsend and Robert V. Sharp, pp. 57-71. The Art Institute of Chicago and Yale University Press, New Haven.

1995 When Words Are Not Enough: Hopewell Interregionalism and the Use of Material Symbols at the GE Mound. In *Native American Interactions: Multiscalar Analyses and Interpretations in the Eastern Woodlands*, edited by Michael M. Nassaney and Kenneth E. Sassaman, pp. 122-143. University of Tennessee Press, Knoxville.

1986 Adena Houses and Their Implications for Early Woodland Settlement Models in the Ohio Valley. In *Early Woodland Archaeology*, edited by Kenneth B. Farnsworth and Thomas E. Emerson, pp. 564-580. Center for American Archeology Press, Kampsville, Illinois.

1979 The Hopewell Interaction Sphere: The Evidence for Interregional Trade and Structural Complexity. *Indiana Historical Society, Prehistory Research Series* 5 (2).

Seig, Lauren E., and R. Eric Hollinger
2005 Learning from the Past: The History of Ohio Hopewell Taxonomy and Its Implications for Archaeological Practice. In *Woodland Period Systematics in the Middle Ohio Valley*, edited by Darlene Applegate and Robert C. Mainfort Jr., pp. 120-133. University of Alabama Press, Tuscaloosa.

Shane, Orrin C., III
1971 The Scioto Hopewell. *In Adena: The Seeking of an Identity*, edited by B.K. Swartz, pp. 142-157. Indiana: Ball State University, Muncie.

Sherrod, P. Clay, and Martha Ann Rolingson
1987 Surveyors of the Ancient Mississippi Valley: Modules and Alignments in Prehistoric Mound Sites. *Arkansas Archaeological Survey Research Series No. 28*. Arkansas Archaeological Survey, Fayetteville.

Shetrone, Henry C.
1926 Explorations of the Hopewell Group of Prehistoric Earthworks. *Ohio Archaeological and Historical Quarterly* 35:5-227.

Sims, Lionel
2009 Entering, and Returning From, the Underworld: Reconstituting Silbury Hill by Combining a Quantified Landscape Phenomenology with Archaeoastronomy. *Journal of the Royal Anthropological Institute* (n.s.) 15:386-408.

Skinner, Shaune E., and Rae Norris
1984 Excavation of the Connett Mound 4. *Ohio Archaeologist* 34(4):23-26.

Smart, T.L., and Ford, R.I. 1983. Plant Remains. In *Recent Excavations at the Edwin Harness Mound, Liberty Works, Ross County, Ohio*, edited by N'omi Greber. Mid-Continental Journal of Archaeology Special Paper No. 5. Kent, Ohio: Kent State University Press.

Squier, Ephraim G., and Edwin H. Davis
1848 *Ancient Monuments of the Mississippi Valley; Comprising the Results of Extensive Original Surveys and Explorations*. Smithsonian Contributions to Knowledge, Vol. 1. Smithsonian Institution, Washington, D.C.

Star, S. Frederick
1960 The Archaeology of Hamilton County, Ohio. *Journal of the Cincinnati Museum of Natural History* 23(1):1-130.

Swartz, B. K., Jr. (editor)
1971 *Adena: The Seeking of an Identity*. Ball State University, Muncie, Indiana.

Suzuki, David, and Peter Knudson
1992 *Wisdom of the Elders: Honoring Sacred Native Visions of Nature*. Bantam, New York.

Theodoratus, Dorothea J., and Frank LaPena
1994 Wintu Sacred Geography of Northern California. In *Sacred Sites, Sacred Places*, edited by David L. Carmichael, Jane Hubert, Brian Reeves, Audhild Schanche, pp. 20-31. Routledge, New York.

Thomas, Cyrus
1894 Report on the Mound Explorations of the Bureau of Ethnology for the Years 1890-1891. In *Twelfth Annual Report of the Bureau of American Ethnology*. Smithsonian Institution, Washington, D.C.

Thomas, David H.
1999 *Archaeology: Down to Earth*. 2nd ed. Harcourt Brace & Co., Fort Worth, Texas.

1993 The World As It Was. In *The Native Americans: An Illustrated History*, edited by Betty Ballentine and Ian Ballentine. Turner Publishing, Atlanta.

Thomas, Julian
1996 *Time, Culture & Identity: An Interpretive Archaeology*. Routledge, London.

Tight, W. G.
1895 A Preglacial Tributary to Paint Creek and Its Relation to the Beech Flats of Pike County, Ohio. *Bulletin of the Scientific Laboratories of Denison University* IX(1):25-34.

Tilley, Christopher
2008 Phenomenological Approaches to Landscape Archaeology. In *Handbook of Landscape Archaeology*, edited by Bruno David and Julian Thomas, pp. 271-276. Left Coast Press, Walnut Creek, California.

2000 [1989] Interpreting Material Culture. In *Interpretive Archaeology: A Reader*, edited by Julian Thomas, pp. 418-426. Leicester University Press, London.

1994 *A Phenomenology of Landscape: Places, Paths and Monuments*. Berg, Oxford.

Tonetti, Al
2004 Phase I Archaeological Survey for the Camp Sherman Training Site, Chillicothe, Springfield Township, Ross County, Ohio. Report submitted by the ASC Group, Columbus, Ohio to the Adjutant General's Department, Ohio National Guard, Columbus, Ohio.

Tuan, Yi-Fu
1977 *Space and Place: The Perspective of Experience*. University of Minnesota Press, Minneapolis.

Turner, Christopher S.
2011 Ohio Hopewell Archaeoastronomy: A Meeting of Earth, Mind, and Sky. *Time and Mind: The Journal of Archaeology, Consciousness and Culture* 4(3):301-324.

1982 Hopewell Archaeoastronomy. *Archaeoastronomy* 5(3):9.

United States Department of Commerce Coast and Geodetic Survey
1978 Plane Coordinate Projection Tables, Ohio. Special Publication No. 269. US Government Printing Office, Washington, D.C.

Van Dyke, Ruth M.
2007 *The Chaco Experience: Landscape and Ideology at the Center Place*. School for Advanced Research Press, Santa Fe, New Mexico.

2003 Memory and the Construction of Chacoan Society. In *Archaeologies of Memory*, edited by Ruth Van Dyke and Susan E. Alcock, pp. 180-200. Blackwell Publishing, Oxford.

Vecsey, Christopher
1995 Prologue. In *Handbook of Native American Indian Religious Freedom*, edited by Christopher Vecsey, pp. 7-25. Crossroad Publishing, New York.

Walker, Jr., Deward E.
1995 Protection of American Indian Sacred Geography. In *Handbook of Native American Religious Freedom*, edited by Christopher Vecsey, pp. 100-115. Crossroad Publishing, New York.

Watts, Christopher (editor)
2013 *Relational Archaeologies: Humans, Animals, Things.* Routledge, London.

Webb, William S., and Charles E. Snow
1945 *The Adena People*. Publications of the Department of Anthropology and Archaeology, Vol. 6. University of Tennessee Press, Knoxville.

Willey, Gordon R., and Phillip Phillips
1958 *Method and Theory in American Archaeology*. University of Chicago Press, Chicago.

Willoughby, Charles C. (with Ernest A. Hooton)
1922 The Turner Group of Earthworks, Hamilton County, Ohio. Papers of the Peabody Museum, Vol. 8, No. 3. Harvard University, Cambridge.

Wood, John E.
1978 *Sun, Moon, and Standing Stones*. Oxford University Press, Oxford.

Wright, George F.
1890 The Glacial Boundary in Western Pennsylvania, Ohio, Kentucky, Indiana, and Illinois. Bulletin of the United States Geological Survey, No. 58.

Wymer, Dee Anne
1997 Paleoethnobotany in the Licking River Valley, Ohio: Implications for Understanding Ohio Hopewell. In *Ohio Hopewell Community Organization*, edited by William S. Dancey and Paul J. Pacheco, pp. 153-171. Kent State University Press, Kent.

1996. The Ohio Hopewell Econiche: Human-Land Interaction in the Core Area. In *A View From the Core: A Synthesis of Ohio Hopewell Archaeology* edited by Paul J. Pacheco, pp. 36-52. Ohio Archaeological Council, Columbus.

Wyrick, David
1866 Ancient Works Near Newark, Licking County, Ohio. [Map from 1860.] In *Atlas of Licking County, Ohio*, edited by F. W. Beers. Beers + Soula, New York.

Young, Biloine Whiting and Melvin L. Fowler
2000 *Cahokia: The Great Native American Metropolis*. University of Illinois Press, Urbana.

Zaleha, Michael J., and William F. Romain
2014 Geophysical Subsurface Investigation of the Serpent Mound Area (Ohio, USA) Using Electrical Resistivity Ground Imaging (ERGI): Evaluation of Bedrock Controls on Surface Features. Poster presented at the Geological Society of America Annual Meeting, Vancouver, B.C.

2012 Electrical Resistivity Ground Imaging (ERGI) Investigation of Topographic Depressions Near the Serpent Mound Native American Effigy, Adams County, Ohio. Poster presented at the Geological Society of America Annual Meeting, Charlotte, North Carolina.

Zedeño, María N.
2013 Methodological and Analytical Challenges in Relational Archaeologies: A View from the Hunting Ground. In *Relational Archaeologies: Humans, Animals, Things*, edited by Christopher Watts, pp. 117-134. Routledge, London.

2009 Animating by Association: Index Objects and Relational Taxonomies. *Cambridge Archaeological Journal* 19(3):407-417.

2008 Bundled Worlds: The Roles and Interactions of Complex Objects from the North American Plains. *Journal of Archaeological Method and Theory* 15:362-378.

Index

A

Adena culture, *11-12, 23, 39, 288*
Adena Mound, *49, 256, 280, 307, 309*
Adena-Hopewell beliefs, *242*
Alexandersville Works, *293*
Alum Cliffs Gorge, *284*
Anderson Earthworks, *166, 278, 284, 287, 297, 300, 311*
Antares, *289-290*
astronomic north, *32, 268*
Atwater, Caleb, *43, 266-267, 292, 297*
Austin Brown Mound, *276*
axis mundi, *7, 27, 49, 239, 252*

B

Baker Fork, *191-192*
Barnes Works, *282*
Baum Earthworks, *164-166, 285, 287, 295*
Beech Flats, *191-192, 318*
being-in-the-world, *2, 5, 7, 14, 240, 252, 254*
Biggs Mound, *283*
Blackwater Group, *256*
Bourneville Circle, *286-287*
Brush Creek Watershed, *191, 287*
Bunker Hill, *109, 278*

C

Cahokia, *17, 26, 267, 297, 302, 313, 320*
Camp Sherman, *277, 318*
Capitolium Mound, *49, 275-276, 311*
Carlisle Fort, *293-294, 298*
Cedarville Earthworks, *205*
Chenopodium, *165, 243, 260-261*
Chillicothe, *108, 110, 162, 167, 239, 251, 280-281, 284, 287, 297, 304-305, 315, 318*
Chimney Rock, *276*
Circleville Earthworks, *248, 257, 295, 297*
City of the Dead, *108*
Claylick Creek, *46*
Coffman Knob, *48*
concretions, *164*
Connolly, Robert, *301*
Conus Mound, *49, 111, 256, 275, 305*
Copperas Mountain, *164*
Country Club Mound, *162*
Cygnus, *277*

D

Davis Works, *278*
declination, *34-36, 290*
Deneb, *28, 277*
Dunlaps Earthwork, *276*

E

Eagle Mound, *244-245, 303*
East Fork Works, *207, 292-293*
Eastern Woodlands, *1, 17, 21, 26-28, 42, 238, 244, 257, 260, 289-290, 298-299, 304, 308, 314, 316*
Eliade, Mircea, *302*
Ellipse Earthwork, *44, 207*

F

Fairfield Township Works, *294*
Fairmount Mound, *275*
Flint Ridge, *46-48, 274-275, 309*
Fort Ancient, *42, 205-206, 251, 257, 260, 288-289, 291-292, 301, 303, 310, 314*
Fort Hill, *40, 191-192, 251, 287-288, 297, 312*
Frankfort Works, *165, 285*
Fredericktown Earthworks, *39*

G

Geller Hill, *271*
geodetic north, *268*
Glenford Hilltop Enclosure, *48, 310*
Great Circle, *38, 41-42, 45, 249, 251, 256, 261, 270-274, 295, 315*
Great Hopewell Road, *38, 45-46, 111, 270-272, 277, 306, 315*
Great Horned Serpent, *258*
Great Miami River, *219-220, 293-295*
Great Mound, *219, 293, 295*
Great Rift, *28*
Great Serpent, *26, 28, 193, 289, 304-305, 307, 312*
grid north, *32-33, 271*

H

Hall, Robert, *304*
Hancock, John, *304*
Harness Mound, *282, 303, 309, 317*
Hazlett Mound, *46-47, 97, 274, 300*
Hazlett Works, *47*
Heidegger, Martin, *304*
heirophany, *109, 193, 278*
hematite, *22, 98-99*
High Bank Circle, *166, 266-267, 281*
High Bank Earthwork, *166, 266*
Hill-Kinder Mound, *219, 293, 295*
Hively, Ray and Robert Horn, *12-13, 41, 109, 166, 266-267, 270, 278, 280-281, 285, 287*
Hocking River, *18, 97-98, 276*
Holmes, William, *305*
Hopeton Circle, *281*
Hopeton Earthwork, *248, 253, 277-281, 295, 302, 308*
Hopeton Square, *278-280*
Hopewell cultigens, *261*
Hopewell Great Enclosure, *285*
Hopewell Measurement Unit, *12, 249-250*
Hopewell Mound Group, *163, 249, 284-285, 299, 313, 315*
horizon elevation, *33-37, 269-272, 274-277, 279-288, 291-294*
horticultural areas, *260*

J

Jackson Square, *278*
Johnson Works, *40, 278, 282*
Junction Group, *162, 256, 284, 296, 299*

K

Kandel Earthworks, *39-40, 269, 278*
Kokosing River, *39*

L

Land of the Dead, *28-29, 45, 109-110, 258-259, 262, 274*
Lankford, George, *306-307*
Large Stone Mound, *275*
Liberty Circle, *249*
Liberty Earthworks, *163, 165-166, 249, 253, 257, 260, 280-282, 285, 287, 295, 303, 317*
Licking River, *40, 42, 44, 46, 320*
LiDAR, *14-15, 17, 29-33, 37, 46, 48, 165, 207, 269, 271-272, 281, 283-284, 288-289, 291, 313, 315*
Little Miami River, *205-207, 291, 293*
Locke, John, *308*
Lowerworld, *7, 26-28, 193, 240, 242, 252-253, 257, 272*
Lynott, Mark, *308*

M

Marietta Conus mound, *256*
Marietta Earthworks, *15, 33, 39, 48-49, 249, 251, 253-254, 256, 260, 275-276, 295, 305, 311, 314*
Marietta Large Square, *249*
Marshall, James, *266-267, 308*
Massie's Creek, *205*
Miamisburg Mound, *111, 219-220, 251, 295*
Milford Earthworks, *205-207, 248, 292, 295, 309*
Milford West, *206, 292*
Milky Way, *28, 45-46, 240, 272, 274, 277, 289, 291*
Milky Way Path, *28, 45-46, 272, 274, 277, 289*
Mills Square, *280-281, 285, 287*
Mills, William C., *309*
Moats, Richard, *310*
Mound City, *108-111, 251, 257, 259, 277-278, 283-284, 298, 308*
Mount Logan, *109, 278*
Mount Prospect, *287*
Murphy, James, *310*
Muskingum River, *18, 39-40, 48-49, 97, 254, 269, 275*

N

Newark Eagle mound, *244-245*
Newark Earthworks, *38, 40-43, 46-47, 49, 109, 251-252, 270-271, 274, 305, 308, 314-315*
Newark Ellipse, *257*
Newark Great Circle, *42, 249, 251, 256, 261, 271, 273, 295*
Newark Observatory Circle, *253, 266, 281*
Newark Octagon, *46-47, 166, 245, 247, 251, 261, 266, 270, 273, 282, 295, 314*
Newark Octagon and Observatory Circle, *166, 261, 266, 270, 282, 295*
Newark Valley, *40-41, 251, 270*
North Fork Licking River, *42*
North Fork Paint Creek, *162, 165*

O

Ohio Brush Creek, *191*
Old Fort Earthworks, *283*
orthogonal relationships, *245, 249*
Otherworld, *44-46, 108, 110, 238, 254-255, 257-258, 290, 314*

P

Paint Creek, *108, 112, 162-166, 191, 280-281, 284, 305, 318*
Paint Creek Valley, *163-165, 191, 280, 305*
parallax, *35-37, 268, 270-271, 278, 280-283, 289*
parallel walls, *111, 206-207, 253-254, 273, 283-284, 292*
Pollock Wilson Mounds, *294*
portals, *255, 257*
Porter Mounds, *285*
Portsmouth Earthworks, *110-111*
Portsmouth Group A, *283, 295*
Portsmouth Group B, *112, 283, 295*
Portsmouth Group C, *283*
precession, *34, 290*
Pricer Mound, *111, 286-287*
Prufer, Olaf, *312*
Putnam, Frederic W., *312*

Q

Quadranaou Mound, *275*

R

Raccoon Creek, *42, 253, 308*
refraction, *33, 35-37, 268-271, 274-289, 291-294, 316*
relational field, *2, 6-7, 263*
Reservoir Mound, *275*
Riordan, Robert, *312*
Roberts Mound, *48, 275*
Rock Mill Works, *97, 276*
Rowley Mound, *39, 270*
Ruby, Bret, *315*
Ruggles, Clive, *316*

S

Sacra Via, *251, 254, 276*
sacred circles, *23, 256*
Salisbury map, *44, 48, 273*
Scioto River, *108, 111-112, 162-163, 165-166, 219, 239, 254, 276, 280-281, 283, 294*
Scioto River Valley, *111, 162-163, 165, 219, 280, 294*
Scioto River Watershed, *108, 276*
Scorpius, *28, 289-291*
Seal Township Works, *282*
Seeman, Mark, *316*
Seip Earthworks, *111, 164-166, 257, 278, 286-287, 295, 297*
Seip-Pricer Mound, *111*
Serpent Mound Impact Crater, *192, 309, 313*
Shriver Circle, *278*
signal mounds, *219, 260*
Snake Den Group, *164*
South Fork Licking River, *42, 44*
Spirit House, *110, 278*
spirit realm, *7-9, 254*
Spruce Hill, *164, 251, 286-287, 315*
Squier and Davis, *49, 111, 164, 191, 219, 266, 278-282, 284-289, 291-294, 299, 308, 311, 315*
Stackhouse Earthworks, *39, 269, 278*
standard unit of length, *12*
State Plane Coordinate systems, *32, 302*
Steel Group, *162-163, 266, 284*
Stitt Mound, *276*
Stonehenge, *1, 26*
Story Mound, *162*
Stubbs, *205, 295*
Sugarloaf, *109, 239, 272, 278-279, 291*
Sunfish Creek, *98*

T

theta angle, *32-33*
This World, *7, 26, 108-110, 252, 256-257, 263, 290*
Thunderbirds, *28*
Till Plains, *108, 205*
Tilley, Christopher, *318*
Trefoil Works, *191-192*
Tremper Earthworks, *257, 283, 309*
true north, *32-34, 268, 271, 282, 289*
Turner Earthworks, *205-207, 257, 267, 279, 292, 295, 318-319*
Twin Mounds, *294*
Two Mounds, *34, 162, 220, 295*

U

U-shaped earthworks, *111-112*
Upham Mound, *272*
Upperworld, *7, 26-28, 252-253*

V

Vecsey, Christopher, *319*

W

West Milford Earthworks, *207, 292, 309*
Whiskey Hollow, *191-192*
Williamson Mound, *205, 291*
Willoughby, Charles, *319*
Wilson Mound, *48, 295*
Wolf Plains Group, *98*
Works East Earthworks, *163, 165-166, 248, 253, 280-281, 285, 287, 295*
Wright Square, *44, 46, 249, 251, 270, 272, 295*

Y

Yost Earthworks, *47-48, 245, 274-275*

www.ingramcontent.com/pod-product-compliance
Lightning Source LLC
Chambersburg PA
CBHW081149290426
44108CB00018B/2492